AF411436

Visual and Camera Sensors

Visual and Camera Sensors

Editors

Kang Ryoung Park
Sangyoun Lee
Euntai Kim

MDPI • Basel • Beijing • Wuhan • Barcelona • Belgrade • Manchester • Tokyo • Cluj • Tianjin

Editors

Kang Ryoung Park
Division of Electronics and
Electrical Engineering
Dongguk University
Seoul
Korea, South

Sangyoun Lee
School of Electrical and
Electronic Engineering
Yonsei University
Seoul
Korea, South

Euntai Kim
School of Electrical and
Electronic Engineering
Yonsei University
Seoul
Korea, South

Editorial Office
MDPI
St. Alban-Anlage 66
4052 Basel, Switzerland

This is a reprint of articles from the Special Issue published online in the open access journal *Sensors* (ISSN 1424-8220) (available at: www.mdpi.com/journal/sensors/special_issues/vcSensors).

For citation purposes, cite each article independently as indicated on the article page online and as indicated below:

LastName, A.A.; LastName, B.B.; LastName, C.C. Article Title. *Journal Name* **Year**, *Volume Number*, Page Range.

ISBN 978-3-0365-1584-7 (Hbk)
ISBN 978-3-0365-1583-0 (PDF)

Contents

About the Editors

Kang Ryoung Park

Kang Ryoung Park received his B.S. and M.S. degrees in electronic engineering from Yonsei University, Seoul, South Korea, in 1994 and 1996, respectively. He received his Ph.D. degree in electrical and computer engineering from Yonsei University in 2000. He has been a professor in the division of electronics and electrical engineering at Dongguk University since March 2013. His research interests include image processing and deep learning.

Sangyoun Lee

Sangyoun Lee received B.S. and M.S. degrees in electrical and electronic engineering from Yonsei University, Seoul, South Korea, in 1987 and 1989, respectively, and a Ph.D. degree in electrical and computer engineering from the Georgia Institute of Technology, Atlanta, GA, USA, in 1999. He is currently a Professor and the Head of Electrical and Electronic Engineering at the Graduate School and the Head of the Image and Video Pattern Recognition Laboratory at Yonsei University. His research interests include all aspects of computer vision, with a special focus on pattern recognition for face detection and recognition, advanced driver-assistance systems, and video codecs.

Euntai Kim

Euntai Kim received B.S., M.S., and Ph.D. degrees in Electronic Engineering, all from Yonsei University, Seoul, Korea, in 1992, 1994, and 1999, respectively. From 1999 to 2002, he was a Full-Time Lecturer in the Department of Control and Instrumentation Engineering, Hankyong National University, Kyonggi-do, Korea. Since 2002, he has been with the faculty of the School of Electrical and Electronic Engineering, Yonsei University, where he is currently a Professor. He was a Visiting Researcher with the Berkeley Initiative in Soft Computing, University of California, Berkeley, CA, USA, in 2008. He was also a Visiting Researcher with Korea Institute of Science and Technology (KIST), Korea, in 2018. His current research interests include computational intelligence, statistical machine learning and deep learning and their application to intelligent robotics, autonomous vehicles, and robot vision.

Preface to "Visual and Camera Sensors"

Recent developments have led to the widespread use of visual and camera sensors, such as visible light, near-infrared (NIR), and thermal camera sensors, in a variety of applications in video surveillance, biometrics, image compression, computer vision, image restoration, etc. While existing technology has matured, its performance is still affected by various environmental conditions, and recent approaches have been attempted to use multimodal camera sensors and fuse deep learning techniques with conventional methods to guarantee higher accuracy. The goal of this Special Issue was to invite high-quality, state-of-the-art research papers that deal with challenging issues in visual and camera sensors. We solicited original unpublished and completed research papers that were not currently under review by any other conference/magazine/journal. Topics of interest included, but were not limited to, the following:

- Image processing, understanding, recognition, compression, reconstruction, and restoration by visible light, NIR, thermal camera, and multimodal camera sensors

- Video processing, understanding, recognition, compression, reconstruction, and restoration by various camera sensors

- Computer vision by various camera sensors

- Biometrics and spoof detection by various camera sensors

- Object detection and tracking by various camera sensors

- Deep learning by various camera sensors

- Approaches that combine deep learning techniques and conventional methods on images by various camera sensors

Kang Ryoung Park, Sangyoun Lee, Euntai Kim
Editors

sensors

MDPI

Article

An Optimum Deployment Algorithm of Camera Networks for Open-Pit Mine Slope Monitoring

Hua Zhang [1,†], Pengjie Tao [1,†], Xiaoliang Meng [1,*], Mengbiao Liu [1] and Xinxia Liu [2]

[1] School of Remote Sensing and Information Engineering, Wuhan University, Wuhan 430079, China; hzhang_rs@whu.edu.cn (H.Z.); pjtao@whu.edu.cn (P.T.); mengbiaoliu@whu.edu.cn (M.L.)
[2] School of Water Conservancy and Electric Power, Hebei University of Engineering, 62#Zhonghua Street, Handan 056038, China; liuxinxia@hebeu.edu.cn
* Correspondence: xmeng@whu.edu.cn; Tel.: +86-27-6877-1236
† These authors contributed equally to this work.

Abstract: With the growth in demand for mineral resources and the increase in open-pit mine safety and production accidents, the intelligent monitoring of open-pit mine safety and production is becoming more and more important. In this paper, we elaborate on the idea of combining the technologies of photogrammetry and camera sensor networks to make full use of open-pit mine video camera resources. We propose the Optimum Camera Deployment algorithm for open-pit mine slope monitoring (OCD4M) to meet the requirements of a high overlap of photogrammetry and full coverage of monitoring. The OCD4M algorithm is validated and analyzed with the simulated conditions of quantity, view angle, and focal length of cameras, at different monitoring distances. To demonstrate the availability and effectiveness of the algorithm, we conducted field tests and developed the mine safety monitoring prototype system which can alert people to slope collapse risks. The simulation's experimental results show that the algorithm can effectively calculate the optimum quantity of cameras and corresponding coordinates with an accuracy of 30 cm at 500 m (for a given camera). Additionally, the field tests show that the algorithm can effectively guide the deployment of mine cameras and carry out 3D inspection tasks.

Keywords: camera networks; open-pit mine slope monitoring; optimum deployment; close range photogrammetry; three-dimensional reconstruction; OCD4M

Citation: Zhang, H.; Tao, P.; Meng, X.; Liu, M.; Liu, X. An Optimum Deployment Algorithm of Camera Networks for Open-Pit Mine Slope Monitoring. *Sensors* **2021**, *21*, 1148. https://doi.org/10.3390/s21041148

Academic Editor: Kang Ryoung Park
Received: 30 December 2020
Accepted: 4 February 2021
Published: 6 February 2021

Publisher's Note: MDPI stays neutral with regard to jurisdictional claims in published maps and institutional affiliations.

1. Introduction

Slope damage results in serious disasters that cause thousands of deaths and injuries and extensive property damage every year [1]. This poses a serious threat to people working in open-pit mines with slopes. There are three scales of slope damage that can occur in open-pit slopes, and they are bench damage, interslope damage and overall damage [2]. With economic development and the rapid growth of the demand for mineral resources, the exploitation of mine enterprises continues to increase. Many hillside open-pit mines are transformed into deep mines, which leads to increasing the overall angles of slopes and consequently, an increased landslide risk [3]. The statistics of industrial accidents that occurred during the 2005–2010 open-pit coal production period in Turkish coal companies indicates that the most likely risks in open-pit mines are related to mine slopes [4]. According to the US Centers for Disease Control and Prevention (CDC) statistics for mine disasters in 2014, mine slope-related accidents were most reported in quarry operations, accounting for 33.3% of all accidents [5]. In total, 40% of Chinese open-pit mines have slope stability problems. According to statistics, there were 240 slope collapses and 369 fatalities from 2013 to 2017, ranking second amongst noncoal mine accidents in China [6]. Consequently, it is especially important to monitor the slope of open-pit mines.

For mine slope safety monitoring, scholars have used geodetic methods, 3S technology, photogrammetry, the Synthetic Aperture Radar (SAR), 3D laser scanning, and other methods in associated research. The geodesic survey methods [7], such as leveling instrument, theodolite, rangefinder, and distance measurement equipment, represent technology which is widely used in the establishment of high-precision control networks of mines [8]. In addition, the high-precision deformation monitor is mature, displays data reliability, and has a high accuracy; however, for open-pit mine slope monitoring, it has the disadvantages of requiring a large degree of manual involvement, being influenced by terrain access and climate, and is not able to automate monitoring, and those disadvantages result in low detection efficiency. Scholars have begun to focus more on other techniques to carry out mine slope safety monitoring. Manconi et al. [9] used surveying robot data to monitor the mine slope and simplify the complex deformation of the slope, which gained support from relevant departments due to good experimental results. Wang et al. [10] integrated Global Positioning System (GPS)/pseudo-satellite (PL) positioning technology to improve the position settlement accuracy and provide a high-precision monitoring model for high-precision slope monitoring in open-pit mines. Akbar et al. [11] employed the integration of GPS, the Geographic Information System (GIS), and Remote Sensing (RS) to map a hill slide disaster map. Zeybek et al. [12] used a long-range terrestrial laser scanner to measure the precision of Taşkent Landslide (Konya, Turkey). Liao et al. [13] applied high-resolution SAR data to monitor landslides in the Three Gorges Reservoir area in China and were able to identify the precise location, deformation, and time range of the landslide more accurately. Tang et al. [14] used new generation SAR satellites (Sentinel-1 and TerraSAR-X) to map surface displacements and slope instability at three open-pit mines in the Rhenish coalfield in Germany, in order to provide a long-term monitoring solution for open pit mining and its operations. Wang et al. [15] employed inclined photogrammetry to generate a mine Digital Surface Model (DSM) and carried out the construction of a Digital Elevation Model (DEM) for an open-pit mine. Alameda-Hernández et al. [16] used ultraclose range terrestrial digital photogrammetry to monitor the stability of soft foliated rocky slopes and analyzed their errors during rock weathering using the example of soft rocks in Alpujarras (Andalusia, Spain). Tong et al. [17] used Unmanned Aerial Vehicle (UAV) photogrammetry and ground-based laser scanning for open-pit mine inspection and three-dimensional (3D) mapping, aligning image data with point cloud data and classifying land cover. González-Díez et al. [18] elaborated on the methods which used digital photogrammetry to accurately measure slope changes caused by landslides.

In consideration of the use of each monitoring method and the research results of the above-mentioned references, the characteristics of each monitoring method and its corresponding scope of application are summarized in Table 1. Among the methods applied for open-pit mine monitoring, the close-up photogrammetry technique is a moderate method which can provide high efficiency and inexpensive measurement, especially compared to the other usual methods of laser scanning, the Interferometric Synthetic Aperture Radar (InSAR), the Laser Radar (LiDAR), etc. [19–21]. The use of photogrammetry has the requirement of capturing images with a degree of overlap, which places demands on the deployment of cameras.

The visual camera network, as a type of sensor network, is a spatially distributed network of smart cameras that collects and processes multimedia information to transform scene images into a more useful form [22]. Visual sensors can perceive more information than ordinary sensors, and visual sensor networks can handle higher-level visual tasks than single vision sensors [23]; for example, Kulkarni et al. [24] designed the multilayer camera network sensEye for object monitoring, identification, and tracking. In addition, coverage is an important aspect when evaluating the quality of detection of multiple regions of interest in visual sensor networks and is an important research direction for camera networks. Related studies on the coverage problem of visual sensor networks have been conducted and relevant algorithms have been designed to obtain the maximum coverage units with the most optimal camera deployment scheme [25,26]. Based on the research of visual sensor networks, it is a good choice to introduce the idea of visual sensors in the field of slope risk monitoring; make full use of the multimedia resources of the camera; and realize functions such as 3D slope monitoring, tramcar positioning, and video monitoring.

Table 1. Characteristics and application scope of the main monitoring methods employed for open-pit mine slopes.

Monitoring Methods	Efficiency	Monitoring Cycle	Expenses (Estimate)	Accuracy
Traditional geodesic methods	Low efficiency (nonautomatic, restricted by terrain access and climate)	Depending on the monitoring task (ranging from one day to six months)	High costs (USD 2000 equipment costs + high labor costs)	High accuracy (millimeter, submillimeter)
GPS technology	High efficiency (automatic)	Real time	High costs (usually USD > 500,000 initial investment)	High accuracy (millimeter)
3D laser scanning technology	Medium efficiency (semiautomatic)	Depending on the monitoring task (ranging from one day to six months)	High costs (USD > 150,000 equipment costs + labor costs)	high accuracy (millimeter, centimeter)
Measuring robot technology	Medium efficiency (automatic, restricted by terrain access conditions)	Near real time	High costs, (USD > 150,000 equipment costs)	High accuracy (millimeter)
RS technology	Medium efficiency (semiautomatic)	Based on the revisit cycle of RS satellites (>4 days)	Low costs (USD > 3000 a view)	Low accuracy (decimeter, meter)
InSAR	Medium efficiency (semiautomatic)	Based on the revisit cycle of RS satellites (>11 days)	Low costs (USD > 7000 a view)	High accuracy (millimeter)
UAV photogrammetry	Medium efficiency (semiautomatic)	Depending on the monitoring task (ranging from one week to six months)	Low costs (USD > 4000 equipment costs + labor costs)	Medium accuracy (centimeter)
Digital close-up photogrammetry	High efficiency (automatic)	Near real time	Low costs (USD 1500 a monitoring camera point)	Medium accuracy (centimeter, decimeter)

The intelligent application of multicamera video data in mining is an important aspect of smart mine constructions, and a number of studies have applied photogrammetry to the digitization of mines [27], reconstructing visual data in three dimensions and measuring parameters such as slope deformation monitoring and the slope gradient. Giacomini et al. [21] used close-up photogrammetry to continuously monitor the rock surface, in order to assess the potential rockfall risk and estimate the area of impact. Aggarwal et al. [28] implemented an Internet of Things (IOT) landslide monitoring system based on Raspberry Pi using a camera. It analyzed the area in real time based on the video stream obtained from the camera and applied computer vision algorithms to detect landslides. This method can only monitor the occurrence of large landslides and cannot provide an early warning, and the distance cannot be too far. A camera can also be mounted on a UAV to reconstruct the mine pit in 3D to obtain a DSM by extracting the comparative elevation, slope, slope direction, surface fluctuations, and surface roughness distribution and performing crack analysis [29–31]. Kromer et al. [32] used a digital Single Lens Reflex (SLR) camera to form a camera system. The use of photogrammetric workflows for mine slope monitoring achieved a high level of accuracy, but the shortcoming of this paper is that it does not describe the camera network deployment options for mine conditions and camera parameters, as well as the corresponding budgets. The intelligent use of video camera data at a mine site for slope collapse risk monitoring is a trend in smart mine constructions, and the rationalization of camera deployment is one of the most important aspects. At the present stage, corresponding optimized camera network deployment rarely occurs in open-pit mines.

To solve the above problem, we propose an optimum deployment algorithm of camera networks for open-pit mine slope monitoring. The remainder of the paper is structured as follows: in Section 2, the optimum deployment algorithm for open-pit mine landslide monitoring is presented; in Section 3, the experimental simulation for the algorithm, field test, and demonstration are described; and finally, in Section 4, the discussions and conclusions are provided.

2. Materials and Methods

This study introduces the idea of combining visual camera observation with digital photogrammetry, and designs the Optimum Camera Deployment algorithm for open-pit mine slope monitoring (OCD4M). For a given mine slope, an observation platform, and camera parameters, the algorithm determines the deployment scheme with the minimum quantity of cameras and the optimal camera positions, in order to meet the need for overlap in 3D monitoring by close-up photogrammetry and the need for full coverage of safety monitoring, as shown in a simple deployment schematic in Figure 1. In this section, we introduce the OCD4M algorithm in terms of monitoring object description, mathematical model, mine surface preprocessing, and deployment algorithm workflow.

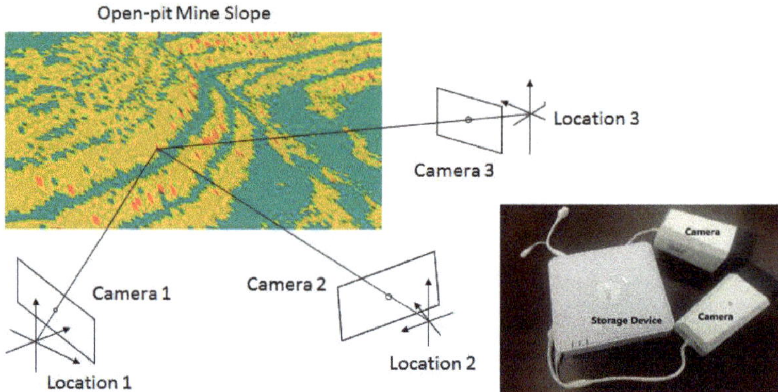

Figure 1. Sketch deployment diagram for open-pit mine slope monitoring. At the top left, the blue area represents the bench work surface, the yellow area represents the bench slope and the red area represents the risk exceeding the threshold.

2.1. Description of Targets and Criteria for Monitoring

Figure 2 shows the composition of the mine's slope. The object of monitoring is to calculate the bench width, bench height, bench slope angle and overall slope angle of the open-pit slope, in order to meet the safety requirements, and to provide early warning if the threshold values are exceeded. The calculation accuracy is required to be decimeter.

2.2. Model Description and Problem Definition

As shown in Figure 3, the camera sensor network contains a number (N) of sensors (S), which are deployed at the observation platform (B) that monitors the target surface (A).

$$S = \{s_1, s_2, \ldots, s_N\}, A : f(x, y, z) = 0, B : g(x, y, z) = 0 \tag{1}$$

From a geometric point of view, for each sensor, its sensing area is defined by tuple $C(i) : (s_i(x, y, z), D, \alpha, \theta)$, where $s_i(x, y, z)$ is the sensor's position, D is the photography distance, α is the sensing angle, and θ is the angle at which the camera deviates from the normal case photography direction. The aims are to ensure that the target surface is fully covered by the sensing area and that the image overlap rate between adjacent sensors is more than 80%. We can define the aims by the following three formulas:

$$A \subseteq C_{s(1)} \cup C_{s(2)} \cup \ldots \cup C_{s(N)}, \tag{2}$$

$$S_{C_{s(i)} \cap C_{s(i+1)}} \geq 0.8 S_{C_{s(i)}}, \tag{3}$$

$$s_i(x, y, z) \subseteq B, \tag{4}$$

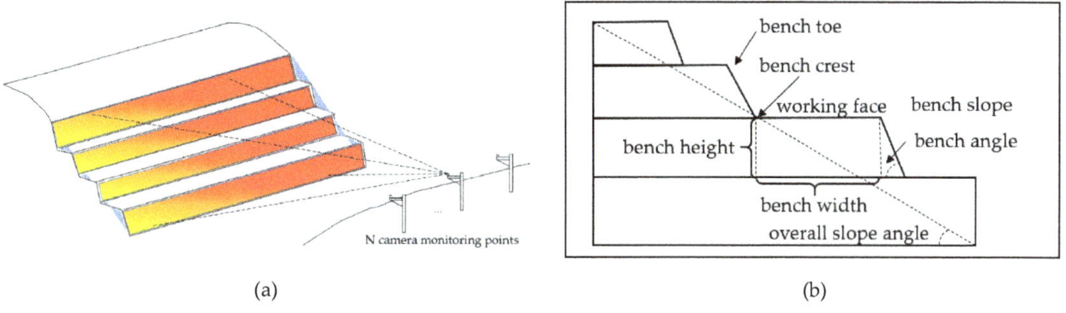

Figure 2. Schematic diagram of the monitoring object. (**a**) Schematic diagram of the monitoring of the open pit slope by cameras. (**b**) Section of the mine slope with the various parts of the open-pit mining.

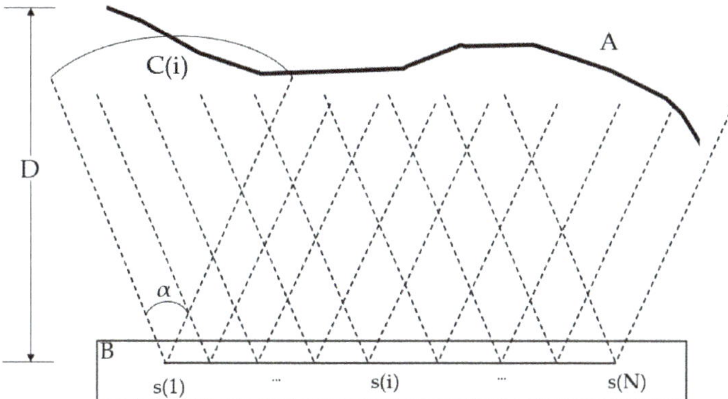

Figure 3. Geometric deployment of the camera sensor network, where curve A is the surface, B is the observation platform, s(i) is the camera sensor, the area corresponding to the two dashed lines is the shooting range C(i), α is the field of view, and D is the shooting distance.

2.3. Discussion of Different Situations

The mine coordinate system is established as shown in Figure 4, with the vertical direction as the Z-axis, the photographic direction as the Y-axis, and the plane perpendicular to the plane formed by Z and Y as the X-axis.

In order to sense the whole Z-direction (Figure 4) area, one or more cameras need to be deployed. As shown in Figure 5a, when the sensing area of a single camera can cover the Z-direction of the mine face, only one camera is sufficient at this point; if not, multiple K cameras as shown in Figure 5b need to be deployed at the same monitoring point to meet the need for sensing the whole coverage of the Z-direction, while the overlap of the monitoring areas of the multiple K deployed cameras also needs to be satisfied to meet the photogrammetry requirements.

Figure 4. Coordinate system established within the mine.

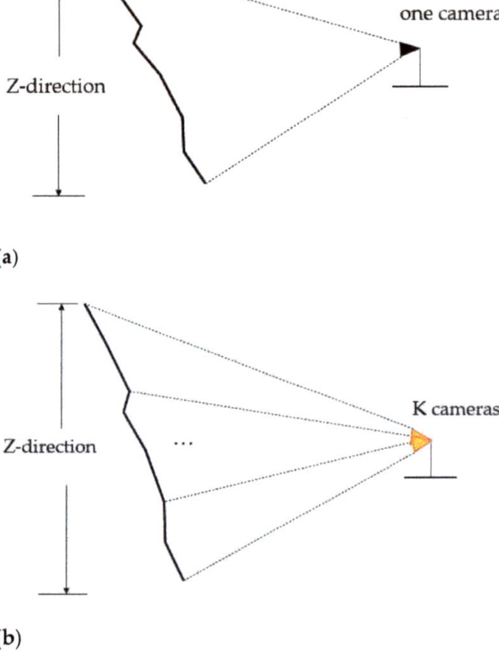

Figure 5. Observation coverage in the Z-direction. (**a**) The case where a single camera can cover the Z-direction of the slope of the open-pit mine, where a single camera is deployed on the right, with the dashed line showing the range of shots and the black curve showing the Z-direction of the mine slope. (**b**) The case where K cameras cover the Z-direction of the broken face of the mine, similar to (**a**), where multiple cameras are deployed on the right, with multiple dashed lines and ellipses indicating the range of the multiple cameras.

In the XY flat area, we simplify the mine surface. In consideration of the complexity of the stereoscopic surface, for the purpose of computational convenience, the mining surface is simplified twice. The slope toe of the mine is closer to the observation platform. In terms of the characteristics of the camera sensors, the closer the observation is, the smaller the observation range. Consequently, we can simplify the surface at Figure 6 to form a curve which is located at the bottom of the mine.

Figure 6. Mine surface simplification: in the XY flat, closest to the camera is the slope line where the red curve in the diagram is located. The closer the distance, the smaller the camera's sensing range. If this line can be fully covered, other heights of the slope must be covered, so we have chosen this line to represent the situation of the mine surface in the XY plane.

In order to reduce the complexity of the calculation, we use a straight line instead of the curve obtained by simplifying Figure 7. The specific approach was carried out using least squares to find the best-fitting straight line. This is modeled by selecting several points on the curve at certain intervals, which are used to solve the linear equation:

$$y = ax + b, \tag{5}$$

for the parameters a and b, where a and b are calculated by the following equations:

$$a = \frac{\text{Num} \sum x_i y_i - \sum x_i \sum y_i}{\text{Num} \sum x_i{}^2 - \left(\sum x_i\right)^2}, \; b = \frac{\sum y_i \sum x_i{}^2 - \sum x_i \sum x_i y_i}{\text{Num} \sum x_i{}^2 - \left(\sum x_i\right)^2}, \; i = 1, 2, \ldots, \text{Num}, \tag{6}$$

where Num is the number of data points (red points in Figure 7).

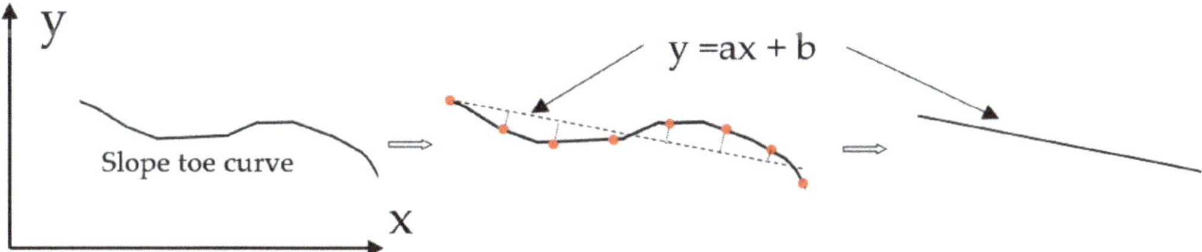

Figure 7. Slope toe simplification: the diagram shows how the broken line is finally reduced to a straight line by line fitting.

According to the observation platform and mine surface, it can be divided into two situations, as Figure 8 shows. In the X-direction, when the length of the observation is longer than the mine surface, we adopt normal case photography; otherwise, we adopt convergent photography, which has a larger coverage area, in order to meet the requirement.

2.4. The Optimum Deployment Algorithm

The OCD4M algorithm is used to solve the camera sensor deployment problem. The algorithm starts with the input of camera sensor parameters, mine face parameters, observation platform parameters, and photographic distances. The next step is to simplify the corresponding mine face in conjunction with the mining extent. Next, we calculate the minimum quantity of cameras and the coordinates through the algorithm. Therefore, the workflow of the algorithm is as shown in Figure 9.

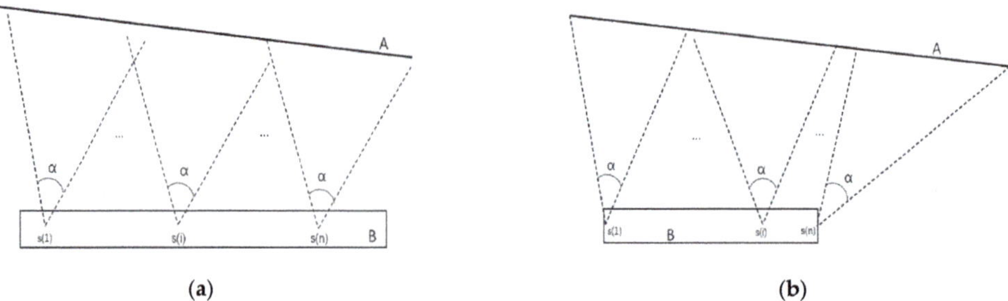

Figure 8. Camera sensor deployment in the X-direction: (**a**) the use of normal case photography and (**b**) the use of convergent photography to cover the entire mine surface. A represents the observed mine surface, the dotted line represents the photographic area, α represents the field of view, s(i) represents the camera monitoring point, and B represents the observation platform.

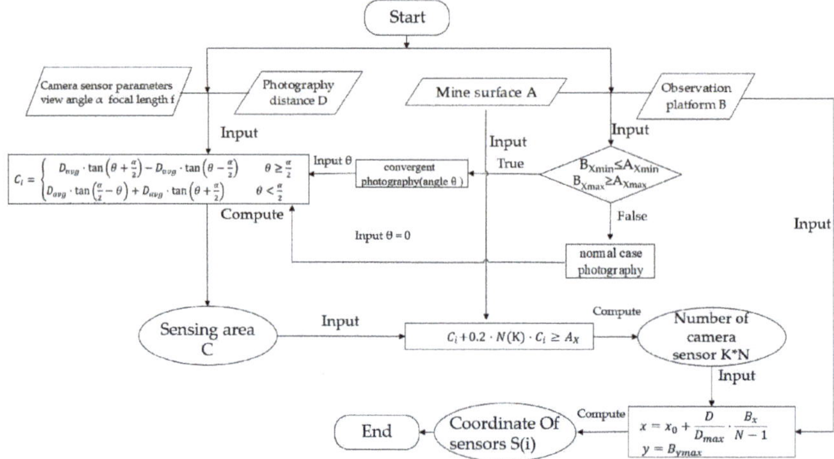

Figure 9. Workflow of the Optimum Camera Deployment algorithm for open-pit mine slope monitoring (OCD4M) algorithm.

The input parameters for the algorithm are the parameters of the camera sensors (mainly the camera's field of view α and focal length f), the shooting distance D, the mine plane A to be observed, and the observation platform B. The sensing range C of the camera sensor can be calculated from the camera parameters and the shooting distance using the following formula:

$$C_i = 2 \cdot D_{avg} \cdot \tan\frac{\alpha}{2}. \tag{7}$$

In the above equation, C_i is the camera area, including C_{ix} and C_{iz}; D_{avg} is the camera distance; and α is the field of view of the camera.

The next step calculates the number of cameras K per surveillance point to achieve full coverage of the open-pit slope in the Z-direction, calculated according to the following formula:

$$C_{iz} + 0.2 \cdot K \cdot C_{iz} \geq A_z \tag{8}$$

where C_{iz} is the length of the photographic area C_i in the Z-direction, K is the quantity of cameras, and A_z is the extent of the mine slope in the Z-direction.

When the X-direction of the open-pit mine slope A is greater than the X-direction range of the observation platform B, normal case photography is used, and the minimum quantity of monitoring points N is calculated according to the following equation:

$$C_{ix} + 0.2 \cdot N \cdot C_{ix} \geq A_X. \tag{9}$$

In the above equation, C_{ix} is the camera area in the X-direction, N is the minimum quantity of monitoring points, and A_x is the extent of the mine slope in the X-direction.

Otherwise, convergent photography is used, and the minimum quantity of cameras N can be obtained from

$$C_i = \begin{cases} D_{avg} \cdot \tan\left(\theta + \frac{\alpha}{2}\right) - D_{avg} \cdot \tan\left(\theta - \frac{\alpha}{2}\right) & \theta \geq \frac{\alpha}{2} \\ D_{avg} \cdot \tan\left(\frac{\alpha}{2} - \theta\right) + D_{avg} \cdot \tan\left(\theta + \frac{\alpha}{2}\right) & \theta < \frac{\alpha}{2} \end{cases}. \tag{10}$$

In the above equation, C_i is the camera area in the X-direction, D_{avg} is the camera distance, α is the field of view of the camera, θ is the angle of deviation with respect to the orthogonal direction, and A_x is the extent of the mine slope in the X-direction. The range of cameras $C_{min} \sim C_{max}$ is calculated to solve $N_{min} \sim N_{max}$ according to Formula (9). Based on the results of the quantity of cameras, the coordinates of each camera on the mining plane can be calculated through the following formula:

$$x = x_0 + \frac{D}{D_{max}} \cdot \frac{B_x}{N-1}, \ y = B_{ymax}, \tag{11}$$

where y is the maximum value employed to achieve greater coverage, and the quantity of cameras is K*N.

The whole workflow of the Algorithm 1 can also be described by the following pseudocodes.

Algorithm 1. OCD4M

Require:

Camera sensor set $S = \{s_1, s_2, \ldots, s_N\}$, target surface $A : f(x, y, z) = 0$, Observation platform $B : g(x, y, z) = 0$, Photography distance D, sensing angle α

Ensure:

A is covered by the set of $C(i) : (xyz(s(i)), D, \alpha, \vec{d_i})$, and the degree of overlap of adjacent $C(i)$ greater than 80%. (1) $A \subseteq C_{s(1)} \cup C_{s(2)} \cup \ldots \cup C_{s(N)}$; (2) $S_{C_{s(i)} \cap C_{s(i+1)}} \geq 0.8 S_{C_{s(i)}}$; (3) $xyz(s_i) \subseteq B$.

Process:

1: Compute the range of A. $A_Z = A_{zmax} - A_{zmin}$; $A_X = A_{xmin} - A_{xmax}$,
2: Compute the range of B. $B_Z = B_{zmax} - B_{zmin}$; $B_X = B_{xmin} - B_{xmax}$,
3: Compute the rang of Ci,
4: Whether to cover the Z direction of Mine Surface A
5: while $C_z * K < A_Z$ do
6:       get the number of camera sensors at each point
7:       K++
8: End while
9: Judge the Length relationship of A and B
10: If ($B_{Xmin} \leq A_{Xmin}$ and $B_{Xmax} \geq A_{Xmax}$) do
11:       normal case photography, compute the minimum of camera sensors by Formula (9):
12:       Compute the coordinate of each camera sensor according to the photography
13:       distance by Equation (11),
14:       $y = B_{ymax}, z$
15: Else
16:       convergent photography (angle θ);
17:       compute the range of camera sensor number by
18:       Equations (9)–(11),
19:       we can get K, N and $(s_i(x, y, z))$
20: End If
21: Return number $(K * N)$ and position of sensor $(s_i(x, y, z))$

---

3. Implementation and Results

3.1. Simulation of Experimental Tests

In this section, the results of the simulation experiments are given. The experimental tests focus on simulating the quantity of cameras and camera accuracy at different distances and the quantity of cameras and camera accuracy at different field of view angles. For small to medium-sized open-pit mines and some large mines, 500 m is relatively adequate. A greater distance means a lower accuracy, so beyond 500 m, it is necessary to improve the quality of the camera or add other means to control the accuracy, which means an increase in cost. This is unacceptable for small to medium-sized mines with low revenues, so we chose 500 m as the camera distance for our simulations. The main test condition parameters are as follows:

- Camera parameters
  Sensor size: 1/3″ inches (4.8 mm × 3.6 mm)
  Focal length: 8 mm
- Ore surface parameters
  $A_Z = 100$ m
  $A_X = 500$ m
- Observation platform parameter
  $B_X = 500$ m

3.1.1. Quantity and Precision Analysis

The camera focal length was 8 mm, the horizontal field of view angle was 32.69°, and the vertical field of view angle was 24.81°. Assuming that the monitoring distance was between 50 and 500 m, the minimum quantity of cameras and the resolutions were calculated as shown in Figure 10.

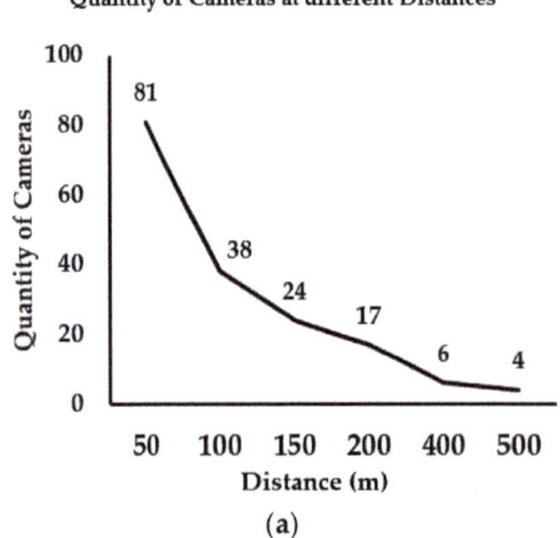

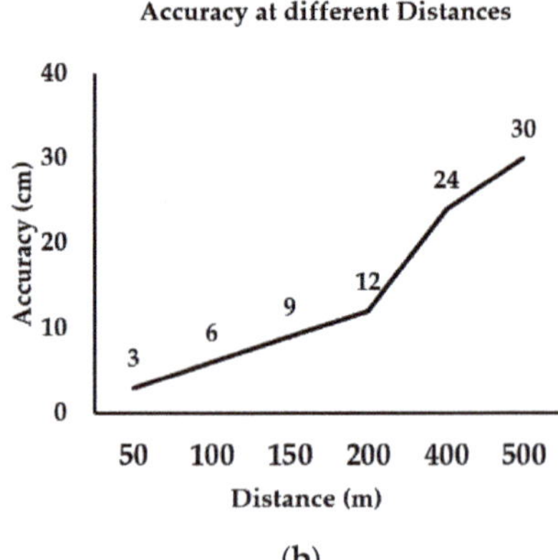

Figure 10. Quantity and accuracy of cameras at different distances. (**a**) The minimum quantity of cameras calculated for different distances. The further the distance, the fewer the cameras needed. (**b**) The distance each pixel represents in the field increases, which means that the accuracy decreases.

As shown in Figure 10a, according to Equation (7), the coverage becomes larger as the distance becomes larger whilst the camera remains the same, which means that the entire mine surface can be covered using fewer cameras. It can be seen from Figure 10b that the resolution of the object decreases as the distance increases, according to Formula (12):

$$\frac{f}{D} = \frac{pixel}{S},$$

$$(12)$$

where f is the focal length, D is the photographic distance, pixel is the size of each image element of the image, and S is the field distance represented by an image element.

As the distance increases, the actual distance of an object represented by a pixel becomes larger, which means that the accuracy decreases. The accuracy can reach 30 cm at 500 m, which meets the requirements for the calculation of slope parameters in open-pit mines and can be used for early warnings on slopes.

3.1.2. Focal length, Field of View Angle and Quantity Analysis

Assuming that the distance was 200 m, we analyzed the quantity of cameras and object resolution results for different focal lengths (2.8–25 mm) and field of view angle conditions.

The calculation result is shown in Table 2. It can be seen that as the focal length increases, the field of view angle decreases accordingly and the quantity of cameras required gradually increases. This is because, as the field of view angle decreases, the sensing range of the camera head decreases, and more camera sensors are needed to meet the coverage requirements. In addition, it can be seen that the accuracy increases as the focal length increases, because, as the focal length increases, the range in which each pixel represents an object becomes smaller at the same distance, allowing the accuracy to increase. With a camera focal length of 2.8 mm, the accuracy can be achieved at a distance of 34 cm at 200 m.

Table 2. Quantity of cameras and accuracy calculated by different focal lengths and view angles.

Focal Length (mm)	Viewing Angle (°)	Photographic Range (m)	Quantity of Cameras	Accuracy (cm)
2.8	79.93	335.22	3	34
4	60.79	234.63	6	24
6	42.72	156.44	11	16
8	32.69	117.31	17	12
12	22.12	78.19	27	8
16	16.68	58.64	38	6
25	10.72	37.50	62	4

Commonly, for open-pit mine slope damage alerts, the resolution of the monitoring should not be less than 50 cm. Most open-pit mines have high-definition cameras which can be fit for this requirement, but only be used for manual monitoring. In addition, the current deployment of their cameras is not suitable for slope monitoring. They need to deploy more cameras facing the mine slope if they want to realize slope damage risk monitoring. Through our OCD4M algorithm, we can make random deployments optimal. The algorithm can calculate the minimum quantity of cameras needed to achieve the large overlap and full coverage required to make the most of the mine's video and multimedia resources.

3.2. Field Testing and Demonstration

The testing field is the Shunxing Quarry in Guangzhou, Guangdong, China, which is located at (113.518859 E, 23.405858 N), as shown in Figure 11. This is a medium and typical open-pit mine. The observed slope length of the mine is 493.22 m. The length of the mining platform is 225.72 m, and the slope height difference is 72.86 m. The distance (that is, the photography distance) is 398.97 m. The proposed camera model is HIKVISION DS-IPC-B12H-I with an 8 mm focal length, 1/2.7″ sensor size, 32.69° field of view angle, and 24.81° vertical angle. Mine parameters, camera parameters, etc. were input into the OCD4M algorithm to calculate the minimum quantity of cameras and the coordinates of the monitoring points. A network of cameras was built in the field according to the coordinates and the 3D monitoring of the mine was automated without human participation by the data processing system we have developed. Given that we were working at the decimeter level of accuracy, we measured the slope of the open-pit, the height and width of the mining benches to be measured, and the volume of mining to be counted.

Figure 11. Study field located in Shunxing Quarry in Guangzhou, Guangdong, China.

In the field, GPS sampling was used to locate camera points and set up the cameras. Multiple photos extracted from camera videos were prepared as the inputs. We used Smart3D software for the data preprocessing, aerial triangulation, dense matching of oblique images, DSM point cloud generation, triangulated irregular network (TIN) construction, texture mapping, model modification, and other processes, in order to produce a realistic 3D model, as Figures 12 and 13 show.

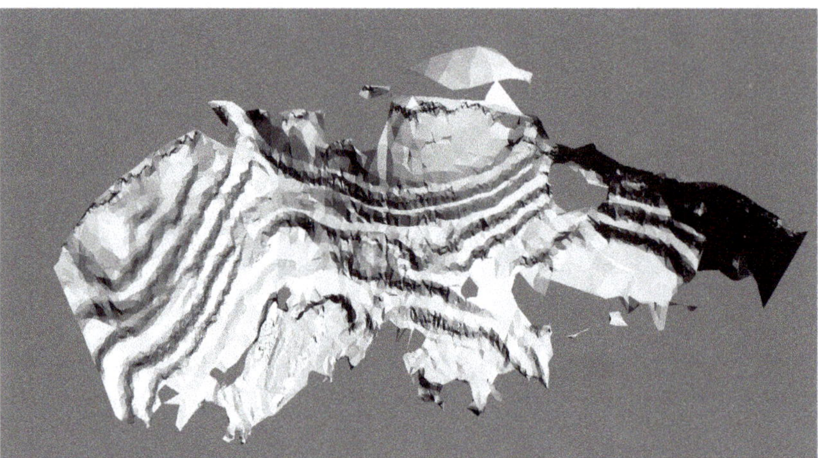

Figure 12. 3D model of the study field.

Figure 13. Schematic diagram of the camera field deployment. ① Schematic view of the slope of the mine. ② Pictures of the field installation, including pole tower transportation, drilling and camera installation, etc. ③ Schematic view of the Shunxing Quarry monitoring.

Figure 14 illustrates the open-pit mine slope monitoring system which we developed for the visualization of the results of the algorithm and the early warnings during the safety monitoring. This can be used to generate the field camera deployment solution, analyze the slope of the mine, and measure the width and height of the working bench to assess and construct warnings about mine slope damage risks. It also integrates 3D reconstruction, 3D monitoring, and 3D visualization of the mine slope.

- The modular ① shows the number of camera points and their coordinates calculated by the algorithm and also indicates the camera status.
- The modular ② shows the result of the 3D reconstruction based on the camera photos and visualizes the mine plane and slope monitoring.
- The modular ③ shows statistical results of the monitoring of the mine plane risks which are related to slope damage indicators.
- The modular ④ shows the parameters of the camera and the monitoring distance.

As calculated by the OCD4M algorithm, the camera sensing range is between 234.62 and 290.58 m, the minimum quantity of cameras is six, and the coordinate results are shown in Table 3. In the given conditions, these six monitoring point coordinates are the best locations for deploying the cameras to monitor their opposite slopes.

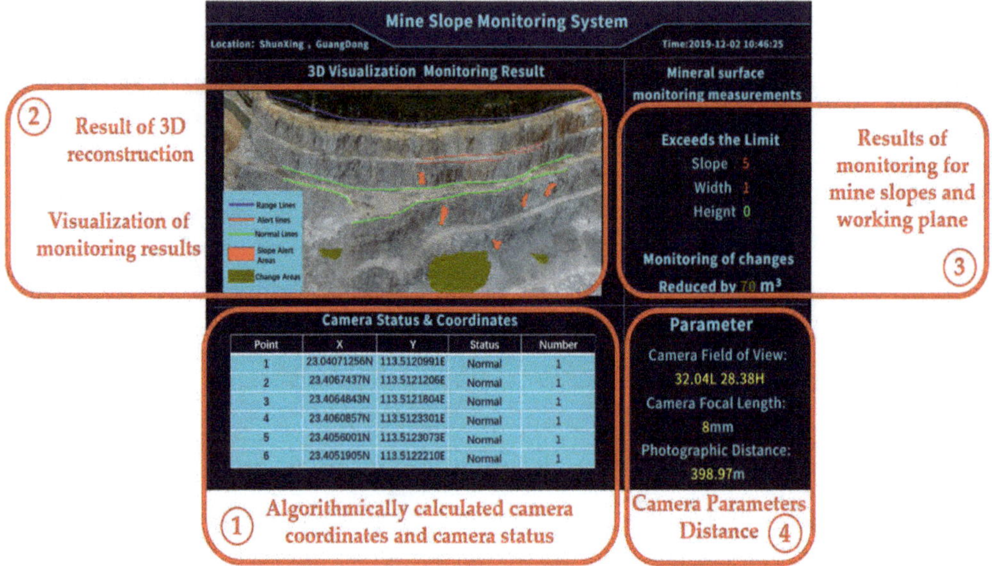

Figure 14. Software interfaces of the demonstration system.

Table 3. Camera coordinates for the 6 different monitoring points shown in Figure 13.

Point	X	Y
1	23.0407125 N	113.5120991 E
2	23.4067437 N	113.5121206 E
3	23.4064843 N	113.5121804 E
4	23.4060857 N	113.5123301 E
5	23.4056001 N	113.5123073 E
6	23.4051905 N	113.5122210 E

Zhang et al. [33] explored the relationship between overlap and accuracy. They assumed that each point is covered by photos five times, as shown in Figure 15. When the overlap is more than 80%, the accuracy is improved, but the speed of accuracy improvement is slowed down.

The red dot(the point on the mine face) is covered five times.

Figure 15. Relative overlap (80%), where each point is covered by five photos at least.

Considering the engineering practice, using more cameras is not conducive to cost control, and 80% overlap is a relatively reasonable choice in terms of the accuracy and engineering practice. Figure 16 shows a comparison of different results when lacking one of the conditions.

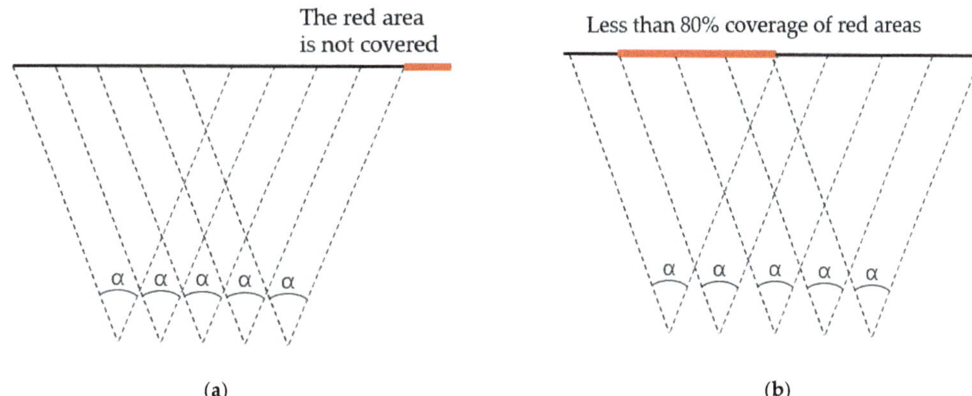

Figure 16. Comparison of different results when lacking one of the conditions: (**a**) 80% overlap at five cameras, no full coverage, and (**b**) full coverage at five cameras, overlap < 80%.

For verifying the result of the algorithm, we conducted a comparison test by reducing and increasing the quantity of cameras. The comparison of overlap and coverage length in conditions of different quantities is shown in Table 4. When the quantity of cameras is less than six, the camera monitoring cannot cover the whole mine surface with regard to overlap, and the overlap between adjacent photos will be less than 80%.

Table 4. Overlap and coverage length for different quantity of cameras.

Quantity of Cameras	Overlap at Full Coverage (%)	Coverage at 80% Overlap(m)
4	72.44	422.31
5	77.96	469.24
6	81.63	516.16
7	84.25	563.08
8	86.22	610.01

As shown in Table 4, meeting both requirements of the 493.2m photography distance and the 80% overlap, six cameras are needed to be deployed at least. Considering the actual engineering costs, using more cameras means higher cost. This proves that the deployment result (in Table 3) calculated by our OCD4M algorithm is relatively reasonable for this study case.

4. Discussion and Conclusions

In summary, the OCD4M algorithm is proposed for the deployment of camera sensor networks for slope monitoring to achieve the minimum quantity of cameras and obtain the deployment location coordinates, in order to optimize the deployment, enabling 3D monitoring capabilities and making full use of the multimedia data obtained from the cameras in the open-pit mine. We have conducted experimental validation with the simulated conditions of quantity, view angle, and focal length of cameras, at different monitoring distances. The OCD4M algorithm was tested in the medium-sized mine field, using Hikvision DS-IPC-B12H-I model 8 mm focal length cameras for mine surfaces photography and reconstructed in Smart3D software. The field test result shows that the accuracy of 30 cm can be achieved at the monitoring distance of 500 m. We also developed the visualization system software, through which the camera deployment scheme for the mine scenario can be generated automatically. According to the result of the algorithm, 3D monitoring of the working platform (e.g., calculating slope angle, height and width of the mine bench) can be realized at the decimeter level.

There are some considerations need to be emphasized in terms of deployment process, application scenario elaboration, engineering costs and limitations of the method. Since the physical deployment of control points is an unstable and costly solution for actual mining work, a high precision calibration of the camera is an important task [34]. Our method guarantees decimeter-level accuracy at 500 m monitoring distance without control points; thus, our method allows for decimeter resolution level safety monitoring in the open-pit mines, such as bench damage and interslope risk monitoring. However, overall landslide monitoring always needs a millimeter scale resolution, which no current camera model on the market can provide, especially when the monitoring slope is further than 500 m away. Due to topographical and other shading issues at the mine site, the use of front-to-surface photography may result in missing images in some areas that cannot be reconstructed in 3D, so the observation platform should be selected with due consideration of whether the area to be observed can be captured by all of the corresponding cameras. Considering the actual cost of the project, the low-cost solution of camera photogrammetry is easy to accept for small and medium-sized mines [7,21]. Additionally, our solution allows for automated monitoring after deployment, which also reduces the investment in manpower costs for the mine. It is more efficient than 3D laser scanning and traditional manual-based measurements. In the case of large mines, where the mines are large, distant, or complex, more cameras are required to ensure coverage and higher quality cameras to ensure accuracy, which can lead to increased costs, which are acceptable for the revenue of large mines.

In addition, the algorithm can be used for the calculation of other slope deployment scenarios (e.g., modeling of cultural heritage objects [35], 3D robot localization [36], monitoring coastal morphology [37], etc.). The algorithm can be improved by considering more photogrammetric geometry factors (e.g., the angle of intersection, the length of the photographic baseline, etc.) to optimize deployment scenarios for obtaining higher measurement accuracy, and by considering the mine topography and the actual deployable location of the mine to perform more complex deployment scenario calculations. The next step of the study will focus on the identification and warn of landslide areas using smart video image recognition based on the deployed system, so that the deployment algorithm can serve both monitoring of slope collapse risk and identifying landslide areas.

Author Contributions: Conceptualization, X.M.; methodology, X.M. and H.Z.; software, H.Z. and M.L.; validation, P.T., H.Z. and M.L.; formal analysis, P.T. and H.Z.; data work, H.Z. and X.L.; writing—original draft preparation, X.M. and H.Z.; writing—review and editing, H.Z, P.T., M.L., X.M. and X.L. All authors have read and agreed to the published version of the manuscript.

Funding: This research was funded by National Natural Science Foundation of China (NSFC) under grant number 41971352 and 42071246, National Key Research and Development Program of China under grant number 2018YFB0505003, Ecological SmartMine Joint Fund of Hebei Natural Science Foundation under grant number E2020402086 and Hebei Natural Science Foundation under grant number E2020402006.

Institutional Review Board Statement: Not applicable.

Informed Consent Statement: Not applicable.

Data Availability Statement: Data sharing not applicable.

Acknowledgments: The authors are very grateful to the many people who helped to comment on the article, and the Large Scale Environment Remote Sensing Platform (Facility No. 16000009, 16000011, 16000012) provided by Wuhan University. Thanks to Shunxing Quarry (Guangdong, China) for providing us with the test field. Special thanks to the editors and reviewers for providing valuable insight into this article.

Conflicts of Interest: The authors declare no conflict of interest.

References

1. Petley, D. Global patterns of loss of life from landslides. *Geology* **2012**, *40*, 927–930. [CrossRef]
2. Dos Santos, T.B.; Lana, M.S.; Pereira, T.M.; Canbulat, I. Quantitative hazard assessment system (Has-Q) for open pit mine slopes. *Int. J. Min. Sci. Technol.* **2019**, *29*, 419–427. [CrossRef]
3. Wei, L.; Feng, Q.; Liu, F.; Mao, Y.; Liu, S.; Yang, T.; Tolomei, C.; Bignami, C.; Wu, L. Precise Topographic Model Assisted Slope Displacement Retrieval from Small Baseline Subsets Results: Case Study over a High and Steep Mining Slope. *Sensors* **2020**, *20*, 6674. [CrossRef] [PubMed]
4. Kasap, Y.; Subaşı, E. Risk assessment of occupational groups working in open pit mining: Analytic Hierarchy Process. *J. Sustain. Min.* **2017**, *16*, 38–46. [CrossRef]
5. CDC–Mining–Mining Facts–2014–NIOSH. Available online: https://www.cdc.gov/niosh/mining/works/statistics/factsheets/miningfacts2014.html (accessed on 4 December 2020).
6. Ehschool. 2017 National Statistical Analysis of Non-Coal Open-Pit Safety Accidents Report. Available online: http://cool.ehsway.cn/news/newsdetail.aspx?pid=2&nid=7379 (accessed on 4 December 2020).
7. Labant, S.; Bindzarova Gergelova, M.; Kuzevicova, Z.; Kuzevic, S.; Fedorko, G.; Molnar, V. Utilization of geodetic methods results in small open-pit mine conditions: A case study from Slovakia. *Minerals* **2020**, *10*, 489. [CrossRef]
8. Can, E.; Erbiyik, H. The Role of Mine Surveying Studies in Preventing Probable Risks that might be emerged in Open Pit Mining Projects. *Int. J. Adv. Sci. Res. Eng.* **2018**, *4*, 207–213. [CrossRef]
9. Manconi, A.; Allasia, P.; Giordan, D.; Baldo, M.; Lollino, G.; Corazza, A.; Albanese, V. Landslide 3D Surface Deformation Model Obtained Via RTS Measurements. In *Landslide Science and Practice: Volume 2: Early Warning, Instrumentation and Monitoring*; Margottini, C., Canuti, P., Sassa, K., Eds.; Springer: Berlin/Heidelberg, Germany, 2013; pp. 431–436. [CrossRef]
10. Wang, J.; Gao, J.; Liu, C.; Wang, J. High precision slope deformation monitoring model based on the GPS/Pseudolites technology in open-pit mine. *Min. Sci. Technol.* **2010**, *20*, 126–132. [CrossRef]
11. Akbar, T.A.; Ha, S.R. Landslide hazard zoning along Himalayan Kaghan Valley of Pakistan-by integration of GPS, GIS, and remote sensing technology. *Landslides* **2011**, *8*, 527–540. [CrossRef]
12. Zeybek, M.; Şanlıoğlu, I. Accurate determination of the Taşkent (Konya, Turkey) landslide using a long-range terrestrial laser scanner. *Bull. Int. Assoc. Eng. Geol.* **2014**, *74*, 61–76. [CrossRef]
13. Liao, M.; Tang, J.; Wang, T.; Balz, T.; Zhang, L. Landslide monitoring with high-resolution SAR data in the Three Gorges region. *Sci. China Earth Sci.* **2011**, *55*, 590–601. [CrossRef]
14. Tang, W.; Motagh, M.; Zhan, W. Monitoring active open-pit mine stability in the Rhenish coalfields of Germany using a coherence-based SBAS method. *Int. J. Appl. Earth Obs. Geoinf.* **2020**, *93*, 102217. [CrossRef]
15. Wang, J.; Wang, L.; Jia, M.; He, Z.; Bi, L. Construction and optimization method of the open-pit mine DEM based on the oblique photogrammetry generated DSM. *Measurement* **2020**, *152*, 107322. [CrossRef]
16. Alameda-Hernández, P.; El Hamdouni, R.; Irigaray, C.; Chacón, J. Weak foliated rock slope stability analysis with ultra-close-range terrestrial digital photogrammetry. *Bull. Int. Assoc. Eng. Geol.* **2017**, *78*, 1157–1171. [CrossRef]
17. Tong, X.; Liu, X.; Chen, P.; Liu, S.; Luan, K.; Li, L.; Liu, S.; Liu, X.; Xie, H.; Jin, Y.; et al. Integration of UAV-Based Photogrammetry and Terrestrial Laser Scanning for the Three-Dimensional Mapping and Monitoring of Open-Pit Mine Areas. *Remote Sens.* **2015**, *7*, 6635–6662. [CrossRef]
18. González-Díez, A.; Fernández-Maroto, G.; Doughty, M.W.; De Terán, J.R.D.; Bruschi, V.M.; Cardenal, J.; Pérez-García, J.L.; Mata, E.; Delgado, J.V. Development of a methodological approach for the accurate measurement of slope changes due to landslides, using digital photogrammetry. *Landslides* **2013**, *11*, 615–628. [CrossRef]
19. Xiang, J.; Chen, J.; Sofia, G.; Tian, Y.; Tarolli, P. Open-pit mine geomorphic changes analysis using multi-temporal UAV survey. *Environ. Earth Sci.* **2018**, *77*, 220. [CrossRef]
20. Slaker, B.A.; Mohamed, K.M. A practical application of photogrammetry to performing rib characterization measurements in an underground coal mine using a DSLR camera. *Int. J. Min. Sci. Technol.* **2017**, *27*, 83–90. [CrossRef]
21. Giacomini, A.; Thoeni, K.; Santise, M.; Diotri, F.; Booth, S.; Fityus, S.; Roncella, R. Temporal-Spatial Frequency Rockfall Data from Open-Pit Highwalls Using a Low-Cost Monitoring System. *Remote Sens.* **2020**, *12*, 2459. [CrossRef]
22. Soro, S.; Heinzelman, W. A Survey of Visual Sensor Networks. *Adv. Multimed.* **2009**, *2009*, 640386. [CrossRef]
23. Akdere, M.; Çetintemel, U.; Crispell, D.; Jannotti, J.; Mao, J.; Taubin, G. Data-Centric Visual Sensor Networks for 3D Sensing. In *Revised Selected and Invited Papers, GeoSensor Networks: Second International Conference, GSN 2006, Boston, MA, USA, 1–3 October 2006*; Nittel, S., Labrinidis, A., Stefanidis, A., Eds.; Springer: Berlin/Heidelberg, Germany, 2008; pp. 131–150. [CrossRef]
24. Kulkarni, P.; Ganesan, D.; Shenoy, P.; Lu, Q. SensEye: A multi-tier camera sensor network. In Proceedings of the 13th annual ACM international conference on Multimedia, Singapore, 6–11 November 2005; pp. 229–238. [CrossRef]
25. He, S.; Shin, D.-H.; Zhang, J.; Chen, J.; Sun, Y. Full-View Area Coverage in Camera Sensor Networks: Dimension Reduction and Near-Optimal Solutions. *IEEE Trans. Veh. Technol.* **2015**, *65*, 7448–7461. [CrossRef]
26. Yaagoubi, R.; El Yarmani, M.; Kamel, A.; Khemiri, W. HybVOR: A Voronoi-Based 3D GIS Approach for Camera Surveillance Network Placement. *ISPRS Int. J. Geo-Inf.* **2015**, *4*, 754–782. [CrossRef]
27. Duarte, J.; Rodrigues, F.; Baptista, J.S. Data Digitalisation in the Open-Pit Mining Industry: A Scoping Review. *Arch. Comput. Methods Eng.* **2020**. [CrossRef]

28. Aggarwal, S.; Mishra, P.K.; Sumakar, K.V.S.; Chaturvedi, P. Landslide Monitoring System Implementing IOT Using Video Camera. In Proceedings of the 2018 3rd International Conference for Convergence in Technology (I2CT), Pune, India, 6–8 April 2018; pp. 1–4. [CrossRef]

29. Yin, Y.; Jiang, C.; Lv, J.; Wang, J.; Ju, X.; Wang, H.; Xing, Y.; Zhang, L. Mining Ground Surface Information Extraction and Topographic Analysis Using UAV Video Data. In Proceedings of the E3S Web of Conferences, Shanghai, China, 18–20 September 2020; p. 05030. [CrossRef]

30. McLeod, T.; Samson, C.; Labrie, M.; Shehata, K.; Mah, J.; Lai, P.; Wang, L.; Elder, J. Using Video Acquired from an Unmanned Aerial Vehicle (UAV) to Measure Fracture Orientation in an Open-Pit Mine. *Geomatica* **2013**, *67*, 173–180. [CrossRef]

31. Cho, S.-J.; Bang, E.-S.; Kang, I.-M. Construction of Precise Digital Terrain Model for Nonmetal Open-pit Mine by Using Unmanned Aerial Photograph. *Econ. Environ. Geol.* **2015**, *48*, 205–212. [CrossRef]

32. Kromer, R.A.; Walton, G.; Gray, B.; Lato, M.J. Robert Group Development and Optimization of an Automated Fixed-Location Time Lapse Photogrammetric Rock Slope Monitoring System. *Remote Sens.* **2019**, *11*, 1890. [CrossRef]

33. Zhang, Y.; Zhang, Y. Analysis of precision of relative orientation and forward intersection with high-overlap images. *Geomat. Inf. Sci. Wuhan Univ.* **2005**, *30*, 126–130.

34. Luhmann, T.; Fraser, C.S.; Maas, H.-G. Sensor modelling and camera calibration for close-range photogrammetry. *ISPRS J. Photogramm. Remote Sens.* **2016**, *115*, 37–46. [CrossRef]

35. Alsadik, B.; Gerke, M.; Vosselman, G.; Daham, A.; Jasim, L. Minimal Camera Networks for 3D Image Based Modeling of Cultural Heritage Objects. *Sensors* **2014**, *14*, 5785–5804. [CrossRef] [PubMed]

36. Feng, S.; Shen, S.; Huang, L.; Champion, A.C.; Yu, S.; Wu, C.; Zhang, Y. Three-dimensional robot localization using cameras in wireless multimedia sensor networks. *J. Netw. Comput. Appl.* **2019**, *146*, 102425. [CrossRef]

37. Godfrey, S.; Cooper, J.; Bezombes, F.; Plater, A. Monitoring coastal morphology: The potential of low-cost fixed array action cameras for 3D reconstruction. *Earth Surf. Process. Landf.* **2020**, *45*, 2478–2494. [CrossRef]

Article

Sky Monitoring System for Flying Object Detection Using 4K Resolution Camera

Takehiro Kashiyama [1,*]**, Hideaki Sobue** [2] **and Yoshihide Sekimoto** [1]

[1] Institute of Industrial Science, The University of Tokyo, 4-6-1 Komaba, Meguro-ku, Tokyo 153-8505, Japan; sekimoto@iis.u-tokyo.ac.jp

[2] School of Engineering, The University of Tokyo, 4-6-1 Komaba, Meguro-ku, Tokyo 153-8505, Japan; hideaki.s.0212@gmail.com

* Correspondence: ksym@iis.u-tokyo.ac.jp

Received: 9 November 2020; Accepted: 8 December 2020; Published: 10 December 2020

Abstract: The use of drones and other unmanned aerial vehicles has expanded rapidly in recent years. These devices are expected to enter practical use in various fields, such as taking measurements through aerial photography and transporting small and lightweight objects. Simultaneously, concerns over these devices being misused for terrorism or other criminal activities have increased. In response, several sensor systems have been developed to monitor drone flights. In particular, with the recent progress of deep neural network technology, the monitoring of systems using image processing has been proposed. This study developed a monitoring system for flying objects using a 4K camera and a state-of-the-art convolutional neural network model to achieve real-time processing. We installed a monitoring system in a high-rise building in an urban area during this study and evaluated the precision with which it could detect flying objects at different distances under different weather conditions. The results obtained provide important information for determining the accuracy of monitoring systems with image processing in practice.

Keywords: flying object detection; drone; convolutional neural network; image processing

---

1. Introduction

In recent years, the use of drones and other unmanned aerial vehicles (UAVs) has expanded rapidly. These devices are currently used in various fields worldwide. In Japan, where natural disasters are a frequent occurrence, these devices are expected to be used for applications such as scanning disaster sites and searching for evacuees. In societies with shrinking populations, especially in areas with declining populations suffering from continuous labor shortages, there is hope that these devices can serve as new means for delivering medical supplies to emergency patients and food supplies to those without easy access to grocery stores.

However, there are growing concerns over these devices being misused for terrorism or other criminal activities. An incident occurred in April 2015 in Japan, when a small drone carrying radioactive material landed on the roof of the Prime Minister's official residence. Moreover, there have been incidents where drones have crashed during large events or at tourist spots. In response to such incidents, legislation has been passed in Japan to impose restrictions on flying UAVs near sensitive locations such as national facilities, airports, and population centers and to set rules and regulations on safely flying these devices. However, this has not stopped the unauthorized use of drones by foreign tourists unaware of these regulations or users who deliberately ignore them. Hence, the number of accidents and problems continues to increase each year.

As the number of drones in use can only increase, it becomes important to ensure that the skies are safe. Therefore, a framework must be developed to prevent drones' malicious use, such as for criminal activities or terrorism. Sky monitoring systems must address this important responsibility. Currently, methods using radar, acoustics, and RF signals to detect UAVs or radio-controlled aircraft [1] have been proposed. However, the systems [2–5] that use radar technology can impact the surrounding environment (such as blocking radio waves from nearby devices) and can only be used in limited areas. The method that uses acoustics [6] cannot detect a drone in a noisy urban environment, limiting the types of environments where the system can be installed. The methods in which drones are detected based on RF signals between the drone and operator [7] face the issue that they cannot detect autonomous control drones that do not transmit RF signals.

Methods employing images [8–11] have likewise been proposed, as we can see from the fact that image processing is used in various fields [12–18]. These methods have gained superiority, as they do not affect the monitoring targets or the surrounding environment and allow for systems to be built at a comparatively lower cost. It is also easy to miniaturize such systems, allowing them to be installed almost anywhere. Another benefit is that it becomes possible to use an indoor monitoring system to monitor an outdoor environment, reducing maintenance costs. Sobue et al. [8] and Seidaliyeva et al. [11] used background differencing to detect flying objects. Schumann et al. [9] used either background differencing or a deep neural network (DNN) to detect flying objects and used a separate convolutional neural network (CNN) model for categorization. Unlu et al. [10] proposed a framework in which the YOLOv3 [19] model was used to scan a wide area captured using a wide-angle camera to detect possible targets, and subsequently, a zoom camera was used to inspect the details of suspicious objects.

Although the aforementioned studies have explained the framework enough, the flying objects detected during the evaluation were large, and factors such as changes in the environment were not considered to a sufficient degree when evaluating their performance. Therefore, it is unclear under which conditions detection would succeed or fail in these methods, thereby making it difficult to determine whether they can be put to actual use. In this study, we propose a system capable of achieving a wide range of observations in real-time at a lower cost compared to existing systems using a single 4K resolution camera and YOLOv3, which is the state-of-the-art CNN. We installed the proposed system in a high-rise building located in the Tokyo metropolitan area, collected images that include flying objects, and evaluated the monitoring precision in changing weather and assuming various distances from the monitoring targets. It should be noted that the monitoring system was tested only during the day.

The remainder of this paper is organized as follows. We describe the proposed sky monitoring system in Section 2. The detection accuracy is presented in Section 3, and the results are discussed in Section 4. Finally, Section 5 concludes the study.

2. Proposed System

2.1. System Overview

We developed a sky monitoring system for use in urban areas that present a substantial risk concerning devices such as UAVs. The following restrictions must be assumed for use in an urban area:

- It is difficult to install equipment outdoors in high-rise buildings.
- It is difficult to use laser-based devices due to the Japanese Radio Act.
- It is difficult to use acoustic-based devices due to the high amount of noise in urban areas.
- It is difficult to monitor a wide area with a single system because of structures such as neighboring buildings.

Due to these restrictions, in this study, we installed a video camera indoors and developed a system that only uses image processing for monitoring. Using a video camera instead of special equipment, such as lasers, allowed the installation environment to be resolved and the system to be built at a reasonable cost. This also allowed us to install multiple systems.

In addition to monitoring nearby UAVs' flight status, we decided to monitor flying objects in general. We assumed that the system would also be used for building management and would need to monitor the sky to prevent damage by birds. The system's objective was to detect objects of a total length of approximately 50 cm or longer, which is the typical size of most general-purpose drones. The system was designed to monitor an area extending 150 m from it. This distance was selected for two reasons. First, UAV regulations within Japan permit flight only at an altitude of 150 m or lower. Second, Amazon has proposed a drone fly zone from 200 to 400 feet (61 to 122 m) for high-speed UAVs, and a fly zone up to 200 feet (61 m) for low-speed UAVs [20]. Although it would also be possible for a UAV used for malicious purposes (such as terrorism or criminal activity) to approach from an altitude of 150 m or higher, we considered that detecting small flying objects at a greater distance is not realistic when using only image processing. In addition, if the goal of drones is surreptitious photographing or dropping hazardous materials, the drones finally need to approach the facility. Therefore, we believe that the proposed system would be suitable for practical use under the aforementioned distance limitations.

2.2. System Configuration

The proposed system was configured from an indoor monitoring system that includes a video camera and a control PC for transmitting the captured video, and a processing server that detects flying objects using a CNN. Figure 1 shows the proposed system configuration.

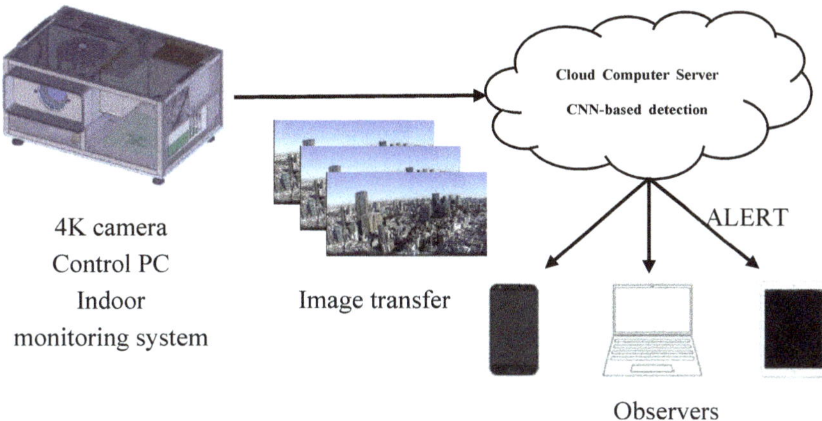

Figure 1. Proposed system configuration.

The monitoring system was equipped with a 4K video camera (which is the highest resolution offered by current consumer video cameras) to detect flying objects up to a distance of 150 m, and the widest angle is used to capture the footage of the sky to cover a wide area. It was assumed that the system would be installed indoors. Hence, we cannot ignore the fact that light would often be reflected from windows in typical office buildings, as blinds are opened and closed, or multiple lights are switched on. A covering hood was installed between the camera and window to reduce the impact of window reflections. The system was painted mostly black to reduce reflected light. This is a simple physical measure against reflected light, but it is essential for indoor installation.

A c5.4xlarge instance type processing server with 16 virtual CPUs was selected from Amazon EC2 as the processing server. In our proposed system, one 4K frame per second was sent from the server's monitoring system for processing. This instance type was selected as it would be sufficient to process 4K images in real-time.

2.3. CNN-Based Detection

Recently, CNN, which represents a deep learning approach, has been applied in numerous computer vision tasks, such as image classification and object detection [21]. Many models [19,22,23] have been proposed and compared in terms of accuracy on the COCO [24] and KITTI [25] benchmark datasets. In this study, we use the YOLOv3 [19] model to detect flying objects. YOLOv3 extracts the features of an image by down-sampling the input image with filters of three sizes of 8, 16, and 32 to detect objects of different sizes. The training process uses the loss that is calculated based on both the objectness score calculated from bounding box coordinates (x, y, w, h) and the class score. The advantage of YOLOv3 is the high balance between processing speed and accuracy. Therefore, we presume that YOLOv3 would be most suitable for use in the security field, which requires both real-time processing and high detection accuracy. In 2020, YOLOv4 [26] and YOLOv5 [27] were developed. However, in consideration of stable operation, we adopted YOLOv3, which boasts abundant practical results.

A collection of images of airplanes, birds, and helicopters was used as data to train the model instead of drone images. This was done for two reasons. First, there is no publicly available dataset for use in CNN training. Second, it would likely be impossible to collect images of all types of drones, because as opposed to automobiles and people, they assume various shapes. Therefore, images of airplanes, birds, and helicopters were collected for model training and evaluation during this study.

The proposed system uses a 4K video camera to capture a wide view of the sky. A drone with a size of approximately 50 cm flying at a maximum distance of 150 m would span approximately 10 pixels in an image. If the system can detect airplanes, birds, and helicopters under such strict conditions, it is expected to be capable of detecting drones.

The monitoring system was installed indoors on the 43rd floor of Roppongi Hills located in the center of Tokyo, to gather training data. Images of external flying objects were captured indoors through a window. Figure 2 shows an example of an image captured.

Figure 2. Image captured by the monitoring system and system installation.

2.4. Input to Model

We determined that compressing 4K images and importing them into the YOLO model would not be an appropriate method to input high-resolution images of distant flying objects into the current model. Therefore, when processing the detection of flying objects on the processing server, 600×600 pixel squares were extracted from 3840×2160 4K images for input into the YOLO model. This is illustrated in Figure 3. These extracted images were partially duplicated, such that the precision of the detected objects would not decrease in the image boundary areas. Detection processes for each image were run in parallel on different virtual CPUs, allowing the detection process for the entire area of a 4K image to be completed within one second (i.e., in real-time).

Figure 3. 4K image input.

2.5. CNN Model Training

Images of flying objects were extracted from videos photographed by the proposed system. Therefore, continuous images were similar among frames, although not completely identical. When such continuous groups of images are randomly split into training and evaluation datasets, the training and evaluation datasets contain almost the same image. In this situation, the trained detection model is not accurately evaluated, resulting in the precision being evaluated highly. Therefore, first, image data were separated by each continuous frame group, as shown in Figure 4. Then, each group was randomly divided into training and evaluation datasets. Notably, 75% of the data was used to train the CNN model, while the remaining 25% was used for evaluation. Furthermore, the size of the flying object and weather were adjusted to be evenly distributed in this process.

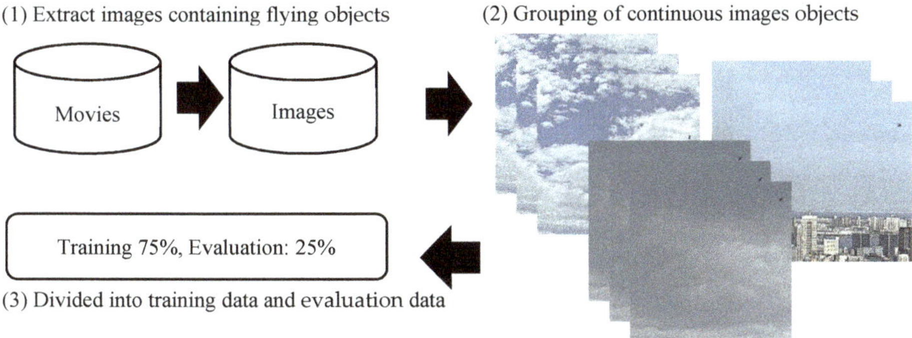

Figure 4. Separation of data for model training.

For hyperparameters, when training the model, we used the default settings published by the developers of YOLOv3. Additional details are provided on the developer's GitHub page [28]. Therefore, when training the model, basic data augmentation, such as flips, was performed. Furthermore, the brightness of the trained images was scaled between 0.3 and 1.2, and the number of trained images was expanded by approximately 10 times to increase robustness to changes in brightness due to changes in the time of day and weather.

3. Evaluation

3.1. Categorization of Data for Evaluation

The purpose of the evaluation process was to determine the effect of differences in the type of weather or sizes of flying objects on flying objects' detection precision. The images gathered, as described in Section 2.3, were categorized by conditions. The collected image data contained 1392 helicopters, 190 airplanes, and 74 birds. It should be noted that the background images did not include buildings and trees, but only the sky, as most flying objects were airplanes and helicopters.

The images were categorized by four types of weather: clear, partly cloudy, hazy, and cloudy/rainy. The images were categorized by eye, based on whether they contained clouds with contours, haze, blue skies, or rain clouds (dark clouds). Figure 5 lists the results of categorizing the images according to the type of weather.

Figure 5. Categorization of data by type of weather.

The images were then sorted into six categories corresponding to the sizes of the flying objects based on the number of horizontal pixels shown in the image: SS (less than 12 pixels), S (12 to 16 pixels), M (17 to 22 pixels), L (23 to 30 pixels), LL (31 to 42 pixels), and 3L (43 pixels or more). These categories can also be expressed as the distance in meters assuming a drone of approximately 50 cm size (SS: 150 m or greater, S: 110 m to less than 150 m, M: 75 m to less than 110 m, L: 55 m to less than 75 m, LL: 38 m to less than 55 m, and 3L: less than 38 m).

Table 1 lists the results of categorizing the images. A few images were categorized as clear, and many were categorized as hazy, with 253 clear, 433 partly cloudy, 601 hazy, and 368 cloudy/rainy images. There were comparatively few images categorized as SS, S, and 3L, but a roughly equivalent number categorized as all other sizes, with 58 SS, 182 S, 368 M, 488 L, 395 LL, and 152 3L images.

Table 1. Number of images by size and weather.

Image Size	Estimated Distance	Clear	Partly Cloudy	Hazy	Cloudy/Rainy
SS (up to 11 pixels)	From 150 m	7	11	19	21
S (up to 16 pixels)	110–150 m	5	57	68	52
M (up to 22 pixels)	75–110 m	12	69	246	41
L (up to 30 pixels)	55–75 m	47	147	149	145
LL (up to 42 pixels)	38–55 m	77	128	99	91
3L (from 43 pixels)	Up to 38 m	105	21	20	19
SUM		253	433	601	369

3.2. Evaluation Indicators

A precision-recall (PR) curve with precision on the vertical axis and recall on the horizontal axis was used for evaluation. Rectangular bounding boxes are generally used for estimation when detecting objects. A fixed threshold was used in numerous studies to evaluate the intersection over union (IoU) calculated from the estimated and actual bounding boxes. However, owing to the extremely small size of the detected objects in this study, a large proportion of measurement errors would be included for the flying objects in the bounding boxes. This is due to the blurriness caused by the lens or misalignment resulting from the human annotation. Therefore, we determined that a flying object was detected during the evaluation process if the actual and estimated bounding boxes overlapped even slightly.

Figures 6 and 7 present the results of evaluating the precision based on the type of weather. In Figure 6, images where a flying object was successfully detected and subsequently successfully categorized as a helicopter, bird, or airplane were considered as true positives. The remainder (where the flying object was successfully detected but categorized incorrectly) was false positive. In Figure 7, images where a flying object was successfully detected (even if categorization failed) were evaluated as true positives, while the remainder was evaluated to be false positive. This was conducted under the assumption that even succeeding only at detection would demonstrate that the system offers the minimum required functionality for monitoring flying objects.

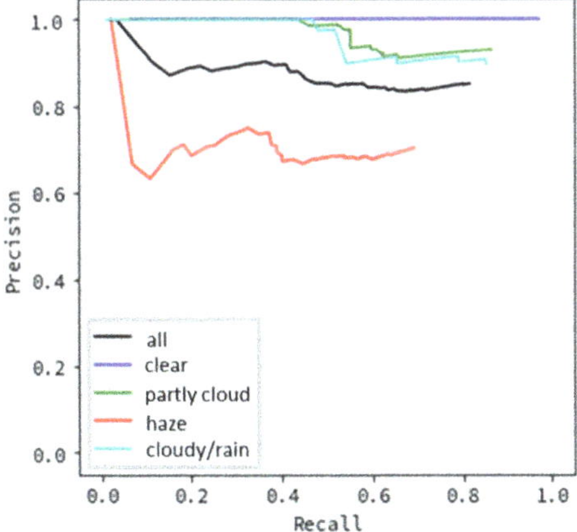

Figure 6. Detection precision by type of weather (including categorization).

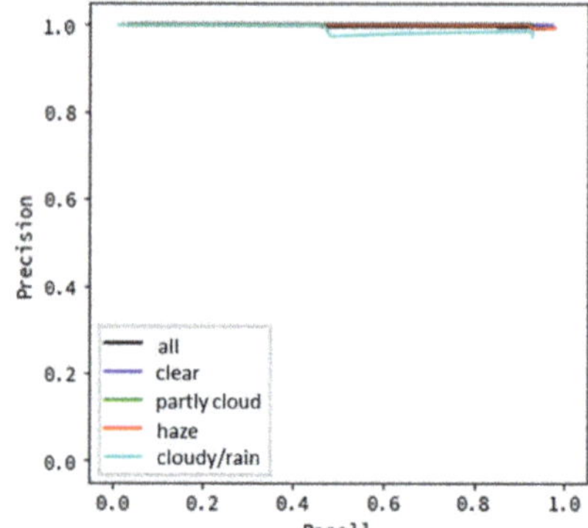

Figure 7. Detection precision by type of weather (not including categorization).

Figures 8 and 9 show the results of evaluating the precision based on the flying object's size. Similar to the evaluation according to the type of weather, in Figure 8, images where a flying object was successfully detected and then successfully categorized as a helicopter, bird, or airplane were evaluated as true positives, while the remainder was considered false positive. Figure 9 shows that images where a flying object was successfully detected (even if categorization failed) were evaluated as true positives, while the remainder was considered false positive.

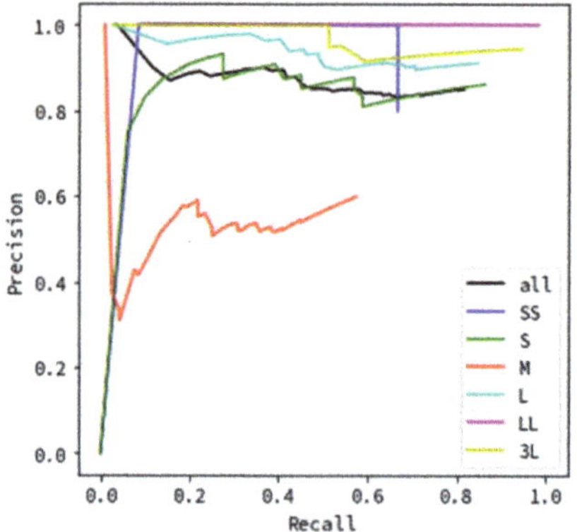

Figure 8. Detection precision by flying object size (including categorization).

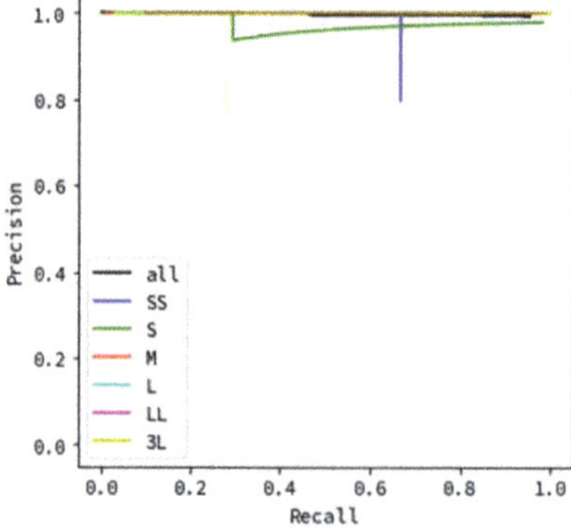

Figure 9. Detection precision by flying object size (not including categorization).

All evaluation data results are depicted as black lines in Figures 6–9, while the results for each type of weather are color-coded.

4. Discussion

We first discuss the results obtained for the precision evaluation based on the type of weather. Figure 6 shows excellent results for clear weather, with almost no false detections. The results for partly cloudy weather (when there are some clouds in the sky) and cloudy/rainy weather (when the entire sky is covered in clouds) are also satisfactory, although not as good as during clear weather. When the recall is 0.8, the results indicate a precision above 0.8. In contrast, the results are worse for hazy weather. When the recall is 0.6, the precision is approximately 0.6. We believe that this is due to the haze blurring the contours of flying objects. In this evaluation, because we targeted detecting flying objects such as helicopters and aircraft, which may be anywhere from several hundred meters to several kilometers away from the monitoring system, the effect may significantly influence detection accuracy. If we consider drones, the flight distance will be remarkably closer than the current targets. Therefore, the accuracy decrease due to hazy weather may be small, which must be investigated in the future. Notably, Figure 6 shows that although clouds and rain lower the overall brightness of images and make it difficult to detect objects by sight, no decrease in precision is observed in the monitoring system. Hence, the noise of the flying object's appearance, and not the change in the brightness of the sky, affects its detection accuracy. In contrast, Figure 7 indicates excellent results in all types of weather, with a precision value of nearly 1.0 for all recall values. This demonstrates that flying object categorization mistakes, rather than detection omissions, have a significant influence in lowering the precision and recall. If the system obtains human support for the classification task, the systems based on image processing can contribute significantly to sky monitoring.

Next, we discuss evaluating the precision based on the sizes of flying objects in the images. The results in Figure 8 indicate excellent results for L, LL, and 3L, which correspond to large sizes; when the recall is 0.8, the precision is above 0.8. In contrast, SS, S, and M, which have relatively small sizes, yield different results for each size. The result of the SS size category is an unnatural value. In the evaluation process, the evaluation data were separated into six categories by size, and the number of images per size category was different (SS: 58, S: 82, M: 368, L: 488, LL: 395, and 3L: 152). Focusing on SS, the number of images is very small. This implies that the precision for SS-sized objects was likely not calculated accurately. However, it also shows the capability of detecting even the smallest flying objects, which should be investigated in the future. The S results in a comparatively good PR curve, comparable to the results for large flying objects (L, LL, 3L). When the recall is 0.8, the precision achieved is above 0.8. This size corresponds to a distance of 110 to 150 m for a drone of maximum length (roughly 50 cm), and it is the minimum size of the monitoring system's observation target. This indicates that the monitoring system using image processing achieves sufficient detection accuracy that satisfies the requirement for a distance of 150 m, as shown in Section 2.1. The M size category shows remarkably poor results that are worse than those of all other size categories. We attribute this to the hazy weather often occurring for this size of an object compared with others. From the values in Table 1, for the M-size images, 246 images, which comprise more than half of the evaluation data of 368 images, are images captured in hazy weather. Hence, it is considered that the accuracy was reduced due to the influence of the weather and not size. Thus, we can conclude that high detection accuracy is achieved for flying object sizes S and above, excluding hazy weather. For the trials without categorizations, similarly to Figure 7, Figure 9 portraying the effect of flying object size shows particularly excellent results with an achieved precision value of nearly 1.0 for all recall values. This indicates that flying object categorization mistakes, rather than detection omissions, significantly impact the lower precision and recall by size, as shown in Figure 8.

In summary, the results indicate that the precision is poor during hazy weather, and the difference in accuracy depending on the object size is not large if it is S or above. Overall, except for hazy weather, the system maintains a precision value of approximately 0.8 when the recall is 0.8, confirming that it offers sufficient precision to function as a monitoring system. Therefore, further improvement of the algorithm to achieve better performance in hazy weather is required for actual use. The results also confirm that the detection precision is affected by flying object categorization mistakes rather than detection omissions. Under the experiment in this study, if we ignore flying objects' classification, almost all flying objects are detected in this study. Assuming that there are not many flying objects in the sky, unlike pedestrian detection in the city, even if the classification accuracy is poor, we believe that it can contribute to real scenes as a monitoring system. However, the images of the flying objects used during this study contain only the sky as a background. Images captured in an urban area are likely to include buildings behind flying objects due to building density in the area. This would create a more complicated background and make it significantly more difficult to detect categories. Therefore, the precision of this system must be verified under a range of conditions for use in urban areas.

Furthermore, although not examined in this study, capturing clear images is an important issue in monitoring systems that use image processing. When installing the system externally, the camera lens becomes dirty due to rain, insects, and dust, and it is not possible to obtain clear images. Even when installed indoors, it is affected by dirt on the glass in front of the monitoring system. If the object is relatively large, such as a pedestrian or a car, this will not have a significant negative impact. However, for flying objects, where the detection target is extremely small, the detection accuracy is substantially affected. Because physical solutions for this challenge are cumbersome, software-based methods are necessary to solve these problems for practical use.

5. Conclusions

We developed a sky monitoring system that uses image processing. This system consists of a monitoring system and a cloud processing server. The monitoring system captures a wide area of the sky at high resolution using a 4K camera and transfers the frame image to the server every second. The processing server uses the YOLOv3 model for flying object detection. In this process, real-time processing is realized by the parallel processing of multiple cropped images. We installed the monitoring system in a high-rise building in the Tokyo metropolitan area and collected CNN model training and evaluation data to evaluate the system's detection precision.

Existing research has employed the CNN model; however, the accuracy was not analyzed for each condition, and it was generally insufficient for practical use. Further, there is no simple system that combines a single 4K camera with the latest CNN to date. We designed a real-time monitoring system prototype using the latest CNN model and 4K resolution camera and demonstrated the detection accuracy of flying objects with various object sizes and weather conditions.

We found that detection accuracy is significantly reduced in hazy weather. From this result, we found that the blurring of the flying object's shape in the haze and the decrease in brightness in rainy or cloudy weather influence the decrease in accuracy. The flying object's size was replaced by the distance of the flying object in this study, and we show that a flying object with a pixel-sized image, assuming a 50 cm drone separated by 150 m, can be detected with high accuracy. Moreover, if we omit the classification accuracy, we find that almost all flying objects can be detected under all weather conditions and sizes considered in this experiment. From these results, the surveillance system using image processing is expected to contribute to sky surveillance, as the cost of human support in classification is low in an environment where there are few objects in the sky.

However, the monitoring system has several limitations: it was verified only during the daytime in this experiment, and we did not consider the effects of dirt and dust on the camera lens on the image. Moreover, the evaluation data targeted long-distance aircraft and helicopters as flying objects and did not include drones that fly relatively short distances. In the future, we plan to evaluate the monitoring accuracy in further detail by collecting evaluation data, including the above conditions. In this research, we will improve the proposed monitoring and develop a system to detect light emitted from LEDs installed on drones and a system that uses a comparatively inexpensive infrared projector for the detection of drones at night.

6. Patents

Japan patent: JP6364101B.

Author Contributions: Conceptualization, T.K. and Y.S.; methodology, T.K.; software, T.K. and H.S.; validation, T.K. and H.S.; formal analysis, T.K. and H.S.; investigation, H.S.; resources, T.K.; data curation, H.S.; writing—original draft preparation, T.K.; writing—review and editing, T.K. and Y.S.; visualization, T.K.; supervision, Y.S.; project administration, Y.S.; funding acquisition, Y.S. All authors have read and agreed to the published version of the manuscript.

Funding: This research received no external funding.

Acknowledgments: MORI Building Co., Ltd. supported this research.

Conflicts of Interest: The authors declare no conflict of interest.

References

1. Taha, B.; Shoufan, A. Machine Learning-Based Drone Detection and Classification: State-of-the-Art in Research. *IEEE Access* **2019**, *7*, 138669–138682. [CrossRef]
2. Samaras, S.; Diamantidou, E.; Ataloglou, D.; Sakellariou, N.; Vafeiadis, A.; Magoulianitis, V.; Lalas, A.; Dimou, A.; Zarpalas, D.; Votis, K.; et al. Deep Learning on Multi Sensor Data for Counter UAV Applications-A Systematic Review. *Sensors* **2019**, *19*, 4837. [CrossRef] [PubMed]
3. Mendis, G.J.; Randeny, T.; Wei, J.; Madanayake, A. Deep learning based doppler radar for micro UAS detection and classification. In Proceedings of the MILCOM 2016—2016 IEEE Military Communications Conference, Baltimore, MD, USA, 1–3 November 2016; pp. 924–929.
4. Ganti, S.R.; Kim, Y. Implementation of detection and tracking mechanism for small UAS. In Proceedings of the International Conference on Unmanned Aircraft Systems (ICUAS), Arlington, VA, USA, 7–10 June 2016; pp. 1254–1260.
5. Kwag, Y.; Woo, I.; Kwak, H.; Jung, Y. Multi-mode SDR radar platform for small air-vehicle Drone detection. In Proceedings of the CIE International Conference on Radar (RADAR), Guangzhou, China, 10–13 October 2016; pp. 1–4.
6. Hommes, A.; Shoykhetbrod, A.; Noetel, D.; Stanko, S.; Laurenzis, M.; Hengy, S.; Christnacher, F. Detection of Acoustic, Electro-Optical and Radar Signatures of Small Unmanned Aerial Vehicles. In Proceedings of the SPIE Security + Defence, Edinburgh, UK, 26–29 September 2016; Volume 9997.
7. Ezuma, M.; Erden, F.; Anjinappa, C.K.; Ozdemir, O.; Guvenc, I. Micro-UAV Detection and Classification from RF Fingerprints Using Machine Learning Techniques. In Proceedings of the IEEE Conference on Aerospace, Big Sky, MT, USA, 2–9 March 2019.
8. Sobue, H.; Fukushima, Y.; Kashiyama, T.; Sekimoto, Y. Flying object detection and classification by monitoring using video images. In Proceedings of the 25th ACM SIGSPATIAL International Conference on Advances in Geographic Information Systems, Redondo Beach, CA, USA, 7–10 November 2017; pp. 1–4.
9. Schumann, A.; Sommer, L.; Klatte, J.; Schuchert, T.; Beyerer, J. Deep cross-domain flying object classification for robust UAV detection. In Proceedings of the IEEE International Conference on Advanced Video and Signal Based Surveillance (AVSS), Lecce, Italy, 29 August–1 September 2017; pp. 1–6.
10. Unlu, E.; Zenou, E.; Riviere, N.; Dupouy, P.E. Deep learning-based strategies for the detection and tracking of drones using several cameras. *IPSJ Trans. Comput. Vis. Appl.* **2019**, *11*, 7. [CrossRef]
11. Seidaliyeva, U.; Akhmetov, D.; Ilipbayeva, L.; Matson, E.T. Real-Time and Accurate Drone Detection in a Video with a Static Background. *Sensors* **2020**, *20*, 3856. [CrossRef]
12. Chato, P.; Chipantasi, D.J.M.; Velasco, N.; Rea, S.; Hallo, V.; Constante, P. Image processing and artificial neural network for counting people inside public transport. In Proceedings of the IEEE Third Ecuador Technical Chapters Meeting, Cuenca, Ecuador, 15–19 October 2018; pp. 1–5.
13. Ahmed, S.; Huda, M.N.; Rajbhandari, S.; Saha, C. Pedestrian and Cyclist Detection and Intent Estimation for Autonomous Vehicles: A Survey. *Appl. Sci.* **2019**, *9*, 2335. [CrossRef]
14. Riyazhussain, S.; Lokesh, C.R.S.; Vamsikrishna, P.; Rohan, G. Raspberry pi controlled traffic density monitoring system. In Proceedings of the International Conference on Wireless Communications, Signal Processing and Networking (WiSPNET), Chennai, India, 23–25 March 2016; pp. 1178–1181.

15. Zhang, F.; Li, C.; Yang, F. Vehicle detection in urban traffic surveillance images based on convolutional neural networks with feature concatenation. *Sensors* **2019**, *19*, 594. [CrossRef]

16. Fedorov, A.; Nikolskaia, K.; Ivanov, S.; Shepelev, V.; Minbaleev, A. Traffic flow estimation with data from a video surveillance camera. *J. Big Data* **2019**, *6*, 73. [CrossRef]

17. Maeda, H.; Sekimoto, Y.; Seto, T.; Kashiyama, T.; Omata, H. Road Damage Detection and Classification Using Deep Neural Networks with Smartphone Images. *Comput.-Aided Civ. Infrastruct. Eng.* **2018**, *33*, 1127–1141. [CrossRef]

18. Arshad, B.; Ogie, R.; Barthelemy, J.; Pradhan, B.; Verstaevel, N.; Perez, P. Computer Vision and IoT-Based Sensors in Flood Monitoring and Mapping: A Systematic Review. *Sensors* **2019**, *19*, 5012. [CrossRef] [PubMed]

19. Redmon, J.; Farhadi, A. Yolov3: An incremental improvement. *arXiv* **2018**, arXiv:1804.02767.

20. Amazon PrimeAir. Available online: https://www.amazon.com/Amazon-Prime-Air/b?ie=UTF8&node=8037720011 (accessed on 19 March 2020).

21. LeCun, Y.; Bengio, Y.; Hinton, G. Deep learning. *Nature* **2015**, *521*, 436–444. [CrossRef] [PubMed]

22. Liu, W.; Anguelov, D.; Erhan, D.; Szegedy, C.; Reed, S.; Fu, C.-Y.; Berg, A.C. Ssd: Single shot multibox detector. In Proceedings of the European Conference on Computer Vision, Amsterdam, The Netherlands, 8–16 October 2016; Springer: Cham, Switzerland, 2016; pp. 21–37.

23. Ren, S.; He, K.; Girshick, R.; Sun, J. Faster r-cnn: Towards real-time object detection with region proposal networks. In Proceedings of the Advances in Neural Information Processing Systems, Montreal, QC, Canada, 7–12 December 2015; pp. 91–99.

24. COCO (Common Objects in Context). Available online: https://cocodataset.org/ (accessed on 19 March 2020).

25. Geiger, A.; Lenz, P.; Stiller, C.; Urtasun, R. Vision meets robotics: The kitti dataset. *Int. J. Robot. Res.* **2013**, *32*, 1231–1237. [CrossRef]

26. Bochkovskiy, A.; Wang, C.Y.; Liao, H.Y.M. YOLOv4: Optimal Speed and Accuracy of Object Detection. *arXiv* **2020**, arXiv:2004.10934.

27. Ultralytics. Available online: https://github.com/ultralytics/yolov5 (accessed on 19 March 2020).

28. YoloV3 GitHub. Available online: https://pjreddie.com/darknet/yolo/ (accessed on 19 March 2020).

Publisher's Note: MDPI stays neutral with regard to jurisdictional claims in published maps and institutional affiliations.

Article

Recognition and Grasping of Disorderly Stacked Wood Planks Using a Local Image Patch and Point Pair Feature Method

Chengyi Xu [1,2]**, Ying Liu** [1,]*****, Fenglong Ding** [1] **and Zilong Zhuang** [1]

[1]   College of Mechanical and Electronic Engineering, Nanjing Forestry University, Nanjing 210037, China; xucy312@163.com (C.X.); dfl@njfu.edu.cn (F.D.); zzl0702@njfu.edu.cn (Z.Z.)

[2]   College of Mechanical Engineering, Nantong Vocational University, Nantong 226007, China

*   Correspondence: liuying@njfu.edu.cn

Received: 27 August 2020; Accepted: 28 October 2020; Published: 31 October 2020

Abstract: Considering the difficult problem of robot recognition and grasping in the scenario of disorderly stacked wooden planks, a recognition and positioning method based on local image features and point pair geometric features is proposed here and we define a local patch point pair feature. First, we used self-developed scanning equipment to collect images of wood boards and a robot to drive a RGB-D camera to collect images of disorderly stacked wooden planks. The image patches cut from these images were input to a convolutional autoencoder to train and obtain a local texture feature descriptor that is robust to changes in perspective. Then, the small image patches around the point pairs of the plank model are extracted, and input into the trained encoder to obtain the feature vector of the image patch, combining the point pair geometric feature information to form a feature description code expressing the characteristics of the plank. After that, the robot drives the RGB-D camera to collect the local image patches of the point pairs in the area to be grasped in the scene of the stacked wooden planks, also obtaining the feature description code of the wooden planks to be grasped. Finally, through the process of point pair feature matching, pose voting and clustering, the pose of the plank to be grasped is determined. The robot grasping experiment here shows that both the recognition rate and grasping success rate of planks are high, reaching 95.3% and 93.8%, respectively. Compared with the traditional point pair feature method (PPF) and other methods, the method present here has obvious advantages and can be applied to stacked wood plank grasping environments.

Keywords: convolutional auto-encoders; local image patch; point pair feature; plank recognition; robotic grasping

1. Introduction

At present, the grasping process of disorderly stacked wooden planks is mainly completed by manual methods, such as the sorting and handling of wooden planks, and the paving of wooden planks. There are existing defects, such as low labor and high cost. With the rapid development and wide application of robotic technology, vision-based robot intelligent grasping technology has high theoretical significance and practical value to complete this work. The visual recognition and positioning of disorderly stacked wooden planks is an important prerequisite for successful robot grasping. The vision sensor is the key component of the implementation. A RGB-D camera can be regarded as a combination of a color monocular camera and a depth camera, and this camera can collect both texture information and depth information at the same time [1], which has obvious application advantages.

The recognition, detection, and location of objects is a research hotspot in both Chinese and overseas contexts. Hinterstoisser et al. [2] extracted features from the color gradient of the target and the normal vector of the surface and matched them to obtain a robust target detection result. Rios-Cabrera et al. [3] used a cluster to find the target and the detection speed was faster. Rusu et al. [4] calculated the angle between the normal vector of a point cloud and the direction of the viewpoint based on the viewpoint feature histogram (VFH), but this was not robust to occlusion problems. The orthogonal viewpoint feature histogram (OVFH) [5] was proposed. Birdal et al. [6–8] used a point cloud to extract point pair features (PPFs), and used shape information to identify and detect targets, but this was mainly for objects with complex shapes. Lowe et al. [9,10] proposed a scale-invariant 2D feature point the scale invariant feature transform method (SIFT) local feature point with good stability, but it was not suitable for multiple similar targets in the same scene. There are also the feature descriptors surf [11], spin image [12], signature of histogram of orientation (SHOT) [13], etc., Choi [14] added color information on the basis of traditional four-dimensional geometric point pair features and obtained better accuracy than the original PPF method. Ye et al. [15] proposed fast hierarchical template Matching Strategy of Texture-Less Objects, which takes less time than the origin method. Muñoz et al. [16] proposed a novel part-based method using an efficient template matching approach where each template independently encodes the similarity function using a forest trained over the templates. In reference to the multi-boundary appearance model, Liu [17] proposed to fit the tangent to the edge point of the model as the direction vector of the point. Through the four-dimensional point pair feature matching and positioning, a good recognition result was obtained. Li [18] proposed a new descriptor curve set feature (CSF), where the descriptor curve set feature describes a point by describing the surface fluctuations around the point and can evaluate the pose. CAD-based pose estimation was also used to solve recognition and grasping problems. Wu et al. [19] proposed constructing 3D CAD models of objects via a virtual camera, which generates a point cloud database for object recognition and pose estimation. Chen et al. [20] proposed a CAD-based multi-view pose estimation algorithm using two depth cameras to capture the 3D scene.

Recently, deep learning has been applied to the recognition and grasping of robot situations. Kehl [21] proposed the use of convolutional neural networks for end-to-end training to obtain the pose of an object. Caldera et al. [22] proposed a novel approach to multi-fingered grasp planning leveraging learned deep neural network models. Kumara et al. [23] proposed a novel robotic grasp detection system that predicts the best grasping pose of a parallel-plate robotic gripper for novel objects using the RGB-D image of the scene. Levine et al. [24] proposed a learning-based approach to hand-eye coordination for robotic grasping from monocular images. Zeng et al. [25] a robotic pick-and-place system that is capable of grasping and recognizing both known and novel objects in cluttered environments. Kehl [26] proposed a 3D object detection method that used regressed descriptors of locally-sampled RGB-D patches for 6D vote casting. Zhang [27] proposed a recognition and positioning method that uses deep learning to combine the overall image and the local image patch. Le et al. [28] proposed applying an instance segmentation-based deep learning approach using 2D image data for classifying and localizing the target object while generating a mask for each instance. Tong [29] proposed a method of target recognition and localization using local edge image patches. Jiang et al. [30] used a deep convolutional neural network (DCNN) model that was trained on 15,000 annotated depth images synthetically generated in a physics simulator to directly predict grasp points without object segmentation.

Even in the current era of deep learning, the point pair feature (PPF) method [7] has a strong vitality in bin-picking problem. Its algorithm performance is still no less than that of deep learning. Many scholars have made a lot of improvements to PPF [6–8,14,15,31] because of its advantages. The general framwork of PPF has not changed significantly at the macro or micro level anyway. Vidal et al. [31] proposed an improved matching method for Point Pair Features with the discriminative value of surface information.

At present, most of the robot visual recognition and grasping scenes are put together with different objects and the target objects that need to be recognized have rich contour feature information. However, the current research methods are not effective for the identification and grasping of disorderly stacked wooden planks. The main reason for this is that the shape of the wooden plank itself is regular and symmetrical and is mainly a large plane. The contour change information is not rich and different planks have no obvious or special features. There are many similar features such as shapes and textures. When these wooden planks are stacked together, it is difficult to identify and locate one of the wooden planks using conventional methods, making it difficult for a robot to grasp a plank when among many. Therefore, we utilize PPF combined other features to recognize and locate unordered stacked planks.

The local image patches of the wooden plank images both from the self-developed scanning equipment and the disorderly stacking plank scene here are taken as a data set. Using the strong fitting ability of a deep convolutional autoencoder, the convolutional autoencoder is trained to obtain stable local texture feature descriptors. The overall algorithm flow realized by robot grasping is shown in Figure 1. In the offline stage, a pair of feature points are randomly selected on the wooden plank image and local image patches are intercepted. These two image patches are sequentially input to the trained feature descriptor to obtain the local feature vector and combine the geometric feature information of the point pair to build the feature code database of the plank model. In the online stage, the disorderly stacking plank scene is segmented and the plank area to be grasped is extracted, and the geometric feature information of the point pair is also extracted and the feature description code of the local image patch is found, similar to the offline stage. Then, point pair matching is performed with the established plank feature database. Finally, the robot is used to realize the positioning and grasping operation after all the point pairs complete pose voting and pose clustering.

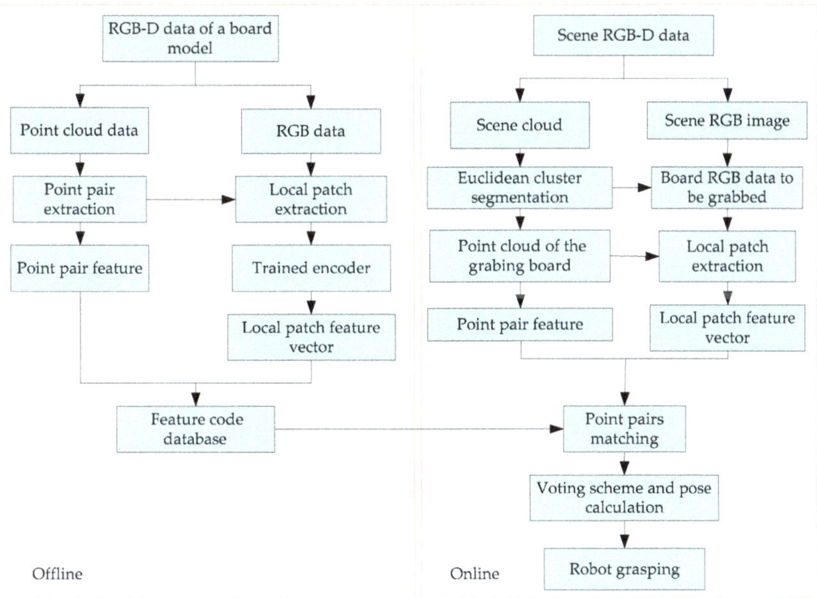

Figure 1. The overall algorithm flow of robot grasping.

2. Methods of the Local Feature Descriptor Based on the Convolutional Autoencoder

Traditional autoencoders [32–34] are generally fully connected, which will generate a large number of redundant parameters. The extracted features are global, local features are ignored, and local features are more important for wood texture recognition. Convolutional neural networks have the characteristics of local connection and weight sharing [35–40], which can accelerate the training of the network and facilitate the extraction of local features. The deep convolutional autoencoder designed in this paper is shown in Figure 2. The robot drives the camera to collect small image patches at different angles around the same feature point in the scene of disorderly stacked wood planks as the input and expected value of output to train the convolutional autoencoder. Its network is mainly composed of two stages, namely, encoding and decoding. The encoding stage has four layers. Each layer implements feature extraction on input data convolution and ReLU activation function operations. There are also four layers in the decoding stage. Each layer implements feature data reconstruction through operations such as transposed convolution and ReLU activation function operations. This network model combines the advantages of traditional autoencoders and convolutional neural networks, in which residual learning connections are added to improve the performance of the network.

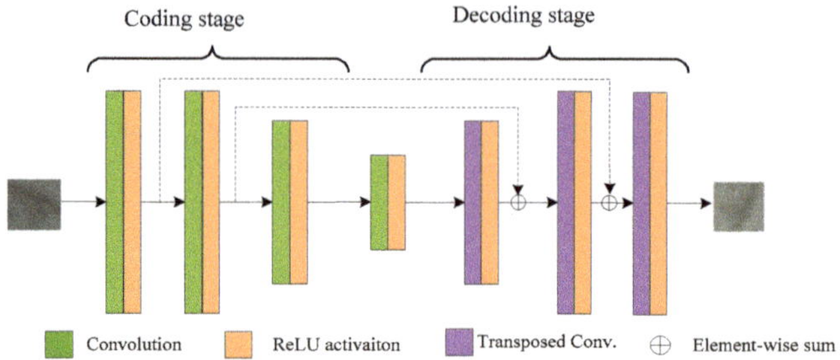

Figure 2. Convolutional auto-encoder mode.

In the encoding stage, the input of the convolutional layer has D groups of feature matrices, and the two-dimensional $M \times M$ feature matrix of the dth group is x^d, $1 \leq d \leq D$. The two-dimensional $N \times N$ convolution kernel of the group of the k'th channel of the output convolution layer is $W^{k,d}$, and the k'th feature map of the output convolution layer is $h^k \in R^{M' \times M'}$, where $M' = M - N + 1$, where h^k can then be expressed as follows:

$$h^k = f\left(\sum_{d=1}^{D} x^d * W^{k,d} + b^k\right) \tag{1}$$

In Equation (1), f is the activation function, $*$ is the two-dimensional convolution, and b^k is the offset of the k'th channel of the convolutional layer.

The decoding stage is designed to reconstruct the feature map obtained in the encoding stage, input a total of K groups of feature matrices, and output the dth feature map, $y^d \in R^{M'' \times M''}$, where $M'' = M' - N + 1$. Below, y^d can be expressed as follows:

$$y^d = f\left(\sum_{k=1}^{K} h^k * \widetilde{W}^{k,d} + c^d\right) \tag{2}$$

In Equation (2), $\widetilde{W}^{k,d}$ is horizontal and vertical flip of the convolution kernel $W^{k,d}$ and $1 \leq k \leq K$. Additionally, c^d is the offset of the d'th channel of the deconvolution layer.

The convolutional autoencoder encodes and decodes the original intercepted image to continuously learn the parameters and offsets of the convolution kernel in the network, so that the output result y is as close as possible to the given output expectation x', and to minimize the reconstruction error function in Equation (3)

$$E = \frac{1}{2n}\sum_{i=1}^{n} \|y_i - x'_i\|^2 \tag{3}$$

where n is the number of samples input into the model training, y_i represents the actual output samples, and x' represents the expected value of the output.

A back propagation (BP) neural network error back propagation algorithm is used to adjust the network parameters. If the training result makes the autoencoder converge, the trained autoencoder encodes the part of the network to obtain a local texture feature descriptor of the wood plank that is robust to changes in perspective, that is, the local image patch is input into the coding part of the convolutional encoder to get the corresponding feature description code.

Only the local image features of the wooden planks are used for the matching and recognition of a single plank. Since there are multiple and similar wooden planks in the visual scene of stacked wooden planks, the algorithm cannot identify whether multiple local image patch features belong to the same wooden plank, and they are likely to be scattered on different wooden planks, as shown in Figure 3. This is very easy to mismatch. We hope local image patches on the same board to be pair relations, as shown in Figure 4.

Figure 3. Local image patches matching.

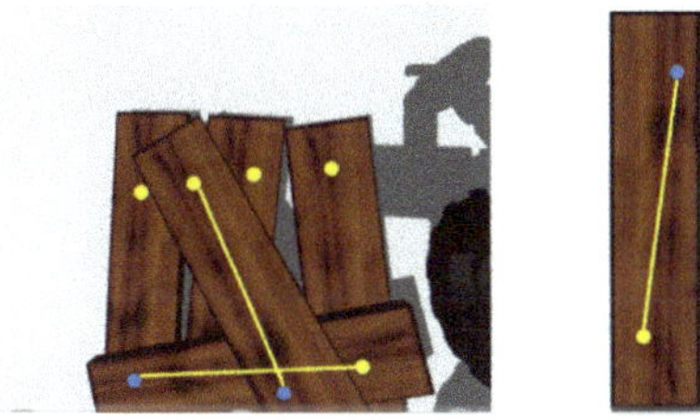

Figure 4. Local image patches of paired relation.

3. Offline Calculation Process: Model Feature Description Using Point Pair Features and Local Image Patch Features

Consider adding a point pair geometric feature constraint relationship to two points on the same wooden plank model, as shown in Figure 5.

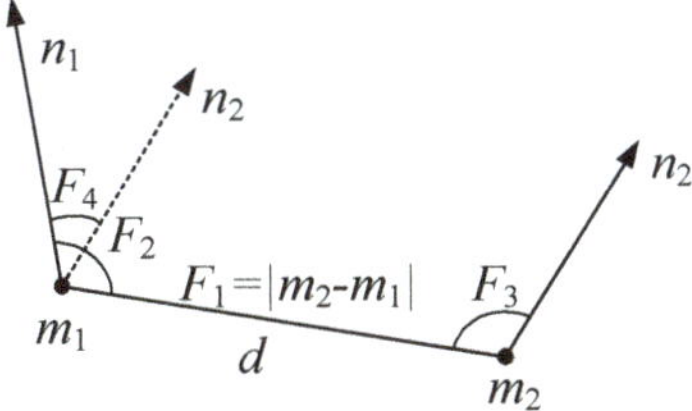

Figure 5. Schematic diagram of point pair features.

For any two points m_1 and m_2 and their respective normal vectors , n_1, n_2, define the constraint relationship describing the local texture feature point pair as shown in Equation (4) [7]:

$$F(m_1, m_2) = [\|d\|, \angle(n_1, d), \angle(n_2, d), \angle(n_1, n_2)] \tag{4}$$

where $\|d\|$ is the distance between the two points, $\angle(n_1, d)$ is the angle between the normal vector n_1 and line d connecting two points, $\angle(n_2, d)$ is the angle between the normal vector n_2 and the line d connecting two points, and $\angle(n_1, n_2)$ is the angle between the normal vector n_1 and the normal vector n_2. Then, namely:

$$\begin{cases} F_1 = \|d\| = |m_2 - m_1| \\ F_2 = \angle(n_1, d) = \arccos\frac{n_1 \cdot d}{|n_1||d|} \\ F_2 = \angle(n_2, d) = \arccos\frac{n_2 \cdot d}{|n_2||d|} \\ F_3 = \angle(n_1, n_2) = \arccos\frac{n_1 \cdot n_2}{|n_1||n_2|} \end{cases} \tag{5}$$

We at first performed point pair feature sampling on the point cloud of a single plank model, that is, we first took a feature point, established a point pair relationship with all other feature points in turn, and then took a new feature point to establish a point pair relationship, repeating the execution until the point pair relationship of all feature points was established, and the characteristic parameter F was calculated for each pair of non-repeated point pairs.

Since the shape of the plank itself is mainly a large rectangular plane, with symmetrical regularity, the corresponding characteristic parameter information of different points on a single plank is the same or similar. Pointpair features alone cannot complete the uniqueness of the plank feature matching, so the feature of the local image patch of the point-pair was added here. We input the intercepted local image patch into the previously trained encoder, where the encoder only needed to take the encoding part of the original convolutional encoder, the feature vector corresponding to the input local image patch can be obtained, and the feature vector of all non-repetitive point pairs of the local image patch can be combined with the point pair feature geometric information.

As shown in Figure 6, we define a more comprehensive description of the characteristic parameters of the wood plank, namely:

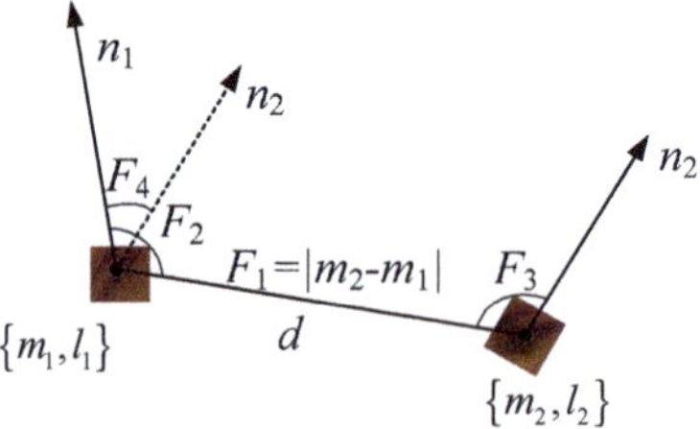

Figure 6. Local patch point pair feature (LPPPF).

$$F_l(m_1, m_2) = [\|d\|, \angle(n_1, d), \angle(n_2, d), \angle(n_1, n_2), l_1, l_2] \tag{6}$$

where l_1, l_2 represent the feature code extracted from the local image patch by the encoder.

The KD-tree (k-dimensional tree) method is used to build a feature code database reflecting the characteristics of the wood plank model, as shown in Figure 7. In summary, the model description method that combines point pair features and image local features can avoid the drawbacks of using one of the methods alone.

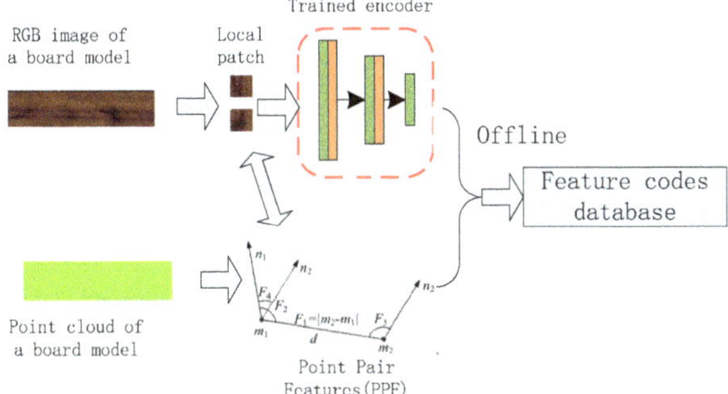

Figure 7. Establishment of feature code database of the wood plank model.

4. Online Calculation Process

4.1. Generating the Feature Code of the Plank to Be Grasped

Since the stacked wooden planks have obvious hierarchical characteristics, the robot's grasping process is generally carried out in order from top to bottom, and the wooden plank grasped each time should be the top layer in the scene at that time. First, the Euclidean distance cluster method [41] was used to segment the scene under robot vision, then calculating the average depth of different clusters and selecting the smallest average depth value as the area to be grasped. As shown in Figure 8, there is no need to match all the wood planks in the scene later, which saves on the computing time. Next, we randomly extracted a part of the scene pointpair feature information and the point pair corresponding local image patches from the area to be grasped, and input the local image patches into the trained encoder to generate local image patch feature vectors. These were combined with the pointpair geometric feature information to form a feature description code.

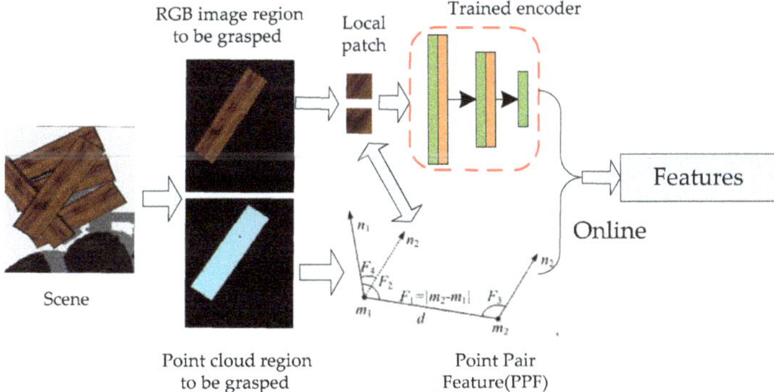

Figure 8. Generation of feature code of the wood plank to be grasped.

4.2. Plank Pose Voting and Pose Clustering

We extracted point pairs that were similar to the feature codes generated in the scene from the feature code database established offline and measured their similarity by the Euclidean distance to complete the point pair matching:

$$dist(F_{off}, F_i) = \|F_{off} - F_i\|_2 \tag{7}$$

where F_i represents the feature code extracted and generated in the scene and F_{off} is the feature code in the feature code database established offline.

We used a local coordinate system to vote in a two-dimensional space to determine the pose here proposed by Drost et al. [7]. We selected a reference point in the scene point cloud to form a point pair with any other point in the scene then calculated the feature value according to Formula (6) and searched for the point pair (m_r, m_i) in the feature code database through Formula (7). Successful matching indicates that feature point s_r is extracted in the scene, where there is a point m_r corresponding to it in the feature code database. We put them in the same coordinate system, as shown in Figure 9. Next, we moved these two points to the origin of coordinates and rotated them so that their normal vectors were aligned with the x-axis. Among them, the transformation matrix that occurs on m_r is $T_{m \to g}$, the transformation matrix that occurs on s_r is $T_{s \to g}$, and, at this time, the other points of their point pairs s_i and s_r are not aligned, and these need to rotate the angle to achieve alignment, with the transformation matrix is $R_x(\alpha)$, which then becomes the following [7]:

$$s_i = T_{s \to g}^{-1} R_x(\alpha) T_{m \to g} m_i \tag{8}$$

where $T_{s \to g}^{-1} R_x(\alpha) T_{m \to g}$ is a temporary plank pose matrix.

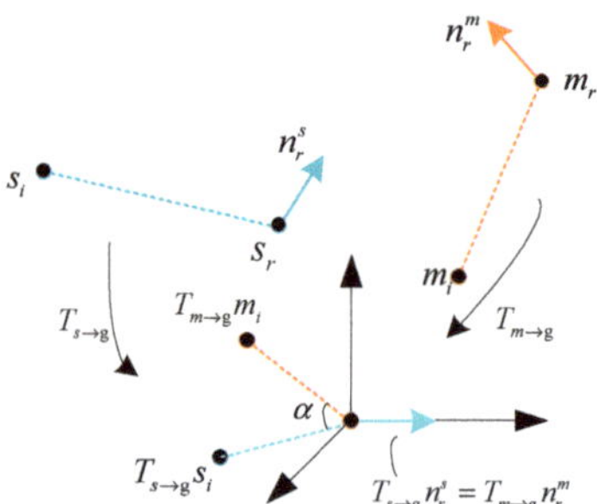

Figure 9. Model and scene coordinate system transformation.

In order to reduce the calculation time and increase the calculation speed, the rotation angle can be calculated by the following formula:

$$\alpha = \alpha_m - \alpha_s \tag{9}$$

where α_m is only determined by the model feature point pair and α_s is only determined by the scene point pair features.

From Equation (8), it can be seen that the complex pose solving problem is transformed into the problem of matching model point pairs and corresponding angles α, so it can be solved by ergodic voting. We created a two-dimensional accumulator, where the number of rows is the number of scene model points M, and the number of columns is the value q after the angle is discretized. When the point pair extracted in the scene matches the point pair of the model correctly, one of the two-dimensional accumulators (m_r, α) corresponds to it, that is, the position is voted.When all the point pairs composed of the scene point s_r and other points s_i in the scene have been processed, the position where the peak vote is obtained in the two-dimensional accumulator is the desired position. An angle α can estimate the posture of the plank, and the position of the model point can estimate the position of the plank.

In order to ensure the accuracy and precision of the pose, multiple non-repetitive reference points in the scene are selected to repeat the above voting. There are also multiple model points in the two-dimensional accumulator used for voting. In this way, there will be multiple voting peaks for different model points, eliminating significantly less incorrect pose votes, which can improve the accuracy of the final result. Multiple voting peaks means that the generated poses need to be clustered.

The highest vote is used as the pose clustering center value. The newly added poses must have the translation and rotation angles corresponding to the pose clustering center pose values set in advance. Within a certain threshold range, when the pose is significantly different from the current pose cluster center, a new cluster is created. The score of a cluster is the sum of the scores of the poses contained in the cluster, and the score of each pose is the sum of the votes obtained in the voting scheme. After the cluster with the largest score is determined, the average value of each pose in the cluster is used as the final pose of the plank to be grasped. Pose clustering improves the stability of the algorithm by excluding other poses with lower scores, and at the same time ensures that only one pose is finalized during each recognition, so that the robot only chooses to grasp one wooden plank at one time. The method of obtaining the maximum score clustering pose average also directly improves the accuracy of the plank's pose. This pose value can be used as the initial value of the iterative closest point method (ICP) [42] to further optimize the plank's pose. In summary, the process to determine the pose of the plank is shown in Figure 10.

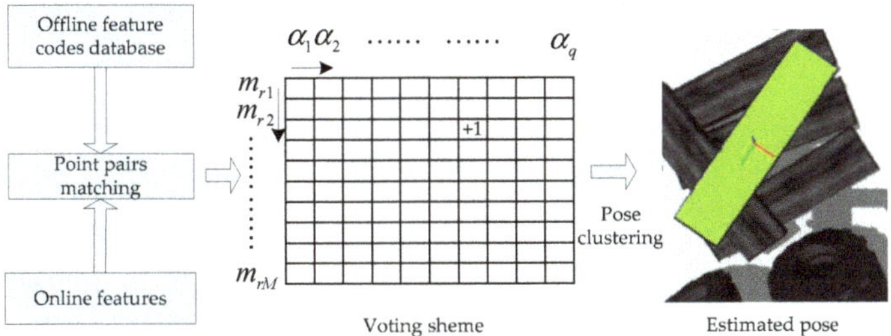

Figure 10. The final pose determination process of the plank.

5. Experiments Results and Discussion

The computer hardware conditions used in the experiment were an Intel Xeon W-2155 3.30 GHz CPU, 16.00 GB of RAM, and a NVIDIA GeForce GTX 1080 Ti GPU. The whole framework is based on C++, OpenCV, Point Cloud Library (PCL) and other open source algorithm libraries. A visual grasp scene model of stacked wooden planks was built on the ROS (robot operating system) with the Gazebo platform. As shown in Figure 11, the RGB-D camera (Xtion Pro Live, Suzhou, China) is fixed at the end of the robot arm and the disorderly stacked wooden boards to be grasped are placed below. Besides that, the robot hand device is a suction gripper with six suction points.

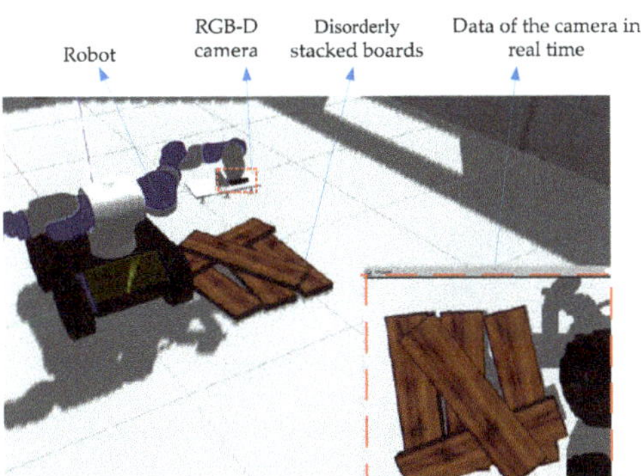

Figure 11. Visual grasping scene model of stacked wooden planks.

5.1. Data Preparation and Convolutional Autoencoder Training

For setting up the data set, we used the self-developed mechatronics equipment to collect the images of wood boards (Figure 12). This device mainly includes a strip light source, a transmission device, a CCD industrial camera (LA-GC-02K05B, DALSA, Waterloo, ON, Canada) and an photoelectric sensor (ES12-D15NK, LanHon, Shanghai, China) mounted on top. When the conveyor belt moves the wood board to the scanning position, the photoelectric sensor will detect the wooden board and start the CCD camera to collect the image of the wood board surface. We collected 100 images of red pine and camphor pine planks (Figure 13) and eventually divide them into small pieces of about 8000 local images (Figure 14).

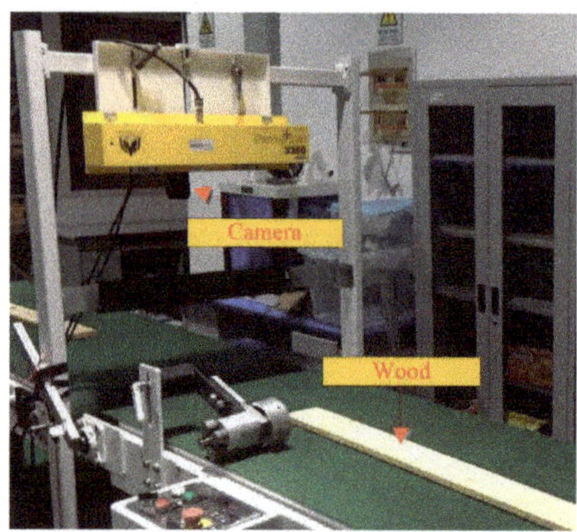

Figure 12. Wood image acquisition equipment.

Figure 13. Collected image of the surface of a wood.

Figure 14. Some local images, which were intercepted from collected wood images.

The data collected by the self-developed equipment accounts for 75% of the whole data set, while the remaining 25% of the data set is collected in the ROS system. The different poses of the end of the robotic arm bring the RGB-D camera to collect scene information from different perspectives to obtain images of the same scene in different perspectives. First, the camera collects feature points at different positions from each image and intercept 22×22 pixels local image patches around the points. From the positive kinematics of the robot and the hand-eye calibration relationship, the pose of the camera can be known. When the scene is fixed, the corresponding points on the image under different perspectives can be known, and the local image patches are intercepted around the same corresponding point on the image under different perspectives. We took a set of two of them as the input end sample and output end expectation value of the training convolutional autoencoder and collected a total of 4000 sets of such local image patches.

The network training used the deep learning autoencoder model designed in this paper. The structure size of each layer is shown in Table 1. The encoding stage contains four convolutional layers, and the decoding stage contains four transposed convolutional layers. The neural network training adopted the form of full batch learning, where the epoch is 160, and the relationship change curve between the training error and iteration number is shown in Figure 15. When the number of epochs was less than 20, the loss value of the network model decreased faster. When the number of epochs was more than 20, the loss value of the network model decreased slowly. When the epoch was 120–160, the loss value of the network model remained basically stable, that is, the model converges. Through training the network model, a 32-dimensional local texture feature descriptor of the wood plank that was stable enough for viewing angle changes was finally obtained. During the experiment, four feature dimensions of local image patches (i.e., 16, 32, 64, and 128) were specifically tested, and recognized 600 planks with different poses. We used the final pose to meet the pose accuracy of the plank to be grasped as the correct pose for recognition. We calculated the recognition rate and pointpair matching time to evaluate the performance of these four feature dimensions, as shown in Figure 16. With the 16-dimensional, 32-dimensional, 64-dimensional, and 128-dimensional feature dimensions increasing sequentially, in other word, the feature expression code was more abundant, so the recognition rate of the wood plank gradually increased, i.e., increases of 83.2%, 95.6%, 95.9%, and 96.4%. After the feature dimension size reached 32 dimensions, the recognition rate was not significantly improved. An increase in feature dimension size was also accompanied by an increase in computing time. When the feature dimensions were 64 dimensions and 128 dimensions, the calculation time of the plank pose increased even more. Considering the recognition rate and computing time, the 32-dimensional local image feature descriptor of the wooden plank was finally selected.

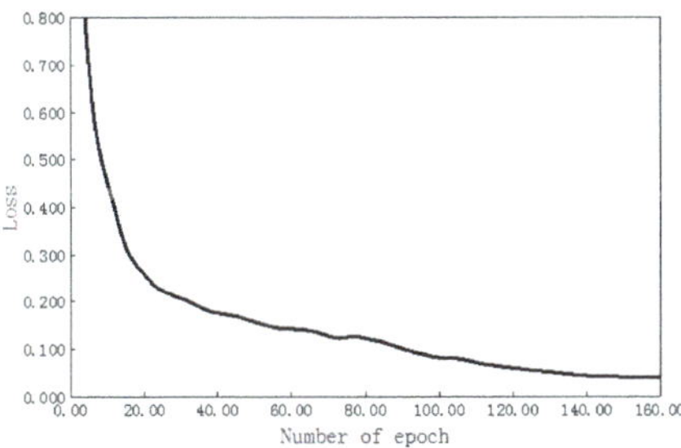

Figure 15. Iterative loss function curve of the deep convolution auto-encoder.

Table 1. Deep convolutional auto-encoder construction.

	Connection Layer	**Filter Size**	**Feature Size**	**Stride Size**	**Activation Function**
	Input layer	——	22×22	——	——
	Convolution layer 1	3×3	$20 \times 20 \times 128$	1	Relu
Coding stage	Convolution layer 2	$3 \times 3 \times 128$	$18 \times 18 \times 128$	1	Relu
	Convolution layer 3	$3 \times 3 \times 128$	$9 \times 9 \times 256$	2	Relu
	Convolution layer 4	$9 \times 9 \times 256$	$1 \times 1 \times 32$	——	Relu
	Transposed convolution 1	$9 \times 9 \times 32$	$9 \times 9 \times 256$	——	Relu
Decoding stage	Transposed convolution 2	$3 \times 3 \times 256$	$18 \times 18 \times 128$	2	Relu
	Transposed convolution 3	$3 \times 3 \times 128$	$20 \times 20 \times 128$	1	Relu
	Transposed convolution 4	$3 \times 3 \times 128$	22×22	1	Relu

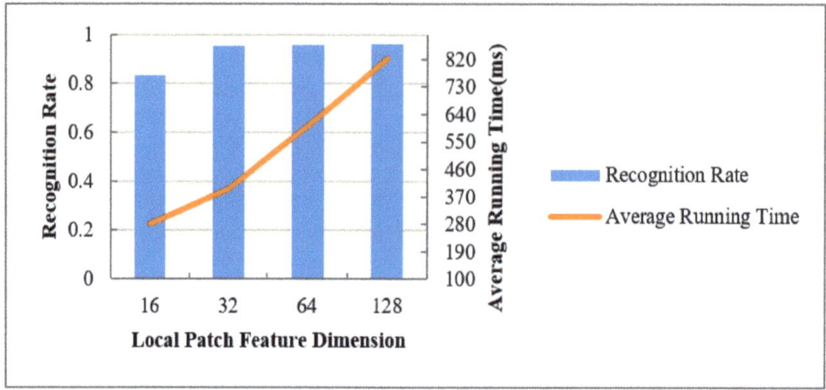

Figure 16. The influence of feature dimensions of the local image patch on recognition performance.

5.2. Grasping of Planks

The robot first grasps the top plank of the stacked plank. The original point cloud visualization result of the stacked plank scene under the RGB-D camera using the Rviz tool is shown in Figure 17. Combining the Euclidean distance clustering method [41], the point cloud under the camera was divided into different areas, so that different planks correspond to different point cloud clusters. The visualization result using the Rviz tool is shown in Figure 18. The cluster with the smallest average depth value was confirmed as the current priority grasping area, and the point pair matching was completed using the aforementioned method based on pointpair features and local image patch features. Then, we performed pose voting and clustering and finally determined the pose $^C_o M$ of the plank to be grasped.

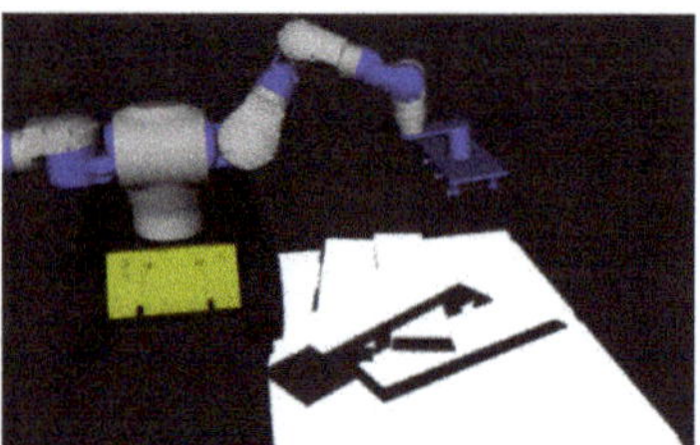

Figure 17. Original point cloud.

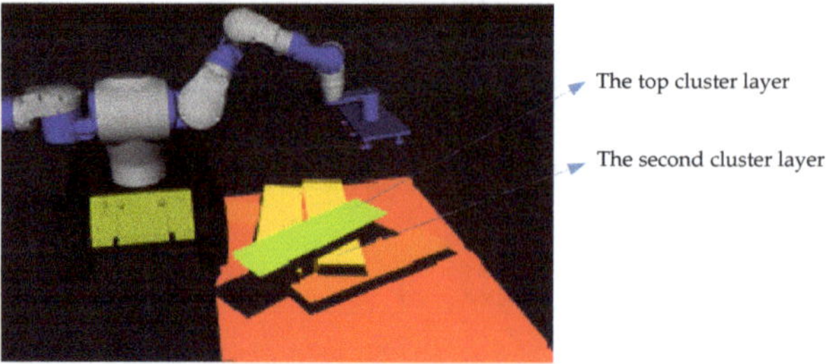

Figure 18. The segmentation result of point cloud after regional clustering.

After obtaining the pose ^c_oM of the plank to be grasped in the camera coordinate system, the current pose ^b_hM of the robot end (by forward kinematic solution of the robot kinematics) and the result ^c_hM of the robot hand-eye calibration, the plank's pose was converted to the robot base coordinate system:

$$^b_oM = {}^b_hM\,{}^c_hM^{-1}\,{}^c_oM \tag{10}$$

Then the grasping operation was realized by driving the end of the robotic arm to move to this position. As shown in the Figure 18, the top cluster has only one piece of wood, which is easy to locate and grasp with our method. When the robot hand has grasped and moved away the top board, the original second cluster layer is now the top position. The cluster contains two planks, since the two planks are close together. As shown in Figure 19, the two planks are not fully presented in camera vision and their respective image dose not include the whole plank, which is similar to the occlusion effect. In this case, our method can also obtain the recognition results of the cluster with multiple boards and select the target to be grasped (Figure 19). The multiple pose voting peak obtained by PPF algorithm get the poses of multiple wood planks in the same cluster and the pose with the highest voting score is selected as the target pose to be grasped by robot. As shown in Figure 20, the recognition and positioning of the plank to be grasped is accurate, and the robot grasping action process is smooth.

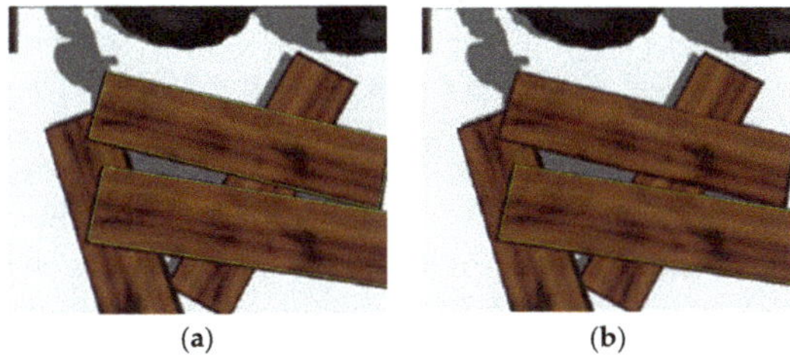

(a)   (b)

Figure 19. Wood plank recognition results: (**a**) Recognition results of some wood planks in the same cluster; (**b**) The board to be grasped.

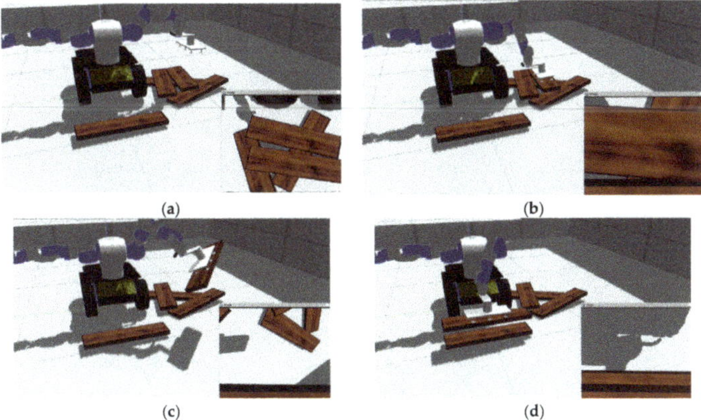

(a)   (b)

(c)   (d)

Figure 20. The result of the robot visually grasping wooden planks: (**a**) Identifying the pose of the plank to be grasped; (**b**) positioning the end of the robot and grasping; (**c**) during the robot handling process; (**d**) preparing to place the grasped wooden plank.

We carried out a grasping experiment 1000 times on randomly placed wooden planks in stacked wooden planks piles, also using other methods to perform the same number of experiments, and then compared the recognition rate, average recognition time, and grasping success rate. If the positioning accuracy of the recognition result is less than 3 mm and the rotation angle error is less than 2°, this situation is good for grasping success. This positioning accuracy is regarded as the correct recognition. As is shown in Table 2, the corresponding recognition performance comparison is shown in Figure 16, where "PPF" [7] was the traditional point pair method; "CPPF" [14] used the pint pair feature added color information. "SSD-6D" [17] used the convolutional neural networks for end-to-end training to obtain the pose of an object. "LPI" only used the local image patch proposed here to match the feature points and ICP calculated the pose, and did not use the point pair method to match; "LPPPF" is a method we proposed to determine the pose based on the local image patch combining point pair feature matching feature points, pose voting, and clustering.

Table 2. Performance of different combination methods in grasping wood planks. PPF: Point pair feature.

	PPF	CPPF	SSD-6D	LPI	LPPPF (Proposed Method)
Recognition rate	0.115	0.753	0.837	0.851	0.953
Average operation time (ms)	327	322	409	563	396
Grasping success rate	0.105	0.746	0.825	0.835	0.938

From Table 2 and Figure 21, it can be seen that the recognition rate of the plank is closely related to the success rate of robot grasping. The higher the recognition rate, the greater the success rate of robot grasping. The feature descriptors that only use PPF methods feature poor description of the surface features of the wood plank, so the recognition rate is significantly lower. Even if color information is added to the traditional point pair feature in CPPF method, color information is only one of the features of the plan and the recognition rate of this method is not high here. SSD-6D using convolutional neural networks for end-to-end training to obtain the pose of an object also does not have a high recognition rate because of the low positioning accuracy. The local image patches used only have certain advantages in describing wood texture features, and the recognition rate has been improved to a certain extent. However, the feature description is not comprehensive enough, resulting in a low recognition rate 85.1%. The LPPPF method we proposed here has a certain improvement in the recognition rate of the wood plank to be grasped compared to other methods, which is about 11 percentage points higher when using deep learning SSD-6D method. Compared with only using local image patch features, it is about 9 percentage points higher. Additionally, the average computing time is also relatively short, i.e., 396 ms. This shows that this method has obvious application advantages in grasping occasions in the scene of disorderly stacking planks. Through the convolutional autoencoder to extract the texture features of the local image patches of the wood, combined with the point pair features, the surface features of the wood can be better expressed. At the same time, in view of the hierarchical nature of the wood stacking, an Euclidean distance clustering method is used for segmentation first, which avoids the entire scene for collecting image patches for matching, greatly reducing the number of local image patches that need to be extracted and ultimately reducing the calculation time for recognition.

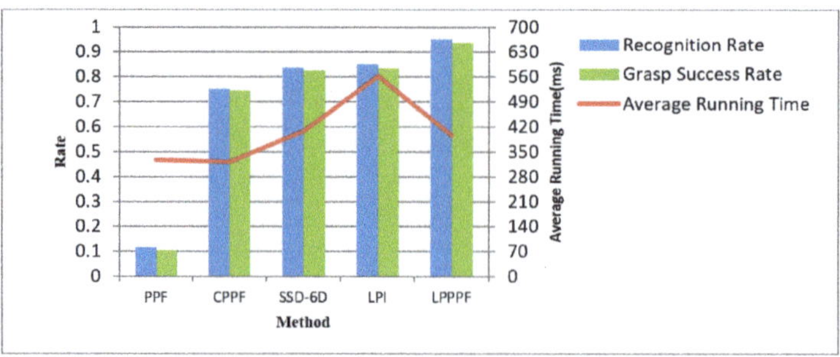

Figure 21. Comparison of the recognition performance of different methods.

6. Conclusions

The main shape of a plank is a large plane that is symmetrical and regular. The current conventional methods make it difficult to identify and locate planks to be grasped in scenes of disorderly stacked planks, which makes it difficult for robots to grasp them. A recognition and positioning method combining local image patches and point pair features was proposed here. Image patches were collected from disorderly stacked wooden boards in the robot vision scene and a convolutional autoencoder was used for training to obtain a 32-dimensional local texture feature descriptor that is robust to viewing angle changes. The local image patches around the point pair from the single-plank model were extracted, the feature code was extracted through the trained encoder, and the point pair geometric features were combined to form a feature code describing the feature of the board. In the stacking plank scene, the area of the plank to be grasped was segmented by a Euclidean distance clustering method and the feature code was extracted, and the plank to be grasped was identified through processes such as matching point pairs, pose voting and clustering. The robot grasping experiment here has proven that the recognition rate of this method is 95.3%, and the grasping success rate is 93.8%. Compared with PPF and other methods, the method presented here has obvious advantages. It is suitable for the grasping of disorderly stacked wood planks. At the same time, it has certain reference significance for recognition and grasping in other similar conditions.

Author Contributions: Methodology, writing—original draft, C.X.; conceptualization, funding acquisition, supervision, Y.L.; validation, F.D., and Z.Z. All authors have read and agreed to the published version of the manuscript.

Funding: Key Research & Development Plan of Jiangsu Province (Industry Foresight and Key Core Technologies) Project (grant no. BE2019112), Jiangsu Province Policy Guidance Program (International Science and Technology Cooperation) Project (grant no. BZ2016028), Qing Lan Project of the Jiangsu Province Higher Education Institutions of China, Natural Science Foundation of Jiangsu Province (grant no. BK20191209), Nantong Science and Technology Plan Fund Project (grant no. JC2019128).

Conflicts of Interest: The authors declare no conflict of interest.

References

1. Wang, Y.; Wei, X.; Shen, H.; Ding, L.; Wan, J. Robust Fusion for Rgb-d Tracking Using Cnn Features. *Appl. Soft Comput.* **2020**, *92*, 106302. [CrossRef]
2. Hinterstoisser, S.; Lepetit, V.; Ilic, S.; Holzer, S.; Bradski, G.; Konolige, K.; Navab, N. Model based training, detection and pose estimation of texture-less 3D objects in heavily cluttered scenes. In Proceedings of the Asian Conference on Computer Vision, Daejeon, Korea, 5–9 November 2012; pp. 548–562.
3. Rios-Cabrera, R.; Tuytelaars, T. Discriminatively Trained Templates for 3D Object Detection: A Real Time Scalable Approach. In Proceedings of the International Conference on Computer Vision (ICCV 2013), Sydney, Australia, 1–8 December 2013; pp. 2048–2055.
4. Rusu, R.; Bradski, G.; Thibaux, R.; HsuRetrievalb, J. Fast 3D recognition and pose using the viewpoint feature histogram. In Proceedings of the 2010 IEEE/RSJ International Conference on Intelligent Robots and Systems, Taipei, Taiwan, 18–22 October 2010; pp. 2155–2162.
5. Wang, F.; Liang, C.; Ru, C.; Cheng, H. An Improved Point Cloud Descriptor for Vision Based Robotic Grasping System. *Sensors* **2019**, *19*, 2225. [CrossRef] [PubMed]
6. Birdal, T.; Ilic, S. Point pair features based object detection and pose estimation revisited. In Proceedings of the 2015 International Conference on 3D Vision (3DV), Lyon, France, 19–22 October 2015; pp. 527–535.
7. Drost, B.; Ulrich, M.; Navab, N.; Ilic, S. Model globally, match locally: Efficient and robust 3D object recognition. In Proceedings of the 2010 IEEE Computer Society Conference on Computer Vision and Pattern Recognition, San Francisco, CA, USA, 13–18 June 2010; pp. 998–1005.
8. Li, D.; Wang, H.; Liu, N.; Wang, X.; Xu, J. 3D Object Recognition and Pose Estimation from Point Cloud Using Stably Observed Point Pair Feature. *IEEE Access* **2020**, *8*, 44335–44345. [CrossRef]
9. Lowe, D.G. Distinctive Image Features from Scale-Invariant Keypoints. *Int. J. Comput. Vis.* **2004**, *60*, 91–110. [CrossRef]

10. Yan, K.; Sukthankar, R. PCA-SIFT: A more distinctive representation for local image descriptors. In Proceedings of the 2004 IEEE Computer Society Conference on Computer Vision and Pattern Recognition, CVPR 2004, Washington, DC, USA, 27 June–2 July 2004; Volume 2, p. 2.

11. Bay, H.; Ess, A.; Tuytelaars, T.; Van Gool, L. Speeded-Up Robust Features (SURF). *Comput. Vis. Image Underst.* **2008**, *110*, 346–359. [CrossRef]

12. Johnson, A.E.; Hebert, M. Using spin images for efficient object recognition in cluttered 3D scenes. *IEEE Trans. Pattern Anal. Mach. Intell.* **1999**, *21*, 433–449. [CrossRef]

13. Salti, S.; Tombari, F.; di Stefano, L. SHOT: Unique signatures of histograms for surface and texture description. *Comput. Vis. Image Underst.* **2014**, *125*, 251–264. [CrossRef]

14. Choi, C.; Christensen, H.I. 3D pose estimation of daily objects using an RGB-D camera. In Proceedings of the 25th IEEE/RSJ International Conference on Robotics and Intelligent Systems, IROS 2012, Vilamoura, Algarve, Portugal, 7–12 October 2012; pp. 3342–3349.

15. Ye, C.; Li, K.; Jia, L.; Zhuang, C.; Xiong, Z. Fast hierarchical template matching strategy for real-time pose estimation of texture-less objects. In Proceedings of the International Conference on Intelligent Robotics and Applications, Hachioji, Japan, 22–24 August 2016; pp. 225–236.

16. Muñoz, E.; Konishi, Y.; Beltran, C.; Murino, V.; Del Bue, A. Fast 6D pose from a single RGB image using Cascaded Forests Templates. In Proceedings of the 2016 IEEE/RSJ International Conference on Intelligent Robots and Systems (IROS), Daejeon, Korea, 9–14 October 2016; pp. 4062–4069.

17. Liu, D.; Arai, S.; Miao, J.; Kinugawa, J.; Wang, Z.; Kosuge, K. Point Pair Feature-Based Pose Estimation with Multiple Edge Appearance Models (PPF-MEAM) for Robotic Bin Picking. *Sensors* **2018**, *18*, 2719. [CrossRef] [PubMed]

18. Li, M.; Hashimoto, K. Curve Set Feature-Based Robust and Fast Pose Estimation Algorithm. *Sensors* **2017**, *17*, 1782.

19. Wu, C.H.; Jiang, S.Y.; Song, K.T. CAD-based pose estimation for random bin-picking of multiple objects using a RGB-D camera. In Proceedings of the 2015 15th International Conference on Control, Automation and Systems (ICCAS), Busan, Korea, 13–16 October 2015; pp. 1645–1649.

20. Chen, Y.K.; Sun, G.J.; Lin, H.Y.; Chen, S.L. Random bin picking with multi-view image acquisition and CAD-based pose estimation. In Proceedings of the 2018 IEEE International Conference on Systems, Man, and Cybernetics (SMC), Miyazaki, Japan, 7–10 October 2018; pp. 2218–2223.

21. Kehl, W.; Manhardt, F.; Tombari, F.; Ilic, S.; Navab, N. SSD-6D: Making RGB-based 3D detection and 6D pose estimation great again. In Proceedings of the IEEE International Conference on Computer Vision (ICCV), Venice, Italy, 22–29 October 2017.

22. Caldera, S.; Rassau, A.; Chai, D. Review of Deep Learning Methods in Robotic Grasp Detection. *Multimodal Technol. Interact.* **2018**, *2*, 57. [CrossRef]

23. Kumra, S.; Kanan, C. Robotic grasp detection using deep convolutional neural networks. In Proceedings of the 2017 IEEE/RSJ International Conference on Intelligent Robots and Systems (IROS), Vancouver, BC, Canada, 24–28 September 2017; pp. 769–776.

24. Levine, S.; Pastor, P.; Krizhevsky, A.; Ibarz, J.; Quillen, D. Learning hand-eye coordination for roboticgrasping with deep learning and large-scale data collection. *Int. J. Robot. Res.* **2018**, *37*, 421–436. [CrossRef]

25. Zeng, A.; Song, S.; Yu, K.T.; Donlon, E.; Hogan, F.R.; Bauza, M.; Ma, D.; Taylor, O.; Liu, M.; Romo, E.; et al. Robotic pick-and-place of novel objects in clutter with multi-affordance grasping and cross-domain image matching. In Proceedings of the 2018 IEEE International Conference on Robotics and Automation (ICRA), Brisbane, QLD, Australia, 21–25 May 2018; pp. 1–8.

26. Kehl, W.; Milletari, F.; Tombari, F.; Ilic, S.; Navab, N. Deep learning of local RGB–D patches for 3D object detection and 6D pose estimation. In Proceedings of the 14th European Conference on Computer Vision (ECCV), Amsterdam, The Netherlands, 11–14 October 2016.

27. Zhang, H.; Cao, Q. Holistic and local patch framework for 6D object pose estimation in RGB-D images. *Comput. Vis. Image Underst.* **2019**, *180*, 59–73. [CrossRef]

28. Le, T.-T.; Lin, C.-Y. Bin-Picking for Planar Objects Based on a Deep Learning Network: A Case Study of USB Packs. *Sensors* **2019**, *19*, 3602. [CrossRef] [PubMed]

29. Tong, X.; Li, R.; Ge, L.; Zhao, L.; Wang, K. A New Edge Patch with Rotation Invariance for Object Detection and Pose Estimation. *Sensors* **2020**, *20*, 887. [CrossRef]

30. Jiang, P.; Ishihara, Y.; Sugiyama, N.; Oaki, J.; Tokura, S.; Sugahara, A.; Ogawa, A. Depth Image–Based Deep Learning of Grasp Planning for Textureless Planar-Faced Objects in Vision-Guided Robotic Bin-Picking. *Sensors* **2020**, *20*, 706. [CrossRef] [PubMed]
31. Vidal, J.; Lin, C.-Y.; Lladó, X.; Martí, R. A Method for 6D Pose Estimation of Free-Form Rigid Objects Using Point Pair Features on Range Data. *Sensors* **2018**, *18*, 2678. [CrossRef]
32. Ni, C.; Zhang, Y.; Wang, D. Moisture Content Quantization of Masson Pine Seedling Leaf Based on StackedAutoencoder with Near-Infrared Spectroscopy. *J. Electr. Comput. Eng.* **2018**, *2018*, 8696202.
33. Shen, L.; Wang, H.; Liu, Y.; Liu, Y.; Zhang, X.; Fei, Y. Prediction of Soluble Solids Content in Green Plum by Using a Sparse Autoencoder. *Appl. Sci.* **2020**, *10*, 3769. [CrossRef]
34. Ni, C.; Li, Z.; Zhang, X.; Sun, X.; Huang, Y.; Zhao, L.; Zhu, T.; Wang, D. Online Sorting of the Film on CottonBased on Deep Learning and Hyperspectral Imaging. *IEEE Access.* **2020**, *8*, 93028–93038. [CrossRef]
35. Li, Y.; Hu, W.; Dong, H.; Zhang, X. Building Damage Detection from Post-Event Aerial Imagery Using Single Shot Multibox Detector. *Appl. Sci.* **2019**, *9*, 1128. [CrossRef]
36. Ranjan, R.; Patel, V.M.; Chellappa, R. Hyperface: A deep multi-task learning framework for face detection, landmark localization, pose estimation, and gender recognition. *IEEE Trans. Pattern Anal. Mach. Intell.* **2017**, *41*, 121–135. [CrossRef]
37. Zhao, W.; Jia, Z.; Wei, X.; Wang, H. An FPGA Implementation of a Convolutional Auto-Encoder. *Appl. Sci.* **2018**, *8*, 504. [CrossRef]
38. Ni, C.; Wang, D.; Vinson, R.; Holmes, M. Automatic inspection machine for maize kernels based on deep convolutional neural networks. *Biosyst. Eng.* **2019**, *178*, 131–144. [CrossRef]
39. Ni, C.; Wang, D.; Tao, Y. Variable Weighted Convolutional Neural Network for the Nitrogen Content Quantization of Masson Pine Seedling Leaves with Near-Infrared Spectroscopy. *Spectrochim. Acta Part A Mol. Biomol. Spectrosc.* **2019**, *209*, 32–39. [CrossRef] [PubMed]
40. Gallego, A.-J.; Gil, P.; Pertusa, A.; Fisher, R.B. Semantic Segmentation of SLAR Imagery with Convolutional LSTM Selectional AutoEncoders. *Remote Sens.* **2019**, *11*, 1402. [CrossRef]
41. Aloise, D.; Deshpande, A.; Hansen, P.; Popat, P. NP-hardness of Euclidean sum-of-squares clustering. *Mach. Learn.* **2009**, *75*, 245–248. [CrossRef]
42. Dong, J.; Peng, Y.; Ying, S.; Hu, Z. LieTrICP: An improvement of trimmed iterative closest point algorithm. *Neurocomputing* **2014**, *140*, 67–76. [CrossRef]

Publisher's Note: MDPI stays neutral with regard to jurisdictional claims in published maps and institutional affiliations.

Article

Enhanced Image-Based Endoscopic Pathological Site Classification Using an Ensemble of Deep Learning Models

Dat Tien Nguyen, Min Beom Lee, Tuyen Danh Pham, Ganbayar Batchuluun *, Muhammad Arsalan  and Kang Ryoung Park

Division of Electronics and Electrical Engineering, Dongguk University, 30 Pildong-ro 1-gil, Jung-gu, Seoul 04620, Korea; nguyentiendat@dongguk.edu (D.T.N.); mblee@dongguk.edu (M.B.L.); phamdanhtuyen@dongguk.edu (T.D.P.); arsal@dongguk.edu (M.A.); parkgr@dongguk.edu (K.R.P.)
* Correspondence: ganabata87@dongguk.edu; Tel.: +82-10-9948-3771; Fax: +82-2-2277-8735

Received: 9 September 2020; Accepted: 21 October 2020; Published: 22 October 2020

Abstract: In vivo diseases such as colorectal cancer and gastric cancer are increasingly occurring in humans. These are two of the most common types of cancer that cause death worldwide. Therefore, the early detection and treatment of these types of cancer are crucial for saving lives. With the advances in technology and image processing techniques, computer-aided diagnosis (CAD) systems have been developed and applied in several medical systems to assist doctors in diagnosing diseases using imaging technology. In this study, we propose a CAD method to preclassify the in vivo endoscopic images into negative (images without evidence of a disease) and positive (images that possibly include pathological sites such as a polyp or suspected regions including complex vascular information) cases. The goal of our study is to assist doctors to focus on the positive frames of endoscopic sequence rather than the negative frames. Consequently, we can help in enhancing the performance and mitigating the efforts of doctors in the diagnosis procedure. Although previous studies were conducted to solve this problem, they were mostly based on a single classification model, thus limiting the classification performance. Thus, we propose the use of multiple classification models based on ensemble learning techniques to enhance the performance of pathological site classification. Through experiments with an open database, we confirmed that the ensemble of multiple deep learning-based models with different network architectures is more efficient for enhancing the performance of pathological site classification using a CAD system as compared to the state-of-the-art methods.

Keywords: pathological site classification; in vivo endoscopy; computer-aided diagnosis; artificial intelligence; ensemble learning

---

1. Introduction

Currently, cancer is a leading cause of human death worldwide [1]. There are several types of cancer such as lung [2,3], breast [4,5], skin [6,7], stomach [8–10], colorectal (also known as colon cancer) [11–16], thyroid [17–19], and brain [20–22] cancers. Among these, stomach and colorectal cancer (CRC) are two of the most common types causing death in humans. To diagnose these types of cancer, in vivo endoscopy is widely used. This technique allows for a detailed visualization of the in vivo structure of the colon or stomach, which is significantly useful to doctors for examining the evidence of disease. However, conventional medical imaging-based diagnostic techniques are still predominantly dependent on the personal knowledge and experiences of doctors (radiologists). Consequently, the diagnosis results have a large variance. To reduce this variance, a double-screening process can be invoked, in which two or more experts (radiologists) are required to read the captured medical images of a single case. Although this method is more efficient than the conventional diagnosis method, it is costly and time-consuming. Recently, a computer-aided diagnosis (CAD) has been widely employed to assist doctors in the diagnosis process, and it is now becoming important for enhancing the performance of the diagnosis process using medical imaging techniques.

Depending to the stage of the disease, several signatures may appear on the colon or stomach, such as polyp or gastritis regions. These regions are called pathological sites. Previous studies on pathological site detection/classification mostly used a single convolutional neural network (CNN). In a previous study [11], Patino-Barrientoo et al. proposed the use of a Visual Geometry Group (VGG)-based network for classifying endoscopic colon images into malignant and nonmalignant (benign) cases. Similarly, Ribeiro et al. [13] used the CNN for automated classification of polyp images. Experimental results with a relatively shallow network (five convolution layers and three fully connected layers) showed that the CNN is useful for polyp classification problems and outperforms other handcrafted feature extraction-based methods. Instead of training a new classification network, as shown in the studies by Patino-Barrientoo et al. [11] and Ribeiro et al. [13], Fonolla et al. [16] used a pretrained CNN model based on the residual network [23] that was trained on ImageNet image dataset as an image feature extractor, and classified input images into malignant and nonmalignant categories using conventional classification methods. Zhu et al. [9] also used a deep residual network and the transfer learning technique to determine the invasion depth of gastric cancer as well as the classification accuracy. They showed that the CNN-based method achieves significantly higher accuracy than human endoscopists. Similarly, Li et al. [10] used a CNN model, named GastricNet, for gastric cancer identification. As reported by their study, deep CNN-based methods are efficient for gastric cancer identification problems.

Although these studies showed that CNN-based methods have been successfully applied to solve the pathological site (polyp) classification problem, they all have a common limitation, which is the use of a single CNN model for the problem. As indicated by several previous studies on CNNs, the performance of the CNN-based method is highly dependent on several factors such as the depth and width of the network, number of network parameters, and architecture of the network. Among these factors, the design of network architecture performs an important role, especially when the depth of the network increases. All these previous studies used relatively shallow networks [11,13] or extracted image features at the last convolution layer of a deep residual network for classification problems [16]. As indicated by previous studies, the depth of a deep learning-based system can significantly enhance the performance of a detection/classification system, while a small number of network parameters help to prevent over/underfitting problems, ensuring that the network is easy to train [23–27]. Therefore, the performance of a classification system is limited because of the use of a shallow network or the extraction of image features using a pretrained model. In addition, it is significantly difficult to recognize pathological sites as benign or malignant cases at an early stage. Therefore, the classification of polyps into benign or malignant cases as performed by previous studies can yield incorrect results at the early stage of a disease.

In a recent study, Kowsari et al. [28] proposed a new ensemble, deep learning approach for classification, namely random multi-model deep learning (RMDL). The RMDL approach solves the problem of finding the best deep learning structure and architecture to improve the classification. As a result of their study, the author showed that the ensemble learning method is efficient in enhancing the classification performance of various systems such as image-based system and text-based classification system. Inspired by the work by Kowsari et al., we proposed the use of multiple CNNs that differ in network architecture and depth to efficiently extract image texture features from input images to overcome these limitations of previous studies.

In contrast to previous studies that classify pathological site images into malignant and nonmalignant cases, our study classifies an input endoscopy image into one of two categories, with or without the appearance of pathological sites. Although this task can be accomplished by using a method to detect pathological sites in endoscopic images, the training process of a detection method requires strong efforts to accurately localize the pathological sites in the training dataset. However, at the early stage of the disease, the pathological sites are small and/or unclear, which can cause difficulties in creating ground-truth labels as well as in the detection process. Therefore, we simplified this task by simply classifying the input images into two classes (with or without the appearance of pathological sites) without the requirement for correct labeling of pathological site in input training images. This approach helps doctors focus on images with pathological sites rather than the other normal images during the diagnosis process. Table 1 lists comparative summaries of the proposed and previous studies. Our proposed method is novel in the following four ways:

- In a variation from the previous research focusing on only the polyp classification, our research is the first to classify pathological sites including both polyp and complex vascular information.
- To overcome the limitations of previous methods, whose performance is limited due to the use of a single model for classification, our study uses multiple deep learning-based models for pathological site classification. By employing an ensemble of multiple deep learning-based models, we can enhance the classification performance of pathological site classification.
- We employed and trained three different CNN model architectures, including VGG-, inception-, and densely connected convolutional network (DenseNet)-based networks, for ensemble learning purposes. Consequently, each model has its own strengths and weaknesses, and we can combine three models to enhance the classification performance using three combination methods, i.e., MAX, AVERAGE, and VOTING.
- Our algorithm is available to the public through [29] so that other researchers can impartially compare with our method.

The remainder of this paper is organized as follows. In Section 2, we propose an image-based pathological site classification method based on an ensemble of deep learning models. In Section 3, we present a validation of the performance of the method proposed in Section 2 using a public in vivo gastrointestinal (GI) endoscopic images, namely the in vivo GI endoscopy dataset [30], and compare it with previous studies and discuss our results. Finally, we present the conclusion of our study in Section 4.

Table 1. Comparative summaries of proposed and previous studies on image-based polyp or pathological site classification.

Category	Method	Strength	Weakness
Handcrafted feature-based	- Extracts image features using handcrafted image features - Classification based on the extracted image features and classification methods, such as Support Vector Machine (SVM) and k-Nearest Neighbor (k-NN) [11,13]	Easy to implement	- Low accuracy - Only focuses on polyp classification
Deep feature-based	- Trains a single CNN model for classification problem. [9–11,13] - Extracts image features using a pretrained convolutional neural network (CNN) and classifies using classification methods such as SVM, k-NN [16]	High accuracy when compared to handcrafted-based method	- More complex than the handcrafted-based method - Requires large amount of training data, strong hardware etc. - Only focuses on polyp classification
	Ensemble of multiple CNNs with different network architectures (Proposed method)	- Focuses on the classification of pathological sites including both polyp and complex vascular information - Extracts rich image features using different architectures of CNN - Combines and takes advantage of single CNN model to enhance classification accuracy	- More complex than previous studies - Requires large amount of training data, strong hardware etc.

2. Proposed Method

2.1. Overview of Proposed Method

Figure 1 shows the overall procedure of our proposed method for enhancing the performance of the pathological site classification system. As explained in Section 1, previous studies predominantly classified the pathological sites using conventional classification methods [11], or used a single CNN model [9–11,13,16]. Consequently, the classification performance is limited. These studies have a common drawback that they used single network architectures for classification problems. Therefore, the extracted image features are dependent on the network architecture. As stated in previous studies [23–26], the architecture of the CNN performs an important role in the performance of deep learning-based systems. For example, conventional CNNs, which are a linear stack of convolution layers, are suitable for a shallow design to solve a simple problem, whereas the residual or dense network is used to increase the depth of conventional CNNs; consequently, they can easily train a complex problem; alternatively, the inception network is used to extract richer features in a single network when compared to conventional CNNs. Based on these observations, we designed our CNN for the pathological site classification problem by incorporating these observations into a single classification network, as shown in Figure 1.

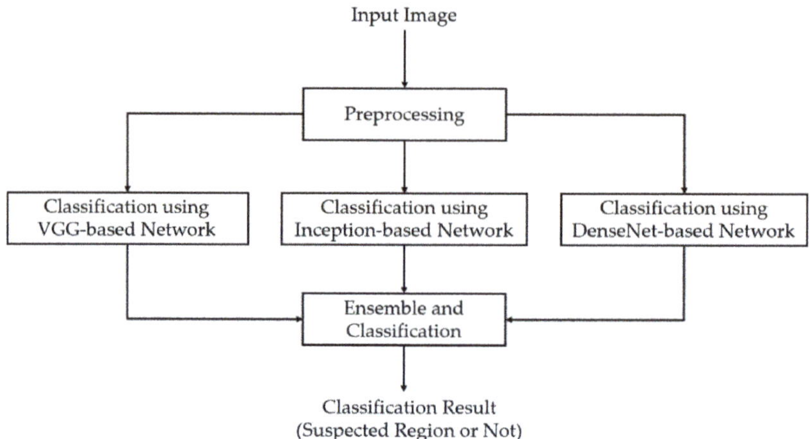

Figure 1. Overview of the proposed method.

As shown in Figure 1, our proposed method first preprocesses the input endoscopic images to reduce noise and prepares them for inputting to the subsequent stages, which are based on deep learning techniques for the classification problem. A detailed explanation of this step is presented in Section 2.2. Our proposed method comprises of the following three main steps for classification. The first step is the classification using a conventional CNN based on VGG16 network architecture [25]. The second step is the classification using an inception-based network to extract richer features according to the different sizes of objects [26]; finally, the third step is based on a DenseNet architecture to exploit the effect of a significantly deep network [31]. From the results of these three networks, we can obtain three classification results. We believe that these three branches are equivalent to the previous studies that are based on a single simple CNN for classification problems. To enhance the performance of the pathological site classification problem, our study further combines these classification results using the ensemble learning technique and performs classification based on the combined results. These explanations are provided in detail in Section 2.3.

2.2. Preprocessing of Captured Endoscopic Images

As explained in Section 2.1, the first step in our proposed method is the preprocessing step to eliminate redundant information from the endoscopic images before inputting them to CNNs. In Figure 2a, we show an example of a captured GI endoscopic image. As shown in this figure, the captured image normally contains two parts, including the background region with low illumination, and the foreground region with higher illumination. It can be observed that the background contains no useful information for our classification steps. Therefore, it should be removed before using the classification steps. This step is useful because the background region not only contains no information about the pathological sites but also presents noise that can consequently decrease the performance of further classification networks.

As shown in Figure 2a, the background region appears with a significantly low illumination when compared to the foreground region. Based on this characteristic, we implemented a simple method for background removal. A graphical explanation using a GI endoscopic image is shown in Figure 2a where the accumulated histogram-like features of pixels are represented in the horizontal direction. As shown in this figure, we first accumulate the histogram-like features of the input image by projecting the gray-level of the image pixel in the horizontal and vertical directions. Because of the low illumination characteristics of the background regions when compared to the foreground regions, we can arbitrarily set a threshold for separating the background and foreground. An example of the experimental result of this step using the image in Figure 2a is shown in Figure 2b. We can observe that, although Figure 2b still contains certain small background regions owing to the characteristics of the capturing devices, most of the background region was removed from the input image and the image is ready for further use in our proposed method.

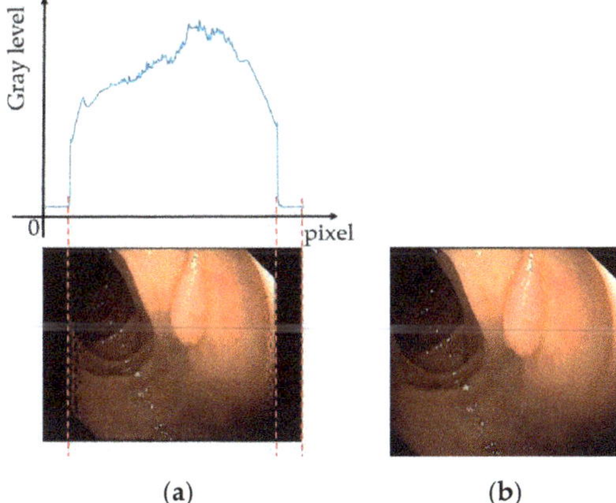

(a)   (b)

Figure 2. (**a**) Example of a captured gastrointestinal endoscopic image and (**b**) the corresponding result of the preprocessing step.

2.3. Pathological Site Classification Method

2.3.1. Deep Learning Framework

Recently, with the development of learning-based techniques, deep learning has been widely used in computer vision research. This technique has been successfully applied to various computer vision/pattern recognition problems such as image classification [23–27], object detection [32,33], and image generation [34–36]. The success of deep learning comes from the fact that this technique simulates the way in which the human brain processes information (images). Originally, a deep learning network was constructed by using several neural network layers to create a deep network that is used to process incoming information (images, voice, text, etc.). For computer vision research, an efficient type of deep learning technique, called CNN, has received significant attention. The idea of this type of deep learning technique is the use of convolution operations to extract useful information from input images, and the use of a neural network to efficiently solve a classification/regression problem. Figure 3 shows a graphical representation of a conventional CNN. As shown in this figure, a conventional CNN is composed of two primary parts, i.e., multiple convolution and classification/regression layers. The primary purpose of the convolution layers is to extract abstract and useful texture features that satisfactorily represent the characteristics/content of input images. The classification/regression layers are used to classify the input images (image classification problem), or regress continuous values such as height and width position of an object (object detection problem) based on the extracted image features produced by the first part of the CNN. Because of the use of convolution operation with the weight sharing scheme, the number of network parameters is significantly reduced when compared to a fully connected network with the same number of network layers. Consequently, it can help to successfully train a deep network and reduce the over/underfitting problem that normally occurs while using deep neural networks because of the large number of network parameters. However, with the increase in the depth and width of modern networks, the number of network parameters is still large, which prevents the successful training of a significantly deep network.

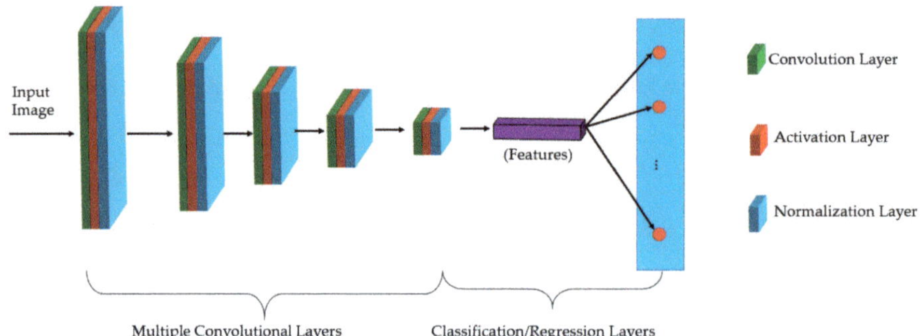

Figure 3. Conventional architecture of convolutional neural networks.

2.3.2. Enhanced Convolutional Neural Network (CNN) Structures for Efficient Feature Extraction

As explained in Section 2.3.1, although the CNN method is efficient for many computer vision tasks, it still has several drawbacks that are caused by the presence of a large number of layers (depth) and network parameters. As the number of layers in the CNN increases, it increases difficulty in training the network and presents the gradient vanishing problem. In addition, training a network with a large number of network parameters requires a large amount of training data to prevent the over/underfitting problem. Further, extracting efficient texture features is also crucial for enhancing the performance of a CNN-based system. To reduce the effects of these problems and enhance the performance of CNN-based systems, several network architectures were proposed. In our study, we used two popular network architectures, including the inception and dense networks that are designed for this purpose.

First, the conventional CNN is constructed by linearly stacking layers to create a deep network. Further, each convolution layer uses a fixed convolution kernel (such as 3×3, 5×5, or 7×7 kernel), as shown in Figure 3. However, when the problem becomes complex with the complex texture structure of input images and/or different sizes of objects, the use of fixed and single convolution kernels is insufficient for extracting efficient features for the classification/detection problem. To solve this problem, Szegedy et al. [26] proposed a new network architecture named inception block to extract richer information from input images than the conventional convolution layers. The concept of the inception block is depicted in Figure 4. In this figure, the input from the previous layer is shown as a red box, while the output and different feature maps obtained using various convolution kernels are marked as blue, yellow, green, and purple boxes. As shown in Figure 4, the inception block is constructed by using multiple convolution layers with different convolution kernels such as 1×1, 3×3, 5×5, and a max-pooling layer. These operations are performed in a parallel manner, and the results of all operations are concatenated to form the final input of an inception layer. It can be easily observed from Figure 4 that the inception layer can extract more useful texture information than the conventional convolution layer because of the use of different convolution kernels. A small convolution kernel can extract small texture features (small object texture). By increasing the convolution kernel, larger receptive fields are used to extract the texture information in the input images. By concatenating the results of all single convolution operations, the final feature maps are expected to contain a large amount of information at various texture levels when compared to the conventional convolution layer. Thus, more useful and efficient information is extracted using the inception layer. As demonstrated by the author of the inception method, it is significantly more efficient than a conventional CNN for the classification problem using the ImageNet dataset [26].

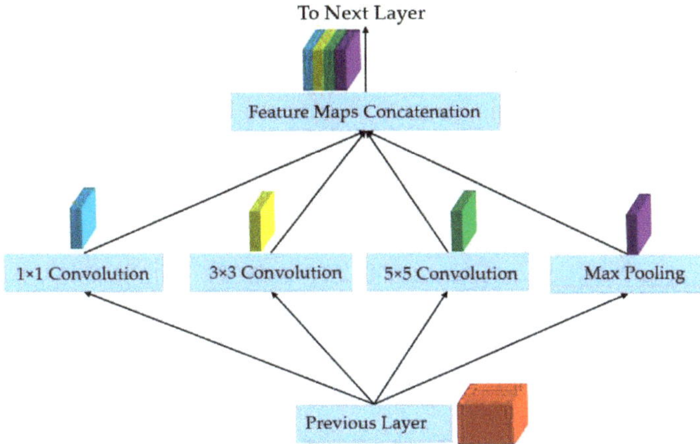

Figure 4. Naïve inception architecture for rich feature extraction used in inception network.

The second CNN architecture used in our study, the dense connection, was proposed by Huang et al. [31], which was designed to alleviate the vanishing gradient problem, reuse features, make the network easier to train, and reduce the number of network parameters. In Figure 5, we show the difference between the conventional CNN architecture (Figure 5a) and the dense-connection network architecture (Figure 5b). In this figure, Conv-ReLU-BN indicates the sequence of convolution (Conv), activation (rectified linear unit (ReLU)), and batch normalization (BN) blocks that are used to manipulate the input feature maps. While the conventional CNN architecture only has a connection between a single parent/child layer, the dense-connection architecture uses all the outputs of the preceding layers as inputs to the current layer. This design helps to reuse the features from early layers and reduces the effects of the vanishing gradient problem. By using dense connections, the network can be thinner and more compact. Consequently, it helps to reduce the number of network parameters that normally cause the over/underfitting problem in CNNs.

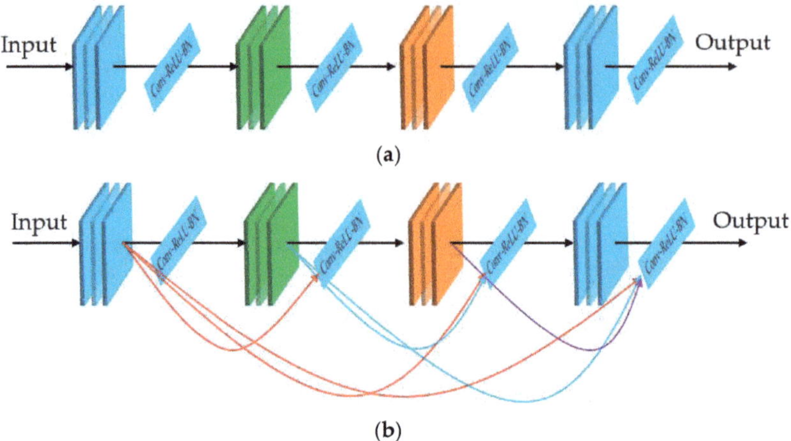

Figure 5. Comparison between (**a**) plain convolutional blocks and (**b**) dense-connection block architecture.

Inspired by the advantages of the inception and dense-connection networks, we used these two networks in addition to the conventional CNN based on the VGG architecture in our proposed method and experiments. Using these three networks with three different architectures, we processed input images in a different manner to enhance the performance of our classification system when combining the strength of each network.

2.3.3. Ensemble of CNN Models for Pathological Site Classification

Based on the success of the CNN in the image classification problem, we propose a method for pathological site image classification, as shown in Figure 1 and Section 2.1. As explained in Section 2.3.1, although CNNs were successfully applied to various computer vision problems, this technique still has several drawbacks that prevent us from obtaining high-performance classification systems. First, the large number of network parameters can increase the difficulty in training the network and can also result in the over/underfitting problem. Second, the difference of the size of objects (texture features) that appear in the input images can affect the classification performance. To reduce the effects of these drawbacks and enhance the classification performance of the pathological site classification system, we propose the use of an ensemble learning technique that combines the classification results of three different CNN architectures for the problem, as shown in Figure 1.

In our study, we used three CNN architectures with different characteristics and depths for the classification problem, including VGG-based, DenseNet-based, and inception-based networks, as shown in Figure 1. Although it is possible to use other network architectures, we selected these network architectures because of our purpose for ensemble learning, that is, to combine the classification results of different network architectures in which each network classifies input images in a specific way. In our study, the VGG-based network serves as a conventional deep CNN for classification problems, which is composed of a linear stack of convolution layers. The DenseNet-based network serves as a very deep CNN with a short-cut path, which helps to easily train the network and extract more abstract and efficient image features. Further, the inception-based network helps extract image features with rich texture features and different sizes of objects. The detailed descriptions of these network architectures are listed in Table 2.

Table 2. Detail description of the convolutional neural networks (CNNs) used in our study (N/A means "not available").

Network	Layer	Input Shape	Output Shape	Number of Network Parameter
VGG-based Network	Input Layer	$224 \times 224 \times 3$	N/A	0
	Main Convolution Layers in VGG16 Network	$224 \times 224 \times 3$	$7 \times 7 \times 512$	14,714,688
	Flatten Layer	$7 \times 7 \times 512$	25,088	0
	Drop-out Layer	25,088	25,088	0
	Dense Layer	25,088	2	50,178
DenseNet-based Network	Input Layer	$224 \times 224 \times 3$	N/A	0
	Main Convolution Layers in DenseNet	$224 \times 224 \times 3$	$7 \times 7 \times 1024$	7,037,504
	Flatten Layer	$7 \times 7 \times 1024$	50,176	0
	Drop-out Layer	50,176	50,176	0
	Dense Layer	50,176	2	100,354
Inception-based Network	Input Layer	$224 \times 224 \times 3$	N/A	0
	Main Convolution Layers in Inception network	$224 \times 224 \times 3$	$5 \times 5 \times 2048$	21,802,784
	Flatten Layer	$5 \times 5 \times 2048$	51,200	0
	Drop-out Layer	51,200	51,200	0
	Dense Layer	51,200	2	102,402

In this table, the convolution layers of the base networks (VGG16 [25], DenseNet121 [31], and inception [26] networks) are marked as "Main Convolution Layers." For the classification part, we modified the number of output neurons from 1000 of the original network (the original networks were designed for the ImageNet classification challenge; therefore, it contains 1000 neurons at the output) to 2 neurons that represent two possible cases of our problem (with/without the appearance of pathological sites).

In the final step in our proposed method, as mentioned in Section 2.1, we attempted to combine the classification results of three different CNN models, i.e., VGG-, inception-, and DenseNet-based networks. In our study, we invoke three combination methods, including the MAX, AVERAGE, and VOTING rules, which are presented in Equations (1)–(3), respectively.

$$\text{MAX rule} = argmax(\max(S_i)) \tag{1}$$

$$\text{AVERAGE rule} = argmax\left(\frac{\sum_{i=1}^{n} S_i}{n}\right) \tag{2}$$

$$\text{VOTING rule} = sign\left(\sum_{i=1}^{n} w_i \times argmax(S_i)\right) \tag{3}$$

In Equations (1)–(3), S_i indicates the classification probability of the ith classifier, and n indicates the number of classifiers. The *"argmax"* operator indicates the selection process of the class label whose classification probability is maximum. In our experiments, we used $n = 3$ as we used three classifiers. As explained in the previous sections, we are dealing with a binary classification problem. Therefore, Si is a vector of two components, i.e., $S_i = (S_{0i}, S_{1i})$, where S_{0i} indicates the probability that the input image belongs to class 0 (negative cases (without pathological sites)), and S_{1i} stands for the probability of the input image belonging to class 1 (positive case (with pathological sites)). In Equation (3), w_i indicates the weight for the ith classifier and $argmax(S_i)$ indicates the classification results of the ith classifier ($argmax(S_i) = 0$ for predicting the input image belonging to class 0; and $argmax(S_i) = 1$ for predicting the input image belonging to class 1. The weight value (w_i) is determined using Equation (4). In our experiments, we will measure the performance of the classification system using all three combination rules and select the rule that is most suitable with our pathological site classification problem based on the experimental results, as shown in Section 3.

$$w_i = \begin{cases} -1 \; if \; argmax(S_i) \; = \; 0 \\ +1 \; if \; argmax(S_i) \; = \; 1 \end{cases} \tag{4}$$

2.4. Classification Performance Measurement

To measure the performance of a CAD system, previous studies used three popular metrics: Sensitivity, specificity, and overall accuracy [37,38]. These metrics are used to measure three different aspects of a binary classification system, i.e., the classification/detection ability of the system with respect to the positive (with the appearance of disease), negative (without the appearance of disease), and overall cases. The definition of these metrics is presented in Equations (5)–(7). In these equations, Sens, Spec, and Accuracy stands for sensitivity, specificity, and overall accuracy; TP indicates the number of true-positive cases (the case in which a positive case is correctly classified as a positive case); FN indicates the number of false-negative cases (the case where a positive case is incorrectly classified as a negative case); TN indicates the number of true-negative cases (the case in which a negative case is correctly classified as a negative case); finally, FP indicates the number of false-positive cases (the case in which a negative case is incorrectly classified as a positive case).

$$Sens = \frac{TP}{TP + FN} \tag{5}$$

$$Spec = \frac{TN}{TN + FP} \tag{6}$$

$$Accuracy = \frac{TP + TN}{TP + TN + FP + FN} \tag{7}$$

As indicated in Equation (5), the sensitivity measurement is the ratio between the TP samples over the total number of positive samples (TP + FN). Therefore, it indicates the ability of the classification system to detect positive cases (distinguish disease cases from all possible disease cases). The specificity is the ratio between the TN samples and the total number of negative samples. Consequently, the specificity is the measurement of the classification performance in detecting negative samples from all possible negative samples. Finally, the overall accuracy is measured by measuring the ratio between all correct classification samples (TP + TN) over all testing samples (TP + TN + FP + FN). Thus, the overall accuracy indicates the ability of a classification system to detect correct samples from the universe of samples.

In our study, we used these three measurements to evaluate the performance of our proposed method. In addition, as indicated by the meaning of the overall accuracy, we used the overall accuracy measurement to compare the performance of our proposed method with that of previous studies.

3. Experimental Results

Using the proposed method mentioned in Section 2, we conducted various experiments using the publicly available in vivo GI endoscopy dataset [30] to measure the classification performance of our proposed method in this section. The detailed experimental results are presented in the following subsections.

3.1. Dataset and Experimental Setups

To evaluate the performance of our proposed method and compare it with previous studies, we conducted experiments using a public dataset, namely the in vivo GI endoscopic dataset [30]. We called this dataset Hamlyn-GI for convenience, as this dataset is collected and provided by the Hamlyn Center for Robotic Surgery [30]. This dataset was originally collected for tracking and retargeting of GI endoscopic pathological sites using Olympus narrow-band imaging and Pentax i-Scan endoscope devices [30]. Specifically, this dataset contains 10 video sequences of GI endoscopic scans. Each video is saved in the format of successive still images. In the study by Ye et al. [30], the authors first manually defined a pathological site (a small polyp or suspected region) at the beginning of the endoscopic image sequence. Then, they tracked and retargeted this region for the remaining sequence of images. The information regarding the selected region and ground-truth tracked-retargeted region is provided for each video sequence in an annotation file. Because these regions are carefully set by experts and the polyp region or possible polyp regions are focused on, we used the information in the annotation file as an indicator of the existence of pathological sites in the still images. In the case of a still image containing a pathological site region, an approximate location of the pathological site is provided in the annotation file; otherwise, a negative value is provided. Based on the provided information in the annotation file, we preclassified the still images into two categories: With and without the existence of the pathological site in the stomach; that is, if the annotation of a still image is provided, then the still image is considered to contain the pathological site and assigned to the "with pathological site" class; otherwise, the still image is assigned to the "without pathological site" class. In Figure 6, we show certain examples of images from the Hamlyn-GI dataset. Figure 6a shows example images without the existence of pathological sites, whereas Figure 6b shows images containing pathological sites, which are marked with white bounding boxes. From Figure 6b, it can be observed that the bounding boxes are approximately provided by the author of the dataset. Therefore, they do not fit the correct location of pathological site regions. Consequently, it is difficult to perform a detection to solve this problem. Instead, we classified the input still images of this dataset into two classes: With and without the existence of pathological sites. In Table 3, we list the detailed statistical information regarding the Hamlyn-GI dataset. In total, the Hamlyn-GI dataset contains 7894 images.

Table 3. Detailed description of images in the Hamlyn-GI dataset.

Sequence Index	1	2	3	4	5	6	7	8	9	10	Total
Images	705	1003	1700	1349	578	336	493	325	266	1139	7894

To measure the performance of the proposed method, we performed two-fold cross-validation. For this purpose, we divided the Hamlyn-GI dataset into two separate parts, namely training and testing datasets. In the first fold, we assigned images of the first five video sequences (video files 1–5 in Table 3) as the training dataset and the images of the remaining five video sequences (video files 6–10 in Table 3) as the testing dataset. In the second fold, we exchanged the training and testing datasets of the first fold, i.e., training dataset contains images of the last five video sequences (video files 6–10 in Table 3) and the testing dataset contains images of the first five video sequences (video files 1–5 in Table 3). This division method ensures that the images of the same person (identity) only exist in either the training or testing dataset. Finally, the overall performance of the dataset with a two-fold cross-validation approach is measured by calculating the average (weighted by the number of testing images) of the two folds. In Table 4, we list a detailed description of the training and testing datasets in our experiments. In this table, "With PS" indicates the existence of a pathological site condition; further, "Without PS" indicates the absence of the existence of a pathological site condition. Although it is possible to use other cross-validation methods such as three-fold, five-fold, or leave-one-out approaches, we decided to use a two-fold approach in our experiments to save the processing time of the experiments.

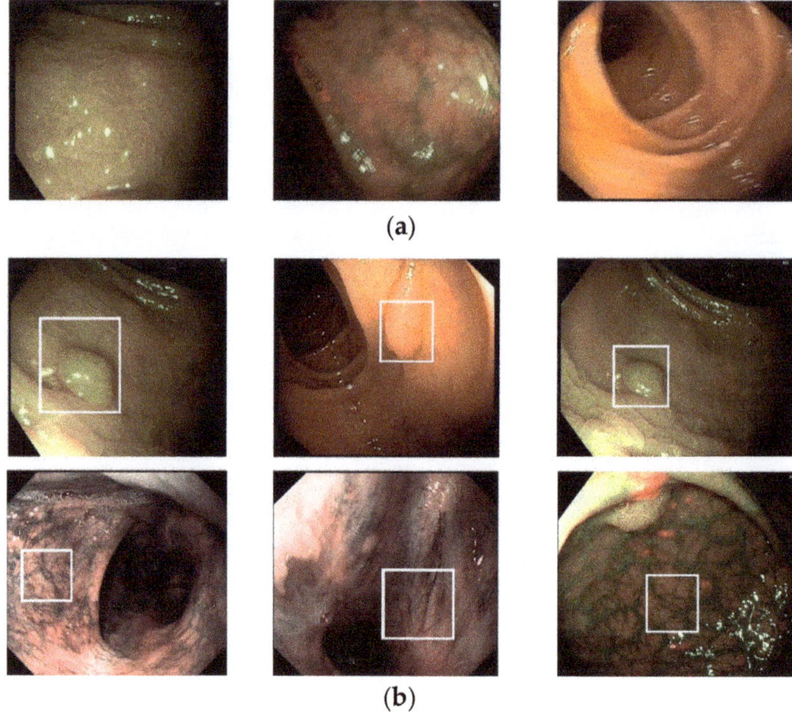

Figure 6. Example of images: (**a**) Gastric images without the existence of pathological sites and (**b**) gastric images with the existence of pathological sites (Upper and lower images indicate those including polyp and complex vascular regions, respectively).

Table 4. Detailed description of the Hamlyn-GI dataset with the two-fold cross-validation scheme used in our study.

Fold Index		Training Dataset		Testing Dataset		Total
First Fold	Number of Videos	5		5		10
	Number of Images	Without PS	With PS	Without PS	With PS	7894
		2036	3299	1639	920	
Second Fold	Number of Videos	5		5		10
	Number of Images	Without PS	With PS	Without PS	With PS	7894
		1639	920	2036	3299	

3.2. Training

In our first experiment, we performed a training process to train the three deep learning-based models which are illustrated in Figure 1 and Table 2. For this experiment, we programmed our network using Python programming language with the help of the Tensorflow library [39] for the implementation of deep learning-based models. A detailed description of the parameters is listed in Table 5. For the training method, we used the adaptive moment estimation (Adam) optimizer with an initial learning rate of 0.0001, and we trained each model with 30 epochs. As the epoch increases, the network parameters become finer; therefore, we continuously reduced the learning rate after every epoch. In addition, a batch size of 32 was used in our experiment.

Table 5. Training parameters used in our study.

Optimizer	Number of Epochs	Initial Learning Rate	Learning Rate Schedule	Batch Size
Adam	30	0.0001	Time decay every epoch	32

In Figure 7, we illustrated the results of the training process using the training datasets. As mentioned in Section 3.1, we used a two-fold cross-validation procedure in our experiments. Therefore, we calculated the average result of the two folds and presented it in Figure 7. In this figure, we show the curves of the loss and training accuracy of the training procedure for all three CNN models. As shown in these curves, the losses continuously decrease while the training accuracies increase with the increase in the training epoch. Thus, we can consider that the training procedures were successful in our experiments.

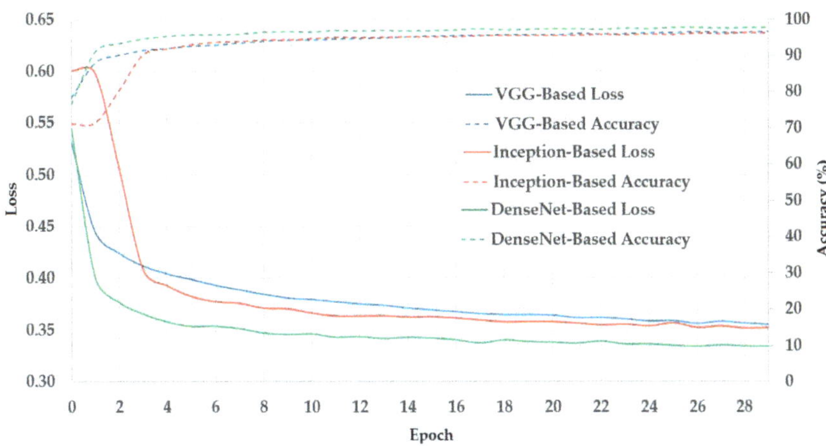

Figure 7. Training results (loss vs. accuracy) of Visual Geometry Group (VGG)-based, Inception-based, and DenseNet-based models in our experiments.

3.3. Testing of Proposed Method (Ablation Studies)

For the ablation studies, we performed the experiments presented in Sections 3.3.1–3.3.3.

3.3.1. Classification Results Based on Individual CNN Models with the Preprocessing Procedure

As explained in Section 2 and Figure 1, our proposed method is composed of three main classification branches. Based on this structure, in our first experiment, we performed experiments using a single CNN-based classification method, as shown in Figure 1, using the three classification networks described in Table 2. For this experiment, we scaled the output images of the preprocessing step to 224 × 224 pixel-sized images and inputted them to each individual network for classification purposes. The detailed experimental results are listed in Table 6. From this table, it can be observed that the VGG-based network achieved an overall classification accuracy of 68.912% with a sensitivity of 66.271% and specificity of 71.946%. Similarly, we obtained an overall classification accuracy of 66.505% with a sensitivity of 87.438% and specificity of 42.476% using the inception-based network; moreover, an overall classification accuracy of 52.609% with a sensitivity of 19.554% and specificity of 90.558% was achieved using the DenseNet-based network. These experimental results show that the CNN is suitable for pathological site classification problems. However, the overall classification result was still low as the largest accurate classification result was obtained using a VGG-based network, which was measured to be 68.912%. This problem is caused by the fact that the dataset used in our experiments was not large and included the complex structure of the GI endoscopic images. In addition, we can observe from these experimental results that the VGG-based network outperforms the inception- and DenseNet-based networks. This is because the inception- and DenseNet-based networks are deeper than the VGG-based network. Consequently, the training of the network is affected because our dataset is not significantly big with images of only five patients.

Table 6. Classification performance using individual CNN-based model (unit: %).

Fold Index	VGG-Based Method			Inception-Based Method			DenseNet-Based Method		
	Sens	Spec	Accuracy	Sens	Spec	Accuracy	Sens	Spec	Accuracy
First Fold	26.304	93.899	69.597	78.370	57.352	64.908	35.978	85.174	67.487
Second Fold	77.417	54.273	68.584	89.967	30.501	67.272	14.974	94.892	45.473
Average	66.271	71.946	68.912	87.438	42.476	66.505	19.554	90.558	52.609

We can also observe from Table 6 that the sensitivity and specificity measurements vary according to each CNN-based model. Using the VGG-based model, we obtained a sensitivity of 66.271% and specificity of 71.946%. This result indicates that the VGG-based network is more efficient in detecting negative images (images without possible pathological sites) than positive images (images with possible pathological sites). Similarly, we obtained a sensitivity of 19.554% and specificity of 90.558% using the DenseNet-based network. As it has a large value of specificity, the DenseNet-based network is more efficient in classifying negative images than positive ones. However, the situation is different when using the inception-based network. Using the inception-based network, we obtained a sensitivity of 87.438% and specificity of 42.476%. The high value of sensitivity indicates that the inception-based network is more efficient for classifying positive images than negative images. In addition, the difference between the sensitivity and specificity using the VGG-based network is approximately 5.675%, which is significantly smaller than the difference of 44.962% obtained using the inception-based network and 71.004% obtained using the DenseNet-based network. This result indicates that the VGG-based network has minimal bias while classifying positive and negative images when compared the other networks. This is because the VGG-based network is significantly shallower than the inception- and DenseNet-based networks. Therefore, the negative effects caused by the over/underfitting problem are less significant than those caused by the inception- and DenseNet-based networks.

3.3.2. Classification Results Based on Individual CNN Models without the Preprocessing Procedure

As shown in Figure 1, our proposed method performs a preprocessing step before applying the classification step by using deep learning-based models to reduce the effect of noise on the classification results. In this experiment, we demonstrated the effects of noise and the advantages of the preprocessing step on our system by measuring the classification performance in a system that does not consider the preprocessing step. For this purpose, we trained and evaluated the performance of the deep learning-based models without considering noise reduction by the preprocessing step. The detailed experimental results are listed in Table 7. As shown in this table, we obtained overall classifications accuracies of 67.392%, 47.745% and 52.356% using the VGG-, inception-, and DenseNet-based methods, respectively. When compared to the experimental results in Table 6, the preprocessing procedure helps to enhance the classification results in the case of all three CNN models. Specifically, the classification accuracy is reduced from 68.912% in the case of using a preprocessing step to 67.392% in the case where a preprocessing step is not used in the experiment with the VGG-based method. In the case of the inception-based method, the overall classification accuracy is significantly reduced from 66.505% to 47.745% for the cases with and without the preprocessing step, respectively. Finally, a marginal reduction is observed in the case of the DenseNet-based method with an overall classification accuracy of 52.609% and 52.356% for the cases with and without the preprocessing step, respectively. Through these experimental results, we can observe that the background and noise have strong negative effects on the classification accuracy of the pathological site classification problem, and the preprocessing step is efficient in reducing these negative effects.

Table 7. Classification performance using an individual CNN-based model (unit: %).

Fold Index	VGG-Based Method			Inception-Based Method			DenseNet-Based Method		
	Sens	Spec	Accuracy	Sens	Spec	Accuracy	Sens	Spec	Accuracy
First Fold	14.565	97.010	67.370	22.065	85.723	62.837	27.065	95.729	71.043
Second Fold	62.412	75.491	67.403	6.790	95.138	40.506	13.641	91.601	43.392
Average	51.979	85.088	67.392	10.121	90.938	47.745	16.567	93.442	52.356

3.3.3. Classification Results of the Proposed Ensemble Model of Three CNNs with the Preprocessing Procedure

As listed in Table 6, the three CNN-based models work differently with the same dataset. This indicates that each network has its own advantages and disadvantages for our classification problem. Based on these characteristics, we considered a combination of the results of these networks to enhance the performance of our classification system. Then, we performed an experiment to evaluate the performance of our proposed method mentioned in Section 2.1 using Equations (1)–(3). The detailed experimental results are listed in Table 8 for all three combination methods.

Table 8. Classification accuracy using our proposed method (unit: %).

Fold Index	MAX Rule			AVERAGE Rule			VOTING Rule		
	Sens	Spec	Accuracy	Sens	Spec	Accuracy	Sens	Spec	Accuracy
First Fold	52.283	82.977	71.942	42.500	88.896	72.215	41.413	88.286	71.434
Second Fold	59.533	63.703	61.124	75.690	61.002	70.084	75.993	59.921	69.859
Average	57.952	72.299	64.630	68.452	73.442	70.775	68.452	72.571	70.369

In the first experiment in this section, we used the MAX combination rule to combine the classification results of the three CNN-based models. As explained in Equation (1), the classification is performed based on the maximum classification scores of all three networks. From Table 8, it can be noted that the overall classification result (the average result of two folds) is 64.630% with a sensitivity of 57.952% and specificity of 72.299%. We can observe that this classification result is higher than 52.609%, which was obtained using only the DenseNet-based model (as listed in Table 6). However, this classification accuracy is lower than 68.912% and 66.505%, which were obtained using the VGG- and inception-based networks, respectively. As the specificity of 72.299% is higher than the sensitivity of 57.952%, we can conclude that the combined system based on the MAX rule is more efficient in recognizing the negative images than the positive images. In addition, the difference between the sensitivity and specificity using the MAX rule is approximately 14.347% (72.299–57.952%), which is smaller than the case of using only the inception- or DenseNet-based network.

In the second experiment, we used the AVERAGE combination rule to combine the classification results of the three CNN-based models. As indicated by its meaning, the classification based on the AVERAGE rule was performed by classifying images based on the average classification scores of the three CNN-based models for negative and positive classes. As listed in Table 8, the AVERAGE combination rule produces an overall classification accuracy of 70.775% with a sensitivity of 68.452% and specificity of 73.442%. When compared to the classification accuracies listed in Table 6, we can observe that the classification accuracy by the AVERAGE rule is significantly better than that of the individual CNN-based models. The overall classification accuracy of 70.775% is significantly higher than the results of 68.912%, 66.505%, and 52.609% obtained using the VGG-, inception-, and DenseNet-based networks, respectively. In addition, the difference between the sensitivity and specificity of the AVERAGE rule is approximately 4.99% (73.442–68.452%), which is better than the value of 5.675% (the smallest difference between sensitivity and specificity of individual CNN-based models, as listed in

Table 6) produced by the VGG-based network. This result indicates that the AVERAGE combination methods have minimal over/underfitting effects when compared to the individual CNNs.

Finally, we performed an experiment using the VOTING combination rule, as indicated using Equations (3)–(4). As listed in Table 8, the VOTING combination rule produces an overall classification accuracy of 70.369% with a sensitivity of 68.452% and specificity of 72.571%. It can be observed that this overall classification accuracy is higher than the overall classification accuracy produced by the individual CNN-based models listed in Table 6. Although the overall classification accuracy produced by the VOTING rule is marginally lower than that produced by the AVERAGE rule (70.369% versus 70.775%), the difference is insignificant (approximately 0.406%). In addition, the difference between the sensitivity and specificity produced by the VOTING rule is 4.119% (72.571–68.452%). This value is the smallest value for the difference between the sensitivity and specificity and it indicates that the VOTING combination rule has fewer effects on the over/underfitting problem among the three combination rules and three individual CNN-based models.

To deeply analyze the obtained experimental results, we obtained the receiver operating characteristic (ROC) curve of various system configurations, including three systems based on three individual CNN networks and three systems based on three combination rules, as shown in Figure 8. ROC curve is one of the most popular measurements used in medical statistic test, which provides a visualization of a classification performance. The ROC curve demonstrates the change of false acceptance rate (FAR) versus false rejection rate (FRR). In our experiment, we use the genuine acceptance rate (GAR) instead of FRR in measuring the ROC curve. That is, GAR is measured by (100 − FRR (%)). Because of this measurement method, a ROC curve looks like the ones in Figure 8, and the higher position on the upper-left side indicate the high performance of a classification system. As we perform a two-fold cross validation procedure, the ROC curves of each system configuration are obtained by taking average of the two ROC curves of two folds. From Figure 8, we can see that the AVERAGE rule outperforms the other system configurations, while the MAX and VOTING rule performs similar with the VGG-based or Inception-based network, and better than DenseNet-based network.

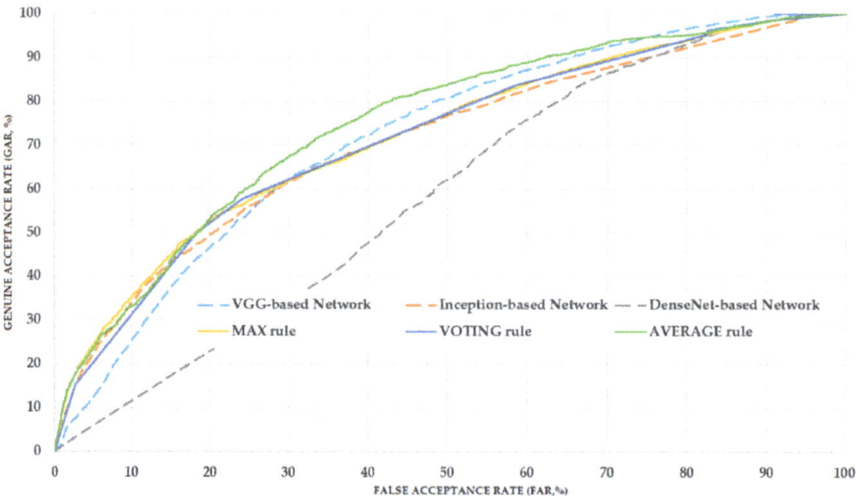

Figure 8. Receiver operating characteristic (ROC) curves of various system configurations in our experiment.

These experimental results show that the combination of the three CNN-based models is efficient in enhancing the classification performance of pathological site classification for a GI endoscopic examination when compared to previous studies. In addition, the AVERAGE and VOTING rules are more efficient than the MAX combination rule. As the best accuracy was obtained using the AVERAGE rule, we can conclude that the AVERAGE rule is the most suitable combination rule for implementation with the Hamlyn-GI dataset. However, as the size of the Hamlyn-GI dataset is relatively small, the applicability of these rules should be further investigated using larger data.

3.4. Comparisons with the State-of-the-Art Methods and Processing Time

Based on the experimental results presented in the above sections, we compared the classification performance of our proposed method with those of previous studies. As explained in Section 1, most of the previous studies used the conventional CNN (training a single shallow network or extracting image features from a pretrained network) for classification purposes. As discussed in Sections 2.3 and 3.3, our proposed method applies the ensemble learning technique to three individual deep CNNs, i.e., VGG-, inception-, and DenseNet-based networks. Therefore, we can consider that the performances of previous studies correspond to the case of using each individual network. Based on this assumption, we conducted a comparison experiment and its results are listed in Table 9. From the table, it can be observed that by using the individual CNNs, we obtained the overall classification accuracies of 68.912%, 66.505% and 52.609% using the VGG-, inception-, and DenseNet-based networks, respectively. The best accuracy of 68.912% was obtained using the VGG-based method. Using our proposed method, we obtained the best accuracy of 70.775% with the AVERAGE combination rule, which is higher than the accuracy produced by the use of an individual network. Moreover, our proposed method with the VOTING combination rule also produced an accuracy of 70.369%, which is also higher than the best value of 68.912% of the individual network. From these experimental results, we can conclude that our proposed method outperforms previous studies in the pathological site classification system. However, as the best accuracy is still low (70.775%), the classification system still requires enhancements in our future works.

Table 9. Comparison of the classification performances between the proposed method and the previous studies (unit: %).

Method	Accuracy
VGG-based Network [11,25]	68.912
Inception-based Network [26]	66.505
DenseNet-based Network [31]	52.609
Proposed Method with MAX rule	64.630
Proposed Method with AVERAGE rule	70.775
Proposed Method with VOTING rule	70.369

In the final experiment, we measured the processing time of our proposed method for pathological site classification problems. For this purpose, we created a deep CNN program in Python programming language using the Tensorflow library [39]. To run the program, we used a desktop computer with an Intel Core i7-6700 central processing unit with a working clock of 3.4 GHz and 64 GB of RAM memory. To increase the speed of the deep learning networks, we used a graphical processing unit, namely GeForce Titan X, to run the inference of the three deep learning models [40]. The detailed experimental results are listed in Table 10. We also performed the comparisons of the processing time between the proposed method and the previous studies. As shown in Table 10, it takes about 37.646 ms, 67.472 ms and 65.901 ms to classify an input image using VGG-based [11,25], Inception-based [26], and DenseNet-based [31] network, respectively. Using our proposed method, it takes 179.440 ms to classify a single input image, and our proposed method can work at a speed of 5.6 frames per second. From these experiment results, we can find that our proposed method takes longer processing time than previous studies. However, it is acceptable in medical image processing applications where the accuracy, but not processing time, is the primary requirement.

Table 10. The comparison of the processing time between the proposed method and the previous studies (unit: ms).

VGG-Based Network [11,25]	Inception-Based Network [26]	DenseNet-Based Network [31]	Proposed Method		
			Preprocessing Step	Ensemble of Three CNNs	Total
37.646	67.472	65.901	8.421	171.019	179.440

3.5. Analysis and Discussion

As shown in our experimental result, the average rule outperforms the max and voting rule in enhancing classification performance. This is caused by the fact that the average rule uses the prediction scores of individual CNN models directly, whereas the max rule is performed based on only the maximum classification score of one CNN model among several CNN models, and the voting rule is performed based on majority decisions of CNN models. As a result, the combined classification score by average rule contains more detailed classification information from all CNN models than the other two combination rules.

We used a public dataset, namely Hamlyn-GI, which showed the large visual differences among the collected images [30]. To measure the performance of the proposed method, we first divided the entire Hamlyn-GI dataset into training and testing sets. The division is done by ensuring that the training and testing datasets are different, i.e., images of the same patient only exist in either training or testing datasets as open world evaluation. As a result, these training and testing set are independent from the other as shown in Figure 6 (for example, the left-bottom and middle-bottom images are the samples of training data whereas the others are those of testing one for the 1st fold validation). We used the training dataset to train the classification models, and testing dataset to measure the performance of the trained models. Therefore, the measured performance is the optimal generalization because the testing set is independent from the training set.

To demonstrate the efficiency of our proposed method, we show certain examples of the classification results in Figure 9. As shown in this figure, images without a pathological site are accurately classified as negative cases (Figure 9a), while images with pathological sites (small polyp regions appear in Figure 9b, complex vascular structure regions in Figure 9c) are classified as positive cases. However, our proposed method also produces certain incorrect classification results, as shown in Figure 10. In Figure 10a, we show certain sample images that were classified as positive cases (images with pathological sites) even if they did not contain pathological sites. The errors were caused either by the fact that the endoscopic images contain complex vascular structures (the left-most image in Figure 10a) or owing to imperfect input images (middle and right-most images in Figure 10a). In Figure 10b,c, we show certain example images in which our classification method incorrectly classified. As we can observe from these figures, certain text boxes in the input images can cause classification errors (right-most image in Figure 10c). In addition, the input images contain small or blurred polyp regions, resulting in classification errors, which can be solved by super-resolution reconstruction or deblurring in our future works.

To provide a deep look inside the actual operation of deep learning-based models for classification problems in our study, we measured the regions of focus in the input images on which the classification models are used for accomplishing their functionalities, and the results are illustrated in Figure 11. For this purpose, we obtained the class activation maps using the gradient-weighted class activation mapping (grad-CAM) method [41]. Grad-CAM is a popular method that explains the working of a deep CNN. In the activation maps of these figures, the brighter regions indicate the regions that are focused on in the feature maps, which our system uses for classification purposes. As shown in Figure 11, our deep learning models focus on pathological sites (possible polyps) or complex vascular structures in classifying input images into negative or positive classes. By providing class activation maps to medical doctors, our system can provide a reasonable explanation about why an input image is classified as positive data, and it can demonstrate the functionality of explainable artificial intelligence. In addition, it is a significantly important characteristic of a high-performance classification system in CAD applications.

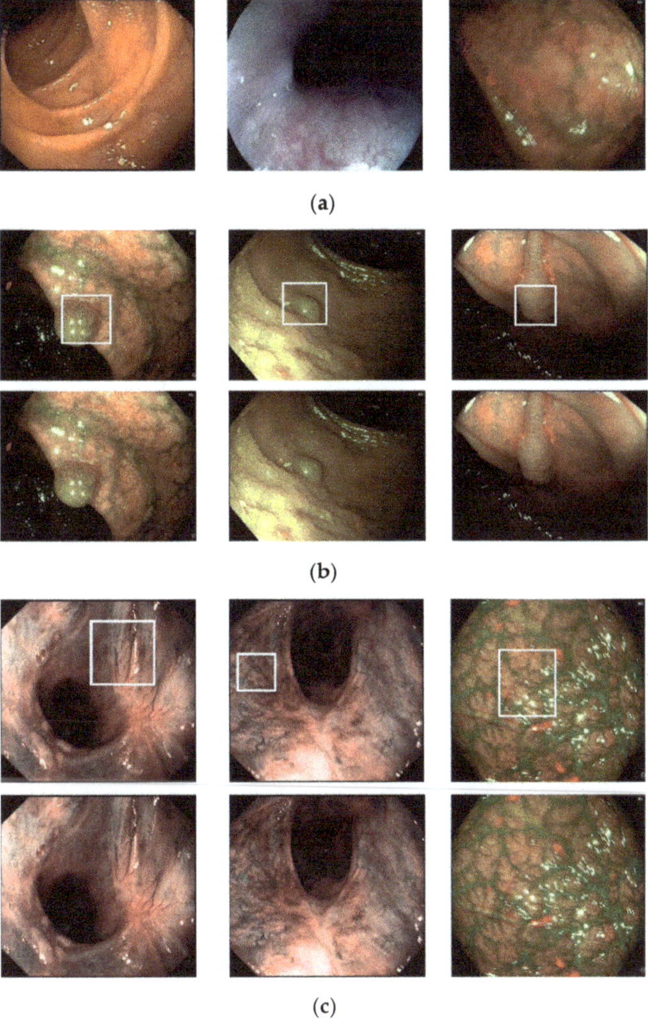

Figure 9. Examples of correct classification results by the proposed method: (**a**) True-negative cases and (**b**,**c**) true-positive cases (Upper and lower images indicate the ground-truth and testing images, respectively).

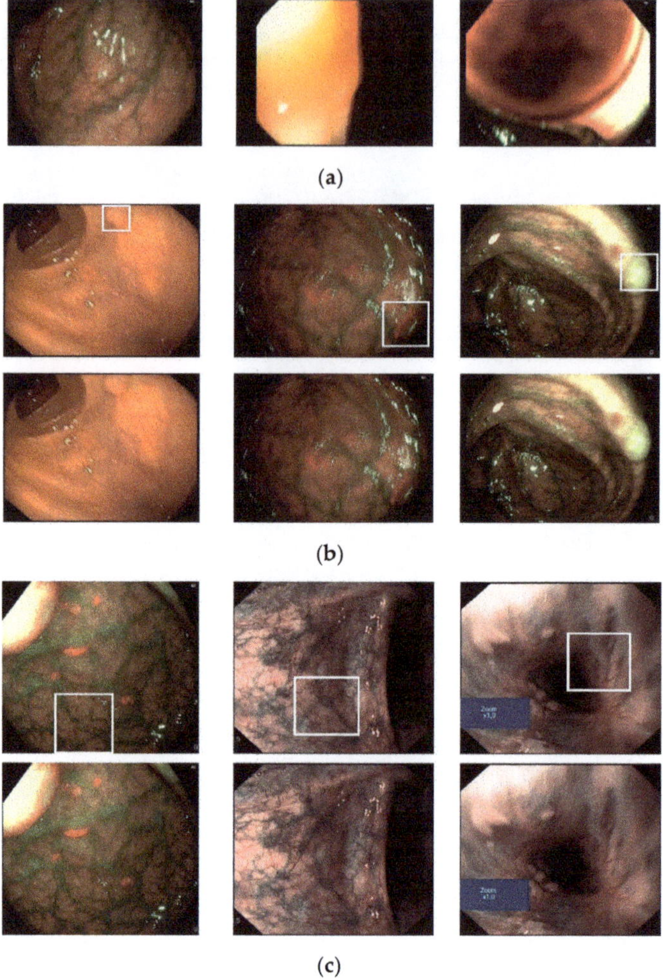

Figure 10. Examples of incorrect classification results by the proposed method: (**a**) False-positive cases and (**b**,**c**) false-negative cases (Upper and lower images indicate the ground-truth and testing images, respectively).

In our experiments, we use only three sub-models for ensemble algorithm because of two reasons. First, these models (VGG-based, Inception-based, and DenseNet-based) are different in network architecture. Therefore, each sub-model has its own advantages and disadvantages that can be compensated by the other sub-models. Second, although we can use more sub-models to perform ensemble algorithm, the complexity and processing time of the proposed method are increased, which can cause the difficulty in training and deployment of model. Because of these reasons, we only used three sub-models in our experiments. However, it is possible to use more sub-models to possibly enhance the system performance. In that case, our work can be seen as a specific example to demonstrate the enhancement possibility of ensemble algorithm in pathological site classification problem. When the number of sub-models increases, it not only increases the complexity and computation of system, but also represents noises to system that caused by error cases of every individual system.

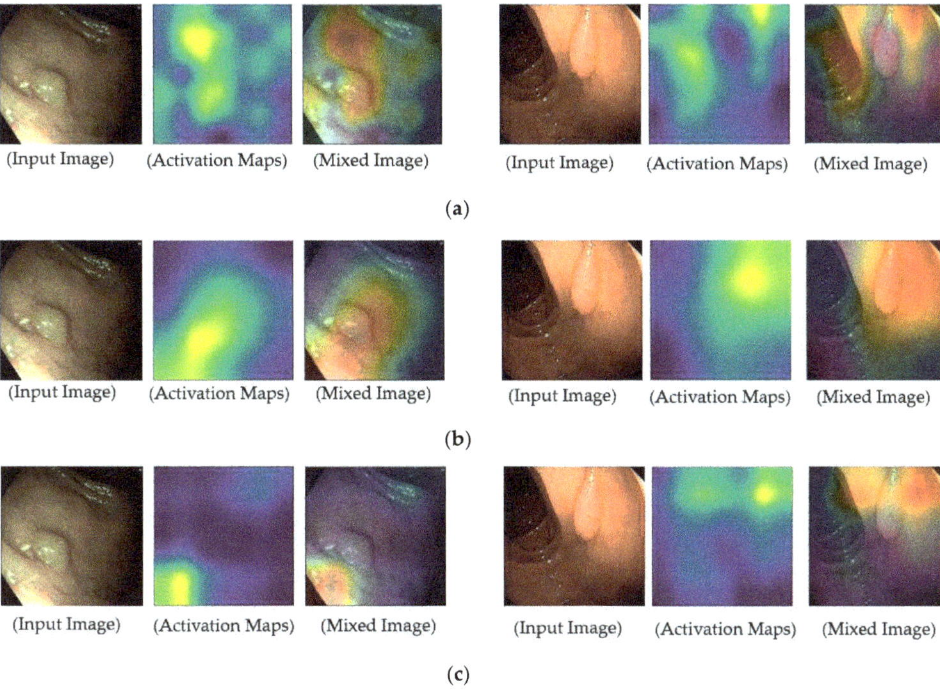

Figure 11. Obtained class activation maps corresponding to the pathological site class using (**a**) VGG-based network, (**b**) inception-based network, and (**c**) DenseNet-based network.

For a demonstration purpose, we additionally performed experiments with more than three sub-models, i.e., ensemble method with four sub-models. For this purpose, we used an additional CNN classifier based on residual network [23]. The experimental results are given in Table 11. From these experimental results, we see that we obtained a highest classification accuracy of 69.686% using the average rule that is little worse than the accuracy of 70.775% obtained using three sub-models mentioned in Section 3.3. The reason is that when the number of models increases, it also represents noises to the ensemble system caused by error cases of the new model. In addition, the residual network also uses the short-cut connection in the similar way as the DenseNet-based network. Therefore, the residual network shares similar characteristics with DenseNet-based network. Because of these reasons, the performance of ensemble system based on four sub-models is little reduced compared to the case of using three sub-models.

Table 11. Classification accuracy using our proposed method with four sub-models (unit: %).

Fold Index	MAX Rule			AVERAGE Rule			VOTING Rule		
	Sens	Spec	Accuracy	Sens	Spec	Accuracy	Sens	Spec	Accuracy
First Fold	57.500	70.165	65.612	54.783	77.181	69.129	55.978	75.290	68.347
Second Fold	59.867	63.114	61.106	82.904	48.969	69.953	79.691	53.978	69.878
Average	59.351	66.286	62.567	76.772	61.551	69.686	74.520	63.483	69.382

4. Conclusions

In this study, we proposed an ensemble learning-based method that combines the classification results of multiple deep learning models to enhance the classification results of the endoscopic pathological site classification system. For this purpose, we first invoked and trained several CNN models, including VGG-, Inception- and DenseNet-based networks for pathological site classification problems. Based on the classification results of these CNN models, we combined them using three combination rules, MAX, AVERAGE and VOTING rules. Using our proposed method, we enhanced the classification performance over that observed in previous studies. In addition, we showed that the AVERAGE rule outperforms the other two combination rules.

Author Contributions: D.T.N. and G.B. designed and implemented the overall system, made the code and performed experiments, and wrote this paper. M.B.L., T.D.P., M.A. and K.R.P. helped with parts of experiments and gave comments during experiments. All authors have read and agreed to the published version of the manuscript.

Funding: This work was supported in part by the National Research Foundation of Korea (NRF) funded by the Ministry of Science and ICT (MSIT) through the Basic Science Research Program under Grant NRF-2020R1A2C1006179, in part by the NRF funded by the MSIT through the Basic Science Research Program under Grant NRF-2019R1F1A1041123, and in part by the NRF funded by the MSIT through the Basic Science Research Program under Grant NRF-2019R1A2C1083813.

Acknowledgments: This work was supported in part by the National Research Foundation of Korea (NRF) funded by the Ministry of Science and ICT (MSIT) through the Basic Science Research Program under Grant NRF-2020R1A2C1006179, in part by the NRF funded by the MSIT through the Basic Science Research Program under Grant NRF-2019R1F1A1041123, and in part by the NRF funded by the MSIT through the Basic Science Research Program under Grant NRF-2019R1A2C1083813.

Conflicts of Interest: The authors declare no conflict of interest.

References

1. World Health Organization–Cancer Report. Available online: https://www.who.int/news-room/fact-sheets/detail/cancer (accessed on 30 July 2020).
2. Bhandary, A.; Prabhu, G.A.; Rajinikanth, V.; Thanaraj, K.P.; Satapathy, S.C.; Robbins, D.E.; Shasky, C.; Zhang, Y.; Tavaresh, J.M.R.S.; Rajac, N.S.M. Deep-learning framework to detect lung abnormality—A study with chest x-ray and lung CT scan images. *Pattern Recognit. Lett.* **2020**, *129*, 271–278. [CrossRef]
3. Huang, W.; Hu, L. Using a noisy U-Net for detecting lung nodule candidates. *IEEE Access* **2019**, *7*, 67905–67915. [CrossRef]
4. Xian, M.; Zhang, Y.; Cheng, H.D.; Xu, F.; Zhang, F.; Ding, J. Automatic breast ultrasound image segmentation: A survey. *Pattern Recognit.* **2018**, *79*, 340–355. [CrossRef]
5. Xu, Y.; Wang, Y.; Yuan, J.; Cheng, Q.; Wang, X.; Carson, P.L. Medical breast ultrasound image segmentation by machine learning. *Ultrasonics* **2019**, *91*, 1–9. [CrossRef]
6. Mirbeik-Sabzevari, A.; Oppelaar, E.; Ashinoff, R.; Tavassolian, N. High-contrast, low-cost, 3-D visualization of skin cancer using ultra-high-resolution millimeter-wave imaging. *IEEE Trans. Med. Imaging* **2019**, *38*, 2188–2197. [CrossRef]
7. Xie, Y.; Zhang, J.; Xia, Y.; Shen, C. A mutual bootstrapping model for automated skin lesion segmentation and classification. *IEEE Trans. Med. Imaging* **2020**, *39*, 2482–2493. [CrossRef]
8. Song, M.; Ang, T.L. Early detection of early gastric cancer using image-enhanced endoscopy: Current trends. *Gastrointest. Interv.* **2014**, *3*, 1–7. [CrossRef]
9. Zhu, Y.; Wang, Q.-C.; Xu, M.-D.; Zhang, Z.; Cheng, J.; Zhong, Y.-S.; Zhang, Y.-Q.; Chen, W.; Yao, L.-Q.; Zhou, P.-H.; et al. Application of convolutional neural network in the diagnosis of the invasion depth of gastric cancer based on conventional endoscopy. *Gastrointest. Endosc.* **2019**, *89*, 806–815. [CrossRef]
10. Li, Y.; Li, X.; Xie, X.; Shen, L. Deep learning based gastric cancer identification. In Proceedings of the IEEE International Symposium on Biomedical Imaging, Washington, DC, USA, 4–7 April 2018; pp. 182–185.
11. Patino-Barrientos, S.; Sierra-Sosa, D.; Garcia-Zapirain, B.; Castillo-Olea, C.; Elmaghraby, A. Kudo's classification for colon polyp assessment using a deep learning approach. *Appl. Sci.* **2020**, *10*, 501. [CrossRef]
12. Hassanpour, S.; Korbar, B.; Olofson, A.M.; Miraflor, A.P.; Nicka, C.M.; Suriawinata, M.A.; Torresani, L.; Suriawinata, A.A. Deep Learning for Classification of Colorectal Polyps on Whole-slide Images. *J. Pathol. Inform.* **2017**, *8*, 30. [CrossRef]

13. Ribeiro, E.; Uhl, A.; Hafner, M. Colonic Polyp Classification with Convolutional Neural Networks. In Proceedings of the IEEE 29th International Symposium on Computer-Based Medical Systems, Dublin, Ireland, 20–24 June 2016.

14. Wickstrom, K.; Kampffmeyer, M.; Jenson, R. Uncertainty and interpretability in convolutional neural networks for sematic segmentation of colorectal polyps. *Med. Image Anal.* **2020**, *60*, 101619. [CrossRef] [PubMed]

15. Ribeiro, E.; Uhl, A.; Wimmer, G.; Hafner, M. Exploring deep learning and transfer learning for colonic polyp classification. *Comput. Math. Method Med.* **2016**, 6584725. [CrossRef] [PubMed]

16. Fonolla, R.; Sommen, F.; Schreuder, R.M.; Shoon, E.J.; With, P. Multi-modal classification of polyp malignancy using CNN features with balanced class augmentation. In Proceedings of the IEEE International Symposium on Biomedical Imaging, Venice, Italy, 8–11 April 2019; pp. 74–78.

17. Ma, J.; Wu, F.; Zhu, J.; Xu, D.; Kong, D. A pretrained convolutional neural network based method for thyroid nodule diagnosis. *Ultrasonics* **2017**, *73*, 221–230. [CrossRef]

18. Nguyen, D.T.; Pham, D.T.; Batchuluun, G.; Yoon, H.S.; Park, K.R. Artificial Intelligence-based thyroid nodule classification using information from spatial and frequency domains. *J. Clin. Med.* **2019**, *8*, 1976. [CrossRef] [PubMed]

19. Chi, J.; Walia, E.; Babyn, P.; Wang, J.; Groot, G.; Eramian, M. Thyroid nodule classification in ultrasound images by fine-tuning deep convolutional neural network. *J. Digit. Imaging* **2017**, *30*, 477–486. [CrossRef]

20. Havaei, M.; Davy, A.; Warde-Farley, D.; Biard, A.; Courville, A.; Bengio, Y.; Pal, C.; Jodoin, P.-M.; Larochelle, H. Brain tumor segmentation with deep neural networks. *Med. Image Anal.* **2017**, *35*, 18–31. [CrossRef]

21. Milletari, F.; Ahmadi, S.-A.; Kroll, C.; Plate, A.; Rozanski, V.; Maiostre, J.; Levin, J.; Dietrich, O.; Ertl-Wagner, B.; Bötzel, K.; et al. Hough-CNN: Deep learning for segmentation of deep brain regions in MRI and ultrasound. *Comput. Vis. Image Underst.* **2017**, *16*, 92–102. [CrossRef]

22. Kamnitsas, K.; Ledig, C.; Newcombe, V.F.J.; Simpson, J.P.; Kane, A.D.; Menon, D.K.; Rueckert, D.; Glocker, B. Efficient multi-scale 3D CNN with fully connected CRF for accurate brain lesion segmentation. *Med. Image Anal.* **2017**, *36*, 61–78. [CrossRef]

23. He, K.; Zhang, X.; Ren, S.; Sun, J. Deep residual learning for image recognition. *arXiv* **2015**, arXiv:1512.03385.

24. Krizhevsky, A.; Sutskever, I.; Hinton, G.E. ImageNet classification with deep convolutional neural networks. In Proceedings of the Neural Information Processing Systems, Lake Tahoe, NV, USA, 3–8 December 2012.

25. Simonyan, K.; Zisserman, A. Very deep convolutional neural networks for large-scale image recognition. *arXiv* **2014**, arXiv:1409.1556.

26. Szegedy, C.; Liu, W.; Jia, Y.; Sermanet, P.; Reed, S.; Anguelov, D.; Erhan, D.; Vanhoucke, V.; Rabinovich, A. Going deeper with convolutions. *arXiv* **2014**, arXiv:1409.4842v1.

27. Srivastava, N.; Hinton, G.; Krizhevsky, A.; Sutskever, I.; Salakhutdinov, R. Dropout: A simple way to prevent neural networks from overfitting. *J. Mach. Learn. Res.* **2014**, *15*, 1929–1958.

28. Kowsari, K.; Heidarysafa, M.; Brown, D.E.; Meimandi, K.J.; Barnes, L.E. RMDL: Random multi-model deep learning for classification. In Proceedings of the 2nd International Conference on Information System and Data Mining, Lakeland, FL, USA, 9–11 April 2018; pp. 19–28.

29. Pathological Site Classification Models with Algorithm. Available online: http://dm.dongguk.edu/link.html (accessed on 1 August 2020).

30. Ye, M.; Giannarou, S.; Meining, A.; Yang, G.-Z. Online tracking and retargeting with applications to optical biopsy in gastrointestinal endocopic examinations. *Med. Image Anal.* **2016**, *30*, 144–157. [CrossRef] [PubMed]

31. Huang, G.; Liu, Z.; Maaten, L.; Weinberger, K.Q. Densely connected convolutional networks. *arXiv* **2016**, arXiv:1608.06993v5.

32. Ren, S.; He, K.; Girshick, R.; Sun, J. Faster R-CNN: Towards real-time object detection with region proposal networks. *arXiv* **2015**, arXiv:1506.01497. [CrossRef]

33. Redmon, J.; Farhadi, A. YOLOv3: An incremental improvement. *arXiv* **2018**, arXiv:1804.02767.

34. Chu, M.P.; Sung, Y.; Cho, K. Generative adversarial network-based method for transforming single RGB image into 3D point cloud. *IEEE Access* **2018**, *7*, 1021–1029. [CrossRef]

35. Nguyen, D.T.; Pham, D.T.; Batchuluun, G.; Noh, K.J.; Park, K.R. Presentation attack face image generation based on a deep generative adversarial network. *Sensors* **2020**, *20*, 1810. [CrossRef]

36. Zhu, J.-Y.; Park, T.; Isola, P.; Efros, A.A. Unpaired image-to-image translation using cycle-consistent adversarial networks. *arXiv* **2017**, arXiv:1703.10593v6.

37. Carvajal, D.N.; Rowe, P.C. Research and statistics: Sensitivity, specificity, predictive values, and likelihood ratios. *Pediatr. Rev.* **2010**, *31*, 511–513. [CrossRef]
38. Nguyen, D.T.; Kang, J.K.; Pham, D.T.; Batchuluun, G.; Park, K.R. Ultrasound image-based diagnosis of malignant thyroid nodule using artificial intelligence. *Sensors* **2020**, *20*, 1822. [CrossRef] [PubMed]
39. Tensorflow Deep-Learning Library. Available online: https://www.tensorflow.org/ (accessed on 20 September 2019).
40. NVIDIA TitanX GPU. Available online: https://www.nvidia.com/en-us/geforce/products/10series/titan-x-pascal/ (accessed on 20 September 2019).
41. Selvaraju, R.R.; Cogswell, M.; Das, A.; Vedantam, R.; Rarikh, D.; Batra, D. Grad-CAM: Visual explanations from deep networks via gradient-based localization. *arXiv* **2016**, arXiv:1610.02391v4.

Publisher's Note: MDPI stays neutral with regard to jurisdictional claims in published maps and institutional affiliations.

 *sensors*

Article

DoF-Dependent and Equal-Partition Based Lens Distortion Modeling and Calibration Method for Close-Range Photogrammetry

Xiao Li [1], Wei Li [1],*, Xin'an Yuan [1], Xiaokang Yin [1] and Xin Ma [2]

[1] School of Mechanical and Electrical Engineering, China University of Petroleum (East China), Huangdao, Qingdao 266580, China; lix2020@upc.edu.cn (X.L.); xinancom@upc.edu.cn (X.Y.); xiaokang.yin@upc.edu.cn (X.Y.)

[2] Polytechnic Institute, Purdue University, West Lafayette, IN 47907, USA; ma633@purdue.edu

* Correspondence: liwei@upc.edu.cn; Tel.: +86-532-86983016

Received: 24 September 2020; Accepted: 16 October 2020; Published: 20 October 2020

Abstract: Lens distortion is closely related to the spatial position of depth of field (DoF), especially in close-range photography. The accurate characterization and precise calibration of DoF-dependent distortion are very important to improve the accuracy of close-range vision measurements. In this paper, to meet the need of short-distance and small-focal-length photography, a DoF-dependent and equal-partition based lens distortion modeling and calibration method is proposed. Firstly, considering the direction along the optical axis, a DoF-dependent yet focusing-state-independent distortion model is proposed. By this method, manual adjustment of the focus and zoom rings is avoided, thus eliminating human errors. Secondly, considering the direction perpendicular to the optical axis, to solve the problem of insufficient distortion representations caused by using only one set of coefficients, a 2D-to-3D equal-increment partitioning method for lens distortion is proposed. Accurate characterization of DoF-dependent distortion is thus realized by fusing the distortion partitioning method and the DoF distortion model. Lastly, a calibration control field is designed. After extracting line segments within a partition, the de-coupling calibration of distortion parameters and other camera model parameters is realized. Experiment results shows that the maximum/average projection and angular reconstruction errors of equal-increment partition based DoF distortion model are 0.11 pixels/0.05 pixels and 0.013°/0.011°, respectively. This demonstrates the validity of the lens distortion model and calibration method proposed in this paper.

Keywords: lens distortion; DoF-dependent; distortion partition; vision measurement

1. Introduction

Vision measurement is a subject that allows quantitative perception of scene information by combining image processing with calibrated camera parameters. Therefore, the calibration accuracy of the parameters is an important determinant of the vision measurement uncertainty. The lens distortion is closely related to the depth of field (DoF), which refers to the distance between the nearest and the farthest objects that are in acceptably sharp focus in an image. For medium- or high-accuracy applications, close-range imaging parameters (e.g., short object distance (<1 m) and small focal length) are often adopted. In such occasions, DoF has a significant influence on lens distortion and, hence, becomes a major cause of vision measurement errors. For instance, to ensure that the vision has a micron-level accuracy when detecting the contouring error of a machine tool [1], the camera is placed 400 mm away from the focal plane to collect and analyze the image sequence of the interpolation trajectory running in the DoF. In this case, measurement errors ranging from dozens to hundreds of microns can be caused by a large lens distortion. Therefore, to improve the vision measurement accuracy in close-range photogrammetry, accurate modeling and calibration of the DoF-dependent lens distortion are urgently needed.

The lens distortion model maps the relation between distorted and undistorted image points. Models to show the relations vary according to the types of optical systems, which include the polynomial distortion model, logarithmic fish-eye distortion model [2], polynomial fish-eye distortion model [2–5], field-of-view (FoV) distortion model [6], division distortion model [7,8], rational function distortion model [9,10], and so on. In 1971, Brown [11,12] proposed the Gaussian polynomial function to express radial and decentering distortion, which is particularly suitable for studying the distortion of a standard lens in high-accuracy measurements [13,14]. Later, researchers noticed that the observed radial and decentering distortion varies with the focal length, the lens focusing state (i.e., focused or defocused), and the DoF position. Since then, researchers have focused on the improvement of distortion calibration and modeling methods to obtain a precise representation of distortion behavior. For the distortion calibration, the study goes in two directions: the coupled-calibration method and the decoupled-calibration method. The former can be generally divided into three types: self-calibration method [15], active calibration method, and traditional calibration method [16]. Among the traditional ones, Zhang's calibration method [17] and its improved method [18–20], used widely in industry and scientific research, are the most popular. In this coupled-calibration method, the distortion parameters are calculated by performing a full-scale optimization for all parameters. Due to the strong coupling effect, the estimated errors of other parameters (i.e., intrinsic and extrinsic parameters) in the camera model would be propagated to that of distortion parameters, thus leading to the failure of getting optimal solutions. By contrast, the decoupled-calibration method does not involve coupling other factors or entail any prior geometric knowledge of the calibration object, and only geometric invariants of some image features, such as straight lines [6,12,21–23], vanishing points [24], or spheres [25], are needed to solve the parameters. Among these features, straight lines can be easily reflected in scenes and extracted from noise images, thus having enormous potential.

Regarding the distortion modeling, some researchers incorporated the DoF into the distortion function. Magill [26] used the distortion of two focal planes at infinity to solve that of an arbitrary focal plane. Then, Brown [12] improved Magill's model by establishing distortion models of any focal plane and any defocused plane (the plane perpendicular to the optical axis in the DoF) on the condition that the distortions of two focal planes are known. Soon after, Fryer [27], based on Brown's model, realized the lens distortion calibration of an underwater camera [28]. Fraser and Shortis [29] introduced an empirical model and solved the Brown model's problem of inaccurate description of large image distortion. Additionally, Dold [30] established a DoF distortion model that is different from Brown's and solved the model parameters through the strategy of bundle adjustment. In 2004, Brakhage [31] characterized the DoF distortion of the telecentric lens in a fringe projection system by using Zernike Polynomials. Moreover, in 2006, the DoF distortion distribution of the grating projection system was experimentally analyzed by Bräuer-Burchardt. In 2008, Hanning [32] introduced depth (object distance) into the spline function to form a distortion model and used the model to calibrate radial distortion.

The above DoF distortion models not only depend on the focusing state but also relate to the distortion coefficients on the focal plane. For these models, on the one hand, the focusing state is usually adjusted by manually twisting the zoom and focus rings, which introduces human errors and changes the camera parameters. On the other hand, the focus distance and distortion parameters on the focal plane cannot be determined accurately. To overcome the problem, Alvarez [33], based on Brown's and Fraser's models, deduced a radial distortion model that is suitable for planar scenarios. With this model, when the focal length is locked, distortion at any image position can be estimated by using two lines in a single photograph. In 2017, Dong [34] proposed a DoF distortion model, by which the researcher accurately calibrated the distortion parameters on arbitrary object planes, and reduced the error from 0.055 mm to 0.028 mm in the measuring volume of 7.0 m × 3.5 m × 2.5 m with the large-object-distance of 6 m. Additionally, in 2019, Ricolfe-Viala [35] proposed a depth-dependent high distortion lens calibration method, by embedding the object distance in the division distortion model, and the highly distorted images can be corrected with only one distortion parameter. However, these researchers only used one set of coefficients, which is not sufficient to accurately represent the distortion. To address this problem, some scholars adopted the idea of partitioning to process image distortion, which uses several sets of distortion coefficients to characterize the distortion. The study, however, which is only applicable to the partitioning of a 2D object plane, fails to take into account the distortion partition within the DoF and the correlation between lens distortion and DoF. Our previous work partitioned the distortion with an equal radius [36]. Although it improved the vision measurement accuracy, the distortion correction accuracy within the partition corresponding to the image edge is still low. Besides, the distortion model we adopted depends on the focusing state of the lens, thus is less practical. In general, the current distortion model and partitioning method cannot accurately reflect the lens DoF distortion behavior in close-range photography, especially for short-distance measurements.

To solve the above problems, the lens distortion model and calibration method for short-distance measurement, which takes into consideration the dimensions of DoF and equal-increment partition of distortion, are proposed in this paper. The rest of this paper is organized as follows. In Section 2, a focusing-state-independent DoF distortion model, which only involves the spatial position of the observed point, is constructed. In Section 3, based on the model in previous section, an equal-increment partitioning DoF distortion model is proposed, which enables a fine representation of the lens distortion in the photographic field. Section 4 details the calibration method for both DoF distortion and camera model parameters, as well as the image processing of the control field for distortion calibration. In Section 5, experimental verification of the proposed lens distortion model and calibration method is carried out. Finally, Section 6 concludes this paper.

2. Focusing-State-Independent DoF Distortion Model

The observed distortion of a point varies with its position within the DoF. Though the close-range imaging configuration increases the visible range, it enlarges the DoF image distortion, consequently affecting the measurement accuracy. To break the limitations of the aforementioned in-plane and DoF distortion model in the vision measurement of short-distance and small-focal-length settings, a DoF-dependent yet focusing-state-independent distortion model is proposed in this paper.

2.1. Pinhole Camera Model with Distortion

As illustrated in Figure 1, the linear pinhole camera model depicts the one-to-one mapping between the 3D points in the object space and its 2D projections in the image. Let $p\left(\begin{array}{cc} u_{li} & v_{li} \end{array} \right)$ be undistorted coordinates mapped from a spatial point in the world coordinate system $O_w X_w Y_w Z_w$ to the image coordinate system ouv through the optical center O_C. Then, camera mapping can be expressed as [17]

$$z' \begin{bmatrix} u_{li} \\ v_{li} \\ 1 \end{bmatrix} = \boldsymbol{K} \cdot \boldsymbol{M} \cdot \begin{bmatrix} X_w \\ Y_w \\ Z_w \\ 1 \end{bmatrix} \tag{1}$$

where z' describes the scaling factor; $\boldsymbol{K}$ is the intrinsic parameter matrix, which quantitatively characterizes the critical parameters of the image sensor (i.e., Charge Coupled Device (CCD) or Complementary Metal-Oxide Semiconductor (CMOS)); Matrix $\boldsymbol{M}$, expressing the transformation between the vision coordinate system (VCS) and the world coordinate system, consists of the rotation matrix $\boldsymbol{R}$ and translation matrix $\boldsymbol{T}$.

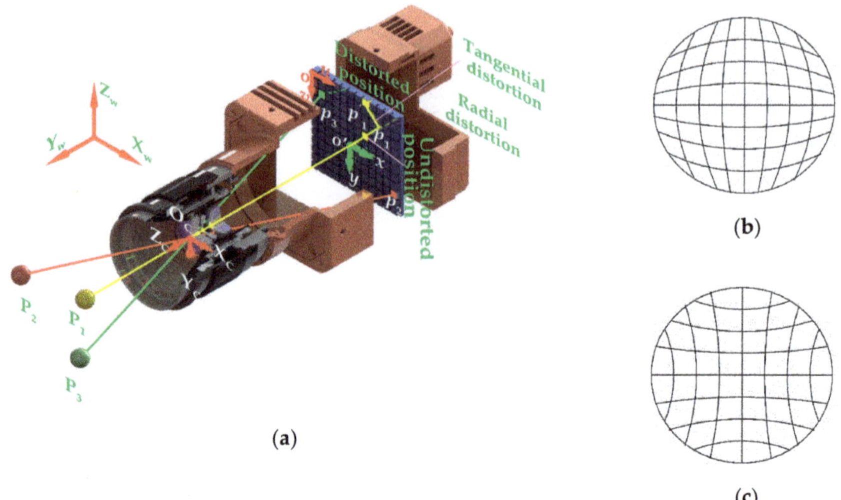

Figure 1. Schematic diagram of camera model and lens distortion: (**a**) camera model; (**b**) barrel distortion; (**c**) pincushion distortion.

However, manufacturing and assembly errors can lead to radial and decentering lens distortion. Consequently, the pinhole assumption does not hold for real camera systems, and the image projection of a straight line would be bent into a curve (Figure 1b,c). To characterize the lens distortion, Brown proposed the distortion model in a polynomial form [11,12]:

$$
\begin{cases}
u_{li} = \overline{u}_{li} + \delta_{u_{li}} \\
v_{li} = \overline{v}_{li} + \delta_{v_{li}} \\
\delta_{u_{li}} = \overline{u}_{li} \cdot (1 + K_1 \cdot r^2 + K_2 \cdot r^4 + \cdots) + \left[P_1 \cdot (r^2 + 2 \cdot \overline{u}_{li}{}^2) + 2P_2 \cdot \overline{u}_{li} \cdot \overline{v}_{li} \right] \cdots \\
\delta_{v_{li}} = \overline{v}_{li} \cdot (1 + K_1 \cdot r^2 + K_2 \cdot r^4 + \cdots) + \left[P_2 \cdot (r^2 + 2 \cdot \overline{v}_{li}{}^2) + 2P_1 \cdot \overline{u}_{li} \cdot \overline{v}_{li} \right] \cdots
\end{cases}
\tag{2}
$$

where $\left(\overline{u}_{li} \quad \overline{v}_{li} \right)$ is the distorted coordinates; $\delta_{u_{li}}$ and $\delta_{v_{li}}$ are the distortion functions of an image point in the u and v direction respectively; $\left(u_0 \quad v_0 \right)$ denotes the distortion center; $r = \sqrt{(\overline{u}_{li} - u_0)^2 + (\overline{v}_{li} - v_0)^2}$ stands for the distortion radius of the image point; K_1 and K_2 are the first and second-order coefficients of radial distortion respectively; while, P_1 and P_2 are the first and second-order coefficients of decentering distortion respectively.

2.2. Distortion Model in the Focal Plane

2.2.1. Radial Distortion Model

Let δr_∞ be the radial distortion for a lens that is focused on plus infinity, while $\delta r_{-\infty}$ on the minus infinity. m_s refers to the vertical magnification in the focal plane at object distance s. According to Magill's model [26], δr_s, the lens radial distortion in the focal plane, can be expressed as

$$
\delta r_s = \delta r_{-\infty} - m_s \cdot \delta r_\infty
\tag{3}
$$

Let δr_{s_m} and δr_{s_k} be the radial distortions in the focal planes when the lens is focused on the distances of s_m and s_k respectively. Then, the distortion function for focal plane δr_s at distance s can be written as

$$
\delta r_s = \alpha_s \cdot \delta r_{s_m} + (1 - \alpha_s) \cdot \delta r_{s_k}
\tag{4}
$$

where, f is the focal length; $\alpha_s = \frac{(s_k - s_m) \cdot (s - f)}{(s_k - s) \cdot (s_m - f)}$. The i-th radial distortion coefficients K_i^s for focused object plane at distance s are

$$K_i^s = \alpha_s \cdot K_i^{s_m} + (1 - \alpha_s) \cdot K_i^{s_k} \quad i = 1, 2. \tag{5}$$

where $K_i^{s_m}$ and $K_i^{s_k}$ are the i-th radial distortion coefficients when the lens is focused on the distances of s_m and s_k respectively. As can be easily noticed in Equation (5), if the radial distortion coefficients of two different focal planes are known, the radial distortion coefficients of any focal plane can be obtained.

2.2.2. Decentering Distortion

As for the decentering distortion, the equations are as follows [12]:

$$\begin{cases} \delta r_u = (1 - \frac{f}{s})(P_1 \cdot (r^2 + 2u^2) + 2P_2 \cdot u \cdot v) \\ \delta r_v = (1 - \frac{f}{s})(P_1 \cdot (r^2 + 2v^2) + 2P_2 \cdot u \cdot v) \\ r^{s,s'} = \frac{s-f}{s'-f} \cdot \frac{s'}{s} \end{cases} \tag{6}$$

where $(1 - \frac{f}{s}) \cdot r^{s,s'}$ is the compensation coefficient; δr_u and δr_v represent the components of the decentering distortion in the u and the v direction respectively; s and s' depict the object distances corresponding to the two focal planes, respectively.

2.3. DoF-Dependent Distortion Model for Arbitrary Defocused Plane

2.3.1. DoF-Dependent Radial Distortion Model

Fraser and Shortis [29] proposed an empirical model for describing the distortion of any object plane (or defocused plane), which solved Brown model's problem of inaccurate description of severe distortion caused by the image configuration of short-distance and small-focal-length settings. The equation is as follows:

$$K^{s,s_p} = K^s + g \cdot (K^{s_p} - K^s) \tag{7}$$

where K^{s,s_p} denotes the radial distortion coefficient in the defocused plane with the depth of s_p when the lens is focused at distance s; g is the empirical coefficient; K^{s_p} and K^s represent the radial distortion coefficients in the focal planes at distances s_p and s respectively. By extending the equation, we can get the radial distortion function $\delta r^{s,s_n}$ at s_n expressed by the $\delta r^{s,s_m}$ at s_m when the lens is focused at the distance of s:

$$\begin{cases} \delta r^{s,s_m} = \delta r^s + g \cdot (\delta r^{s_m} - \delta r^s) \\ \delta r^{s,s_n} - \delta r^{s,s_m} = \delta r^s + g \cdot (\delta r^{s_n} - \delta r^s) - \delta r^s - g \cdot (\delta r^{s_m} - \delta r^s) \end{cases} \tag{8}$$

From the above equation, we can easily obtain $\delta r^{s,s_n} = \delta r^{s,s_m} + \alpha_{s,s_m(s_n)} \cdot (\delta r^s - \delta r^{s,s_m})$. Then, by extending the results to the radial distortion of a point in the defocused plane at distance s_k, the relationship between $\delta r^{s,s_n}$, $\delta r^{s,s_m}$ and $\delta r^{s,s_k}$ can be given by

$$\begin{cases} \delta r^{s,s_n} = \delta r^{s,s_m} + \alpha_{s,s_m(s_n)} \cdot (\delta r^s - \delta r^{s,s_m}) \\ \delta r^{s,s_n} = \delta r^{s,s_k} + \alpha_{s,s_k(s_n)} \cdot (\delta r^s - \delta r^{s,s_k}) \\ \delta r^{s,s_m} = \delta r^{s,s_k} + \alpha_{s,s_k(s_m)} \cdot (\delta r^s - \delta r^{s,s_k}) \end{cases} \tag{9}$$

in which $\alpha_{s,s_m(s_n)} = \frac{s_m - s_n}{s_m - s} \cdot \frac{s - f}{s_n - f}$, $\alpha_{s,s_k(s_n)} = \frac{s_k - s_n}{s_k - s} \cdot \frac{s - f}{s_n - f}$, and $\alpha_{s,s_k(s_m)} = \frac{s_k - s_m}{s_k - s} \cdot \frac{s - f}{s_m - f}$.

After eliminating the focus distance and the distortion in the focal plane, we can obtain the following equation:

$$\begin{cases} \delta r^{s,s_m} = \delta r^{s,s_k} + \frac{s_k - s_m}{s_k - s} \cdot \frac{s - f}{s_m - f} \cdot (\delta r^s - \delta r^{s,s_k}) \\ \delta r^{s,s_n} = \delta r^{s,s_k} + \frac{s_k - s_n}{s_k - s} \cdot \frac{s - f}{s_n - f} \cdot (\delta r^s - \delta r^{s,s_k}) \end{cases} \tag{10}$$

Then, we can have $\frac{s-f}{s_k-s} \cdot (\delta r^s - \delta r^{s,s_k}) = (\delta r^{s,s_m} - \delta r^{s,s_k}) \cdot \frac{s_m-f}{s_k-s_m}$. $\delta r^{s,s_n}$ can be expressed as

$$\delta r^{s,s_n} = \frac{\delta r^{s,s_k} \cdot (s_m - f) \cdot (s_k - s_m) + (s_k - s_n) \cdot (s_k - f) \cdot (\delta r^{s,s_m} - \delta r^{s,s_k})}{(s_m - f) \cdot (s_k - s_m)} \tag{11}$$

Obviously, when the lens is focused at distance s, through two distortions corresponding to object distances s_m and s_k respectively, radial distortion coefficient in any defocused plane with the depth of s_n when the lens is focused at distance s can be obtained:

$$K_i^{s,s_n} = \frac{K_i^{s,s_k} \cdot (s_m - f) \cdot (s_k - s_m) + (s_k - s_n) \cdot (s_k - f) \cdot (K_i^{s,s_m} - K_i^{s,s_k})}{(s_m - f) \cdot (s_k - s_m)}\, i = 1 \cdots 2. \tag{12}$$

When two object planes are set, K_i^{s,s_m}, K_i^{s,s_k}, s_m, s_k and f are known. Thus, K_i^{s,s_n} in Equation (12) is only dependent on s_n, and it is independent of the distortion coefficient K_i^s on the focal plane and the focus distance s.

2.3.2. DoF-Dependent Decentering Distortion Model

In Equation (6), since $\delta P^{s,s_m} = r^{s,s_m} \cdot \delta P^\infty$, the distortion in the focal plane can be written as

$$\begin{cases} \delta P^{s,s_m} = \left(1 - \frac{f}{s}\right) \cdot \frac{s-f}{s_m-f} \cdot \frac{s_m}{s} \cdot \delta P^\infty \\ \delta P^{s,s_k} = \left(1 - \frac{f}{s}\right) \cdot \frac{s-f}{s_k-f} \cdot \frac{s_k}{s} \cdot \delta P^\infty \\ \delta P^{s,s_n} = \left(1 - \frac{f}{s}\right) \cdot \frac{s-f}{s_n-f} \cdot \frac{s_n}{s} \cdot \delta P^\infty \end{cases} \tag{13}$$

where $\delta P^{s,s_m}$, $\delta P^{s,s_k}$ and $\delta P^{s,s_n}$ are the decentering distortion functions in the defocused plane at the object distances of s_m, s_k and s_n when the lens is focused at distance s respectively. From the first two lines of the above equation, we get $\frac{\delta P^{s,s_k}}{\delta P^{s,s_m}} = \frac{s-f}{s_k-f} \cdot \frac{s_m-f}{s-f} \cdot \frac{s_k}{s} \cdot \frac{s}{s_m} = \frac{s_m-f}{s_k-f} \cdot \frac{s_k}{s_m} = M^{s_k,s_m}$, and then

$$\begin{cases} f = \frac{1-M^{s_k,s_m}}{s_k-s_m \cdot M^{s_k,s_m}} \\ \frac{\delta P^{s,s_n}}{\delta P^{s,s_m}} = \frac{s_m-f}{s_n-f} \cdot \frac{s_n}{s_m} \end{cases} \tag{14}$$

Put the first line into the second one, and we obtain

$$\frac{\delta P^{s,s_n}}{\delta P^{s,s_m}} = \frac{s_m - \frac{1-M^{s_k,s_m}}{s_k-s_m \cdot M^{s_k,s_m}}}{s_n - \frac{1-M^{s_k,s_m}}{s_k-s_m \cdot M^{s_k,s_m}}} \cdot \frac{s_n}{s_m} = \frac{s_m \cdot s_k - s_m^2 \cdot M^{s_k,s_m} - 1 + M^{s_k,s_m}}{s_n \cdot s_k - s_m \cdot s_n \cdot M^{s_k,s_m} - 1 + M^{s_k,s_m}} \cdot \frac{s_n}{s_m} \tag{15}$$

Equation (15) can be simplified to

$$\begin{cases} \delta P^{s,s_n} = \frac{M^{s_k,s_m} \cdot (1-s_m^2) + (s_m \cdot s_k - 1)}{M^{s_k,s_m} \cdot (1-s_n \cdot s_m) + s_n \cdot s_k - 1} \cdot \frac{s_n}{s_m} \cdot \delta P^{s,s_m} \\ P_i^{s,s_n} = \frac{M^{s_k,s_m} \cdot (1-s_m^2) + (s_m \cdot s_k - 1)}{M^{s_k,s_m} \cdot (1-s_n \cdot s_m) + s_n \cdot s_k - 1} \cdot \frac{s_n}{s_m} \cdot P_i^{s,s_m}\ i = 1, 2. \end{cases} \tag{16}$$

Put $M^{s_k,s_m} = \frac{P_i^{s,s_k}}{P_i^{s,s_m}}$ $(i = 1, 2)$ into Equation (16), and the following equation is obtained:

$$P_i^{s,s_n} = \frac{P_i^{s,s_k} \cdot (1-s_m^2) + P_i^{s,s_m} \cdot (s_m \cdot s_k - 1)}{P_i^{s,s_k} \cdot (1-s_m \cdot s_n) + P_i^{s,s_m}(s_n \cdot s_k - 1)} \cdot \frac{s_n}{s_m} \cdot P_i^{s,s_m} i = 1, 2. \tag{17}$$

Given the parameters P_i^{s,s_m}, P_i^{s,s_k}, s_m and s_k are known, it can be illustrated from Equation (17) that the decentering distortion coefficient P_i^{s,s_n} in any defocused plane is dependent only on the object distance, s_n, and is independent of the focus distance s and the distortion P_i^s in the focal plane. Moreover, since focal length f is not included in Equation (17), decentering distortion is not affected by this parameter.

Hereto, the DoF-dependent yet focusing-state-independent distortion model suitable for close-range, short-distance measurement scenes is established, which overcomes the limited practicability caused by the way of calibrating DoF distortion by manual adjustment of the focus and zoom rings, and it also solves the problem when the current position and the distortion parameters of the focal plane are not exactly known.

3. Equal-Increment Partition Based DoF Distortion Model

The distortion coefficients are solved by minimizing the straightness error of the observed points. If a set of distortion coefficients is used to describe the distortion in the whole image, the distortion coefficients will be the error balance of all points. However, for each region of the image the error is not the minimum. Hence, an equal-increment partition based DoF distortion model is proposed in this section. The distortion spreads outward from the image center along a circumferential contour, with the characteristics of the image being small in the middle and large on the image edge. In this paper, we first partition the in-plane distortion in an equal-increment way, then the 2D partition strategy is extended to the 3D photographic field.

3.1. Equal-Increment Based Distortion Partitioning Method

Figure 2 presents two distortion partitioning methods. The X axis represents the distance from an image point to the distortion center (u_0, v_0), namely the distortion radius. The Y axis describes the distortion in pixels. The blue curve is the distortion curve calculated by the features in the whole image. As illustrated in Figure 2a, when DoF distortion is partitioned by an equal radius, the distortion increment of each partition is different ($\Delta_1 < \Delta_2 < \Delta_3 < \Delta_4 < \Delta_5$) despite the same distortion radius increment ($R_1 = R_2 = R_3 = R_4 = R_5$) [36]. For a polynomial-based distortion function, it is well known that the more scattered the distorted points and the larger the distortion increments are, the lower the regression accuracy of the function to the distortion is. As a result, the estimated accuracy of the partition's distortion parameters decreases gradually from inside to outside ($\varepsilon_1 > \varepsilon_2 > \varepsilon_3 > \varepsilon_4 > \varepsilon_5$).

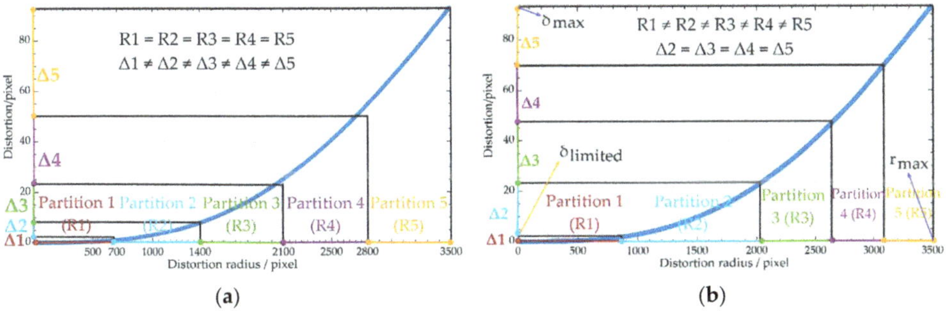

Figure 2. Two distortion partitioning methods: (**a**) equal-radius partition; (**b**) equal-increment partition.

To solve the problem, a DoF distortion model based on the equal-increment partition is proposed in this paper, and the procedures are as follows:

(1) Estimate the distortion curve using all features in the whole image (Figure 2b). Then, determine the maximum value of image distortion δ_{max} according to the maximum distortion radius and distortion curves. The maximum distortion radius of the image is $r_{max} = \frac{1}{2} \cdot \sqrt{(I_l - u_0)^2 + (I_h - v_0)^2}$, where I_l and I_h are the length and height of the image, respectively.

(2) In the central image region, the distortion is so tiny that it cannot converge after iteration, which results in a poorer quality of the undistorted image than that of the original one. Therefore, we use $\delta_{limited}$, the minimum distortion value when the algorithm converges in the central image region, as the threshold to estimate $r_{limited}$, the minimum value of the image distortion radius.

(3) Determine the number of partitions n_p.

(4) Use the maximum distortion δ_{max}, the lower-limit distortion $\delta_{limited}$, and n_p to determine the distortion increment of each partition $\delta_{equ} = (\delta_{max} - \delta_{limited})/(n_p - 1)$, $\delta_{equ} = \Delta_2 = \Delta_3 = \Delta_4 = \Delta_5$ (Figure 2b).

(5) Calculate the radius increment of each partition using δ_{equ} and the distortion curve, $R_1 \neq R_2 \neq R_3 \neq R_4 \neq R_5$ (Figure 2b).

(6) Calibrate the distortion curve of each partition by the features in the corresponding partition of the image utilizing the decoupled-calibration method (see Section 4).

Then the distortion partition of the 2D object plane is extended to the 3D DoF. As can be known from Equation (1), the object-to-image mapping satisfies the following:

$$\begin{cases} x = f \cdot \frac{X_m}{Z_m} = f \cdot \frac{X_k}{Z_k} \\ y = f \cdot \frac{Y_m}{Z_m} = f \cdot \frac{Y_k}{Z_k} \end{cases} \tag{18}$$

where $P_m(\ X_m \quad Y_m \quad Z_m\ )$ and $P_k(\ X_k \quad Y_k \quad Z_k\ )$ are two points in the VCS. The 2D point $p(\ x \quad y\ )$ (in millimeters) is the image projection of the P_m and P_k (P_m, P_k, and O_C are collinear). Let ρ be the partition radius, then $x^2 + y^2 = \rho^2$, and we get

$$\begin{cases} f^2 \cdot \frac{X_m^2}{Z_m^2} + f^2 \cdot \frac{Y_m^2}{Z_m^2} = \rho^2 \\ f^2 \cdot \frac{X_k^2}{Z_k^2} + f^2 \cdot \frac{Y_k^2}{Z_k^2} = \rho^2 \end{cases} \tag{19}$$

From the above equation, we can know that $f \cdot R_m = \rho \cdot Z_m$ and $Z_m \cdot R_k = Z_k \cdot R_m$, where Z_m and Z_k are the depths of the m-th (Π_m) and the k-th (Π_k) object planes in the VCS respectively. R_m and R_k are the partition radius of the two object planes. Let $s_m = Z_m$ and $s_k = Z_k$, and then extend the above distortion partitions to 3D DoF domain. As shown in Figure 3, if the range of the g-th partition in the object plane Π_m is $\left[\ (g-1) \cdot R_m \quad g \cdot R_m\ \right]$, the partition range in object planes Π_k and Π_n are $\left[\ (g-1) \cdot (s_k \cdot R_m/s_m) \quad g \cdot (s_k \cdot R_m/s_m)\ \right]$ and $\left[\ (g-1) \cdot (s_n \cdot R_m/s_m) \quad g \cdot (s_n \cdot R_m/s_m)\ \right]$ respectively. In this way, although the distortion radius in each partition is different, distortion coefficients can be obtained with high accuracy when the image distortion is partitioned by equal distortion increments.

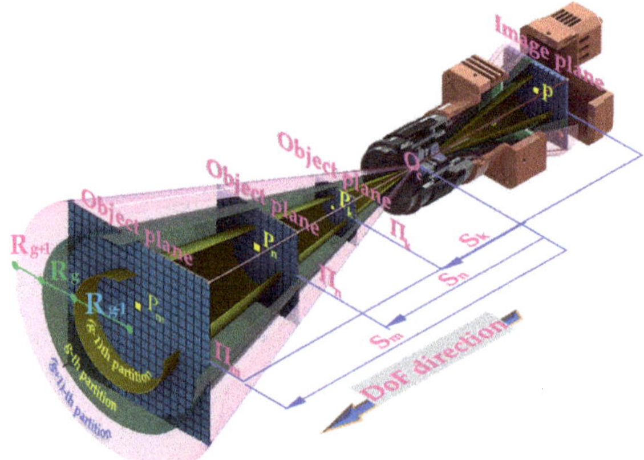

Figure 3. The geometric relationship between the partition radii in different object planes.

3.2. Equal-Increment Partition Based DoF Distortion Model

After partitioning the DoF distortion, we incorporate the partitions into the DoF distortion model. Procedures to solve the partition radius and distortion coefficients on any object distance s_n are as follows:

(1) Partition the distortion in the object plane Π_m using the proposed method, and calculate the i-th order radial and decentering distortion coefficients in the g-th partition. Register the two coefficients as ${}^g K_i^{s,s_m}$ and ${}^g P_i^{s,s_m}$ respectively.

(2) Based on the g-th partition in the object plane Π_m (the object distance is s_m), the corresponding partition radius in the object plane Π_k (the object distance is s_k) is calculated. In addition, the i-th order radial and decentering distortion coefficients can be computed. Register the two coefficients as ${}^g K_i^{s,s_k}$ and ${}^g P_i^{s,s_k}$ respectively.

(3) Based on the partitions in the object plane Π_m, we calculate the partitions in the object distance plane Π_n (the object distance is s_n). Then, for the g-th partition of the object plane Π_n, the radial distortion coefficient ${}^g K_i^{s,s_n}$ and the decentering distortion coefficient ${}^g P_i^{s,s_n}$ can be expressed as

$$\begin{cases} {}^g K_i^{s,s_n} = f({}^g K_i^{s,s_m}, {}^g K_i^{s,s_k}, s_n) \\ {}^g P_i^{s,s_n} = f({}^g P_i^{s,s_m}, {}^g P_i^{s,s_k}, s_n) \end{cases} \tag{20}$$

From the equation, we can know

$$\begin{cases} {}^g K_i^{s,s_n} = \dfrac{{}^g K_i^{s,s_k} \cdot (s_m - f) \cdot (s_k - s_m) + (s_k - s_n) \cdot (s_k - f) \cdot ({}^g K_i^{s,s_m} - {}^g K_i^{s,s_k})}{(s_m - f) \cdot (s_k - s_m)} \\[2ex] {}^g P_i^{s,s_n} = \dfrac{{}^g P_i^{s,s_k} \cdot (1 - s_m^2) + {}^g P_i^{s,s_m} \cdot (s_m \cdot s_k - 1)}{{}^g P_i^{s,s_k} \cdot (1 - s_m \cdot s_n) + {}^g P_i^{s,s_m} (s_n \cdot s_k - 1)} \cdot \dfrac{s_n}{s_m} \cdot {}^g P_i^{s,s_m} \\[2ex] i = 1, 2 \\ g = 1, 2, \cdots, n_p \end{cases} \tag{21}$$

At this point, we have established an equal-increment partition based DoF distortion model for any object plane at s_n when the lens is focused at distance s.

4. Calibration Method for Camera Parameters

In close-range photography, the DoF images are seriously distorted, so the calibration accuracy of the distortion parameters is the decisive factor affecting the vision measurement accuracy. When the coupled-calibration method is used to solve the distortion parameters, the estimated errors of intrinsic and extrinsic parameters will be propagated to distortion parameters. Thus, a two-step method is proposed to calibrate the camera parameters, in which distortion parameters are estimated independently.

4.1. Independent Distortion Calibration Method Based on Linear Conformation

Figure 4 details the experimental system for DoF lens distortion, which consists of a monocular camera, a control field, a light source, an electric control platform, and a multi-axis motion controller. The X, Y, A, and C axes of the platform are in the object space, while the Z-axis is in the image space. A control field, with the features of circle, corner, and line, is used to calibrate the lens distortion, and the geometric relationship between the features is known accurately. On this basis, the pose of the control field relative to the image plane can be adjusted by the Perspective-n-Point (PnP) algorithm.

In this paper, the distortion coefficients can be estimated by the plumb-line method [12] alone. It is defined by Brown (1971) as "a straight line in the object space will be mapped to the image plane in a straight way after a perfect lens, and any change of straightness can be reflected as the lens distortion described by the radial and decentering distortion coefficients."

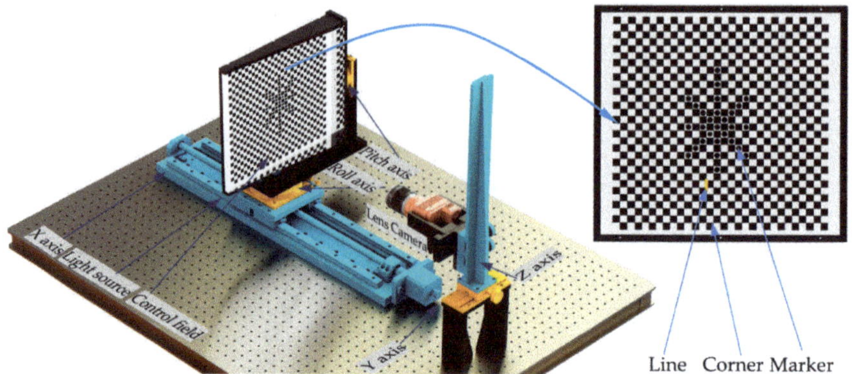

Figure 4. Experimental system for calibrating depth of field (DoF) lens distortion.

As demonstrated in Figure 5, when N edge points $\begin{pmatrix} u_1 & v_1 \end{pmatrix} \cdots \begin{pmatrix} u_N & v_N \end{pmatrix}$ on the same curve are known, the regression line equation determined by the point group is

$$\alpha' u + \beta' v - \gamma' = 0 \tag{22}$$

Let $\alpha' = \sin\theta$, $\beta' = \cos\theta$, $\gamma' = A_u \sin\theta + A_v \cos\theta$, $\tan 2\theta = -\frac{2V_{uv}}{V_{uu}-V_{vv}}$. where θ is the angle between the regression line and the u axis (Figure 5). $A_u = \frac{1}{N}\sum_{i=1}^{N} u_i$, $A_v = \frac{1}{N}\sum_{i=1}^{N} v_i$, $V_{uu} = \frac{1}{N}\sum_{i=1}^{N} (u_i - A_u)^2$, $V_{uv} = \frac{1}{N}\sum_{i=1}^{N} (u_i - A_u)(v_i - A_v)$, $V_{vv} = \frac{1}{N}\sum_{i=1}^{N} (v_i - A_v)^2$.

Given there are L lines and there are N_l points in the l-th line, the average sum of squared distances from the points $\begin{pmatrix} u_{li} & v_{li} \end{pmatrix}$ to all the lines can be written as

$$D = \frac{1}{L} \cdot \sum_{l=1}^{L} \frac{1}{N_l} \cdot \sum_{i=1}^{N_l} (\alpha'_l u_{li} + \beta'_l v_{li} - \gamma'_l)^2 \tag{23}$$

Any distortion of a line's straightness in the image plane can be corrected by a mapping involving radial and decentering distortion. Thus, substitute Equation (2) into Equation (23) and we can get

$$F(\overline{u}_{li}, \overline{v}_{li}; K_1, K_2, P_1, P_2) = 0 \tag{24}$$

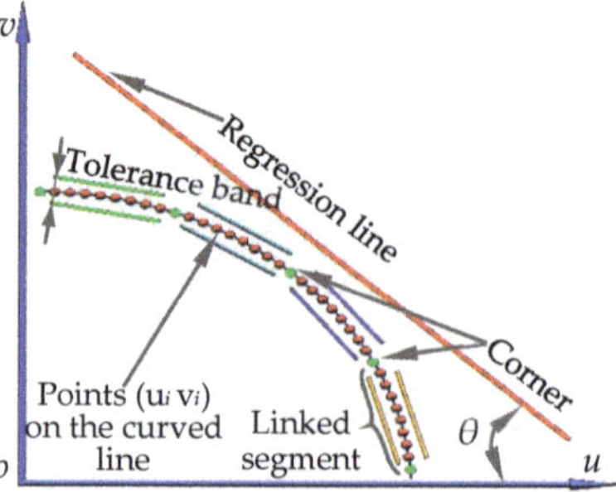

Figure 5. Schematic diagram of distortion calibration based on linear conformation configuration.

If there are L lines in an image and N_l observation points are extracted from each line, we can have $L \cdot N_l$ equations. In these equations, there are $L + 4$ variables (L line coefficients and 4 distortion coefficients). If $L \cdot N_l > L + 4$, the optimal solution of distortion coefficients can be obtained.

After solving the image distortion coefficients, the inverse mapping $imR(u,v) = imD(u_d, v_d)$ between the undistorted image imR and distorted image imD is established by cubic B-spline interpolation. In this way, the image distortion can be corrected. Besides, in this paper, the three straightness indicators of the maximum, average, and root mean square (RMS) $d = \sqrt{D/\sum_{l=1}^{L} N_l}$ of the point-to-line distance, and the Peak Signal-to-Noise Ratio (PSNR) $PSNR = 10 \times \log_{10}((2^n - 1)^2 / MSE$, are used to evaluate the distortion correction effects. D has been defined in Equation (23), and MSE is the mean square error of the image before and after distortion correction.

4.2. Image Processing and Camera Calibration

In this paper, the parameters in the equal-increment partition based DoF distortion model are calculated by using straight lines in a particular area of the control field. To this end, the corner control based method, for extracting line segments within a partition, is proposed. As shown in Figure 6, the image processing procedures include the following:

(1) Image acquisition. Capture the image of the control field using the monocular camera (Figure 6a).
(2) Point detection. Corners of the checkerboard are extracted by the Harris detector (Figure 6b), and the edge points on the curve are detected by the Canny operator with subpixel accuracy.
(3) Point connection. Use the edge points between two adjacent corners to form unit segments (Figure 5). In each segment, David Lowe's method [37] is used to track and connect the edge points in the four-link area from one particular point to the others (Figure 6c). The minimum connection length is set to be greater than 10 pixels.
(4) Point reselection. The distortion is not evenly distributed on the image, with the largest at the image edge, which makes it difficult to remove the noisy point. To solve this problem, the tolerance band of 4 pixels (Figure 5) set in each unit segment is used as the constraint to filter out the outliers. Consequently, the new edge points are determined (Figure 6d).
(5) Line extraction. Any line can be obtained according to the corner position and the predefined distortion radius. Figure 6e,f shows the extraction results of the 19th horizontal line and the lines in different areas of the control field, respectively.

By combining the image processing results with the DoF distortion partition model, distortion parameters at any position of the DoF can be determined. To avoid the coupling effect between the distortion parameters and other parameters in the camera model, the camera's intrinsic and extrinsic parameters are preliminarily calibrated by Zhang's method. Then, we fix the distortion parameters and place the high-precision target in multiple spatial positions to optimize the intrinsic and extrinsic parameters. The cost function to be optimized is

$$\begin{cases} E^q_{depth\_dependent}(\boldsymbol{R}_q) = \sum_{g=1}^{mg} \left(\boldsymbol{H}^{-1}(u_0, v_0, f_x, f_y, {}^g K_i, {}^g P_j, \boldsymbol{R}_q, \boldsymbol{T}_q) \right) \\ \qquad\qquad i = 1, 2 \\ \qquad\qquad j = 1, 2 \end{cases} \tag{25}$$

where $E^q_{depth\_dependent}(\boldsymbol{R}_q)$ describes the cost function when the control field is in the q-th pose. $\boldsymbol{R}_q$ and $\boldsymbol{T}_q$ are the rotation and translation matrices in the q-th pose. ${}^g K_i$ and ${}^g P_j$ are the i-th order radial and j-th order decentering distortion coefficients in the g-th partition of the q-th pose. By using the Levenberg–Marquardt (LM) algorithm, the optimal solution of the camera's intrinsic and extrinsic parameters can be obtained.

Through the above process, the monocular camera calibration can be realized. In practice, the partition where a spatial point is located $\left\lceil \frac{\sqrt{X^2+Y^2}\cdot f}{\rho\cdot Z} \right\rceil$ can be determined after estimating its 3D position $(\ X\ \ Y\ \ Z\ )$. Then, the observed distortion can be corrected by choosing the proper distortion coefficients, thus realizing high-accuracy vision measurements.

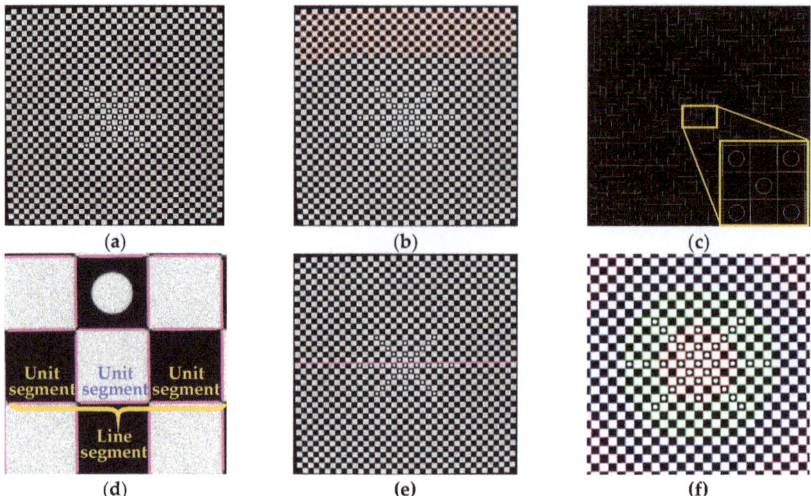

Figure 6. Image processing procedures for linear conformation: (**a**) image of the control field; (**b**) corner detection; (**c**) edge point connection; (**d**) point reselection; (**e**) horizontal line extraction; (**f**) line detection results in different areas.

5. Accuracy Verification Experiments of Both the Distortion Modeling and Calibration Method

5.1. Experimental Verification of the 2D Distortion Partitioning Method

The experimental system is shown in Figure 7. The stroke of the electric control platform along the optical axis of the camera is 500 mm, and the size of the control field is 300 × 300 mm. The SIGMA zoom lens (18–35 mm) and HIK ROBOT camera (MV-CH120-10TM) are selected for imaging. The resolution and focal length are set as 2560 × 2560 pixels and 18 mm respectively. The procedures are as follows:

(1) calibrate the intrinsic and extrinsic parameters of the monocular camera;
(2) make adjustments to ensure that the circle features are distributed symmetrically around the image center;
(3) the pose of the control field is determined and adjusted repeatedly to ensure the object and the image planes are parallel;
(4) the control field is driven by the electronic control platform to move several object planes along the optical axis. The image of the control field in each plane is collected and analyzed by the algorithm on the graphic workstation.

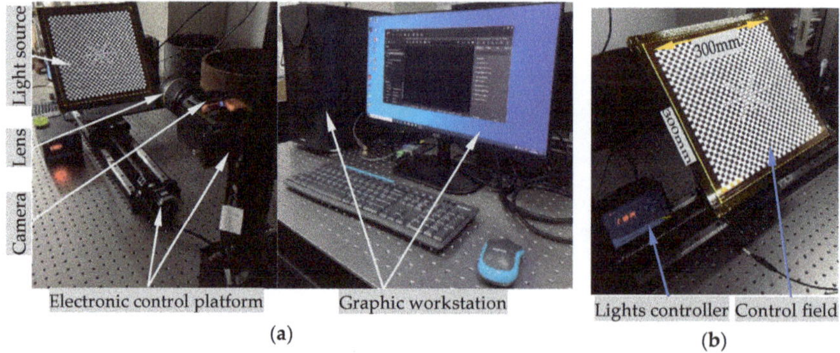

Figure 7. Experimental system for DoF distortion calibration: (**a**) system hardware; (**b**) control field.

First, the accuracy of the 2D distortion partitioning method is verified. The image of the control field at the focus distance is divided into five concentric rings by the equal-radius (Figure 8a–e) and equal-increment (Figure 9a–e) distortion partition models, respectively. In each partition, the corresponding lines (green ones) are selected to solve the distortion coefficients and correct the image distortion. For each of the two partitioning methods, five corrected images can be obtained (i.e., Figures 8f–j and 9f–j). Here, we use Figures 8f and 9f as an example to illustrate the results of distortion correction. The distortion on the image edge solved by the distortion coefficients of the first partition is far beyond the actual distortion here. After distortion correction, distortion is removed overly, thus resulting in the distortion in the opposite direction.

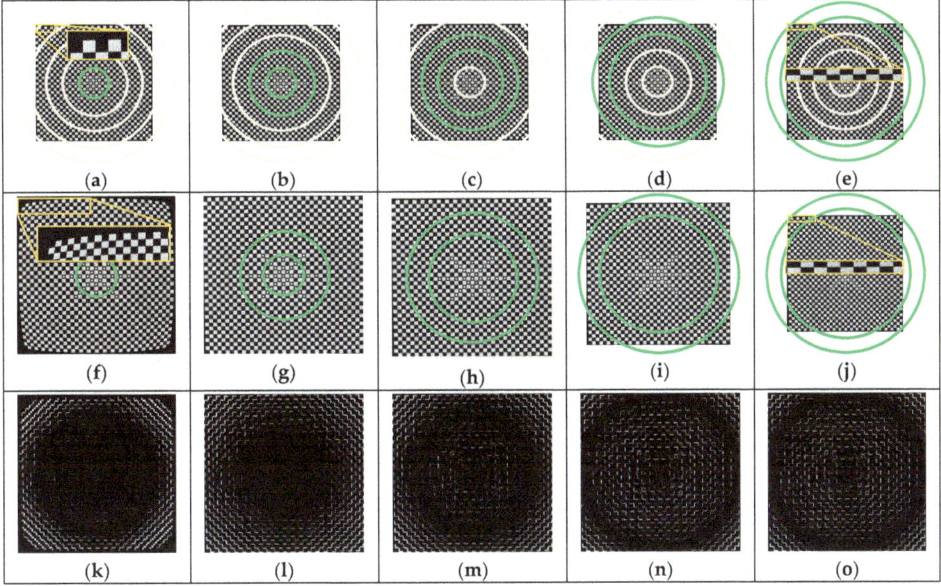

Figure 8. Distortion calibration and correction results based on the equal-radius partition method (f = 18 mm): (**a**) partition 1; (**b**) partition 2; (**c**) partition 3; (**d**) partition 4; (**e**) partition 5; (**f**) distortion correction (partition 1); (**g**) distortion correction (partition 2); (**h**) distortion correction (partition 3); (**i**) distortion correction (partition 4); (**j**) distortion correction (partition 5); (**k**) difference (partition 1); (**l**) difference (partition 2); (**m**) difference (partition 3); (**n**) difference (partition 4); (**o**) difference (partition 5).

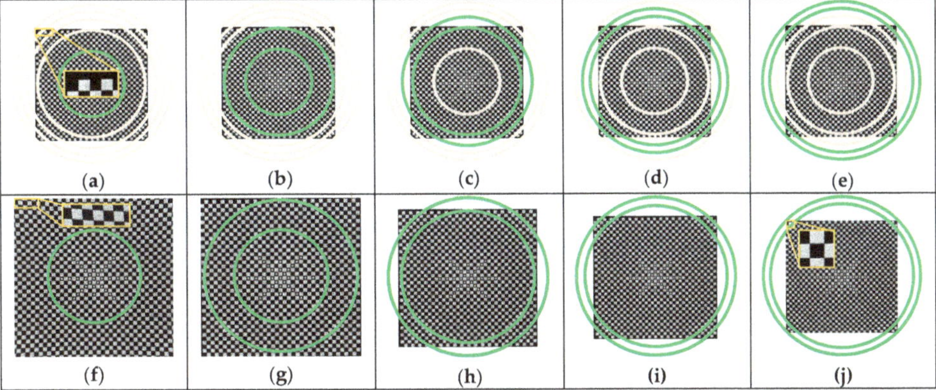

Figure 9. *Cont.*

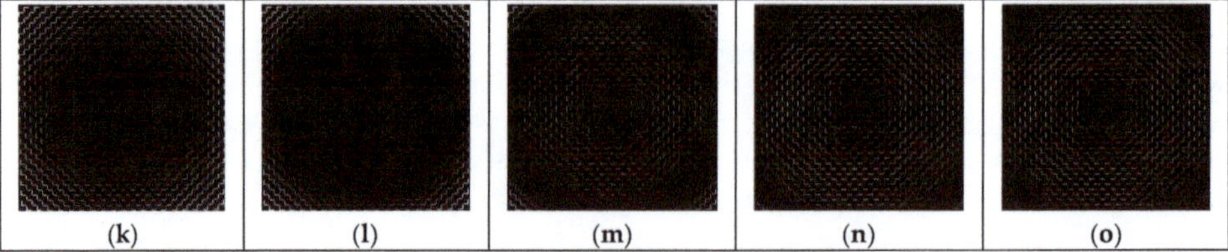

Figure 9. Distortion calibration and correction results based on the proposed partition method (f = 18 mm): (**a**) partition 1; (**b**) partition 2; (**c**) partition 3; (**d**) partition 4; (**e**) partition 5; (**f**) distortion correction (partition 1); (**g**) distortion correction (partition 2); (**h**) distortion correction (partition 3); (**i**) distortion correction (partition 4); (**j**) distortion correction (partition 5); (**k**) difference (partition 1); (**l**) difference (partition 2); (**m**) difference (partition 3); (**n**) difference (partition 4); (**o**) difference (partition 5).

To compare the distortion correction effect of each partition in the image, we subtract the simulated undistorted image with the corrected image, and we get Figures 8k–o and 9k–o. Obviously, the smaller the gray value is, the closer the undistorted image is to the ground truth, and the better the distortion removal effect is. As can be seen from the figures, the distortion correction results of each partition by the equal-radius partitioning method (Figure 8k–m) were not as good as that by the equal-increment partitioning method. Notably, in the fourth partition and the fifth partition located at the edge of the image, the green concentric ring in Figure 8n–o had a larger gray value, while in Figure 9n–o the gray values of pixels in the green concentric ring were approximately 0. This shows that the equal-increment partitioning method had a better performance on eliminating distortion.

Meanwhile, all the lines were used to solve and correct the image distortion as well. Then, the distortion correction effects with and without partition were compared using the aforementioned indexes (Section 4.1). As shown in Table 1, the undistorted images obtained by the two distortion partition methods had a good PSNR of up to 37.61 dB. Compared with the results obtained when without partition, the two partition methods showed a smaller straightness error in each partition. However, compared with the two partitioning methods, the maximum and average errors in the fourth and fifth partitions by the equal-radius partitioning method were at least 4 times and 2 times those by the partitioning method proposed in this paper. That is to say, with the equal-increment partitioning method, each partition can get better distortion correction results. The enlarged image of the best distortion curve for each partition is shown in Figure 10, which validates the effectiveness and accuracy of the proposed partitioning method in 2D settings.

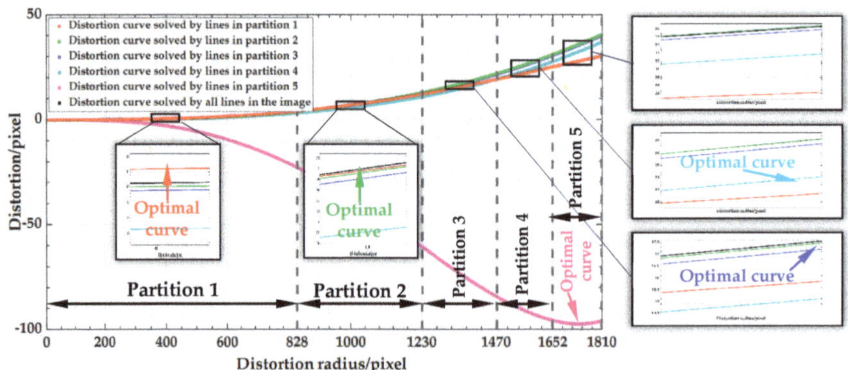

Figure 10. Distortion curves solved by the lines in each partition using the proposed partition method.

Table 1. Comparison of distortion correction of the two partition models.

Indicator	Equal-Radius Partition Model/The Proposed Model					Non-Partitioned Model
	Partition 1	**Partition 2**	**Partition 3**	**Partition 4**	**Partition 5**	
Maximum error/pixel	0.32/0.22	0.62/0.41	0.77/0.53	2.1/0.56	2.7/0.55	7.46
Average error/pixel	0.05/0.03	0.07/0.06	0.10/0.08	0.17/0.08	0.26/0.10	0.52
RMS/pixel	0.04/0.04	0.08/0.06	0.10/0.07	0.11/0.08	0.32/0.09	0.48
PSNR/dB	37.61/37.61	37.26/37.26	37.10/37.30	37.28/37.29	37.34/37.33	37.53

5.2. Accuracy Verification Experiments of DoF Distortion Partitioning Model and Camera Calibration

In this section, the accuracy of the DoF distortion model and camera calibration is verified. The control field is driven to move four different object planes within the DoF, two of which are at the limit positions of the front and rear DoF, and the other two planes are within the DoF. The front object plane was divided into five areas with equal distortion increment of 20.2 pixels. Then, based on the distortion parameters in two object planes with known depths, the distortions in the other two object planes are calculated by the non-partition model, the proposed DoF distortion model with equal-radius partition, and the proposed DoF distortion model with equal-increment distortion partition, respectively. Thereafter, we manually adjusted the ring to focus the lens on the two object planes located at the limit positions of the front and rear DoF. Then, based on the calculated radial and decentering distortion coefficients on the two focal planes, Brown's model [12] with equal-radius partition is used to estimate distortion parameters on the two planes within the DoF.

Furthermore, the results are compared with the distortion directly solved by the lines (the observed value) within the corresponding partition. To compare the accuracy of different DoF distortion models, we took the in-plane point located in the common area (the second column of Table 2) partitioned by the two models at the same object distance (e.g., 400 mm in the first column of Table 2) as an example. As shown in Table 2, for Brown's model [12] with equal-radius partition, the maximum and average absolute differences between the calculated and the observed values were 7.32 µm and 2.81 µm, respectively. Those errors are smaller than that of the traditional Zhang's model without considering the DoF and distortion partition, but much larger than those of the proposed DoF distortion model with equal-radius partition and the proposed DoF distortion model with equal-increment distortion partition, respectively. The maximum and average absolute differences between the calculated and the observed values of the equal-increment distortion partition based DOF distortion model were 1.53 µm and 0.88 µm, respectively. By contrast, the errors of equal-radius partition based DOF distortion model were 4.64 µm and 1.94 µm, which was more than two times those of the proposed model in this paper. The results verified the accuracy of the DoF distortion partitioning model in 3D settings.

Table 2. Accuracy verification for DoF distortion partition model.

Position of Object Plane (mm)	In-Plane Position (Distance from the Distorted Point to the Optical Axis)	Observed Value (μm)	Brown's Distortion Model with Equal-Radius Partition		Equal-Radius Partition Based DOF Distortion Model		Equal-Increment Distortion Partition Based DOF Distortion Model		Zhang's Model	
			Calculated (μm)	Difference $|C-O|$ (μm)	Calculated (μm)	Difference $|C-O|$ (μm)	Calculated (μm)	Difference $|C-O|$ (μm)	Calculated (μm)	Difference $|C-O|$ (μm)
400	Point in partition #1 (56 mm)	86.41	86.33	0.08	86.38	0.03	86.4	0.01	86.27	0.14
	Point in partition #2 (112 mm)	1836.23	1834.12	2.11	1835.82	0.41	1836.09	0.14	1832.42	3.81
	Point in partition #3 (168 mm)	3586.06	3582.81	3.25	3583.7	2.36	3584.84	1.22	3581.81	4.25
	Point in partition #4 (224 mm)	5335.88	5332.01	3.87	5333.06	2.82	5334.57	1.31	5329.22	6.66
	Point in partition #5 (280 mm)	−7085.71	−7093.03	7.32	−7090.35	4.64	−7087.05	1.34	−7095.03	9.32
500	Point in partition #1 (70 mm)	51.3	51.28	0.02	51.28	0.02	51.29	0.01	51.28	0.02
	Point in partition #2 (140 mm)	1468.98	1467.93	1.05	1468.5	0.48	1468.44	0.54	1466.8	2.18
	Point in partition #3 (210 mm)	2886.67	2884.63	2.04	2884.27	2.40	2885.46	1.21	2883.11	3.56
	Point in partition #4 (280 mm)	4304.35	4301.18	3.17	4301.64	2.71	4302.82	1.53	4299.8	4.55
	Point in partition #5 (350 mm)	−5722.03	−5727.29	5.26	−5725.59	4.56	−5723.55	1.52	−5729.84	7.81

Images of circular markers with known precise distance on the planar artifact are collected. The calibration accuracy of the monocular camera is verified by the re-projection errors and the angular reconstruction errors, respectively. Specifically, the planar artifact is driven by the high-accuracy pitch axis to rotate five positions, between two adjacent ones of which are 10°. In each position, the pose matrix between the planar artifact Figure 11a and the calibrated camera is calculated by the OPNP algorithm [38] with the equal-radius and the equal-increment partitioning based DoF distortion models, respectively. Thereafter, 20 markers are projected back to the image via the estimated pose matrix, and the re-projection errors, the image distances between the projected and observed points, are calculated. As shown in Figure 11b, for equal-radius DoF distortion partition model, the maximum and average re-projection errors of the five positions were 0.29 pixels and 0.17 pixels, respectively, while the maximum and average projection errors of the proposed model were 0.11 pixels and 0.05 pixels, respectively. The angle between two adjacent positions of the artifact is reconstructed with the two models as well. As illustrated in Figure 11c, the 3D measurement accuracy of the system is assessed by comparing it with the nominal angle. The results show that the maximum and average angular errors of equal-radius based DoF distortion partition model were 0.48° and 0.30°, respectively, while those of the proposed model were 0.013° and 0.011°, which means that the angular reconstruction errors are effectively reduced. The above results comprehensively verify the accuracy of the DoF distortion partitioning model and the camera calibration method proposed in this paper.

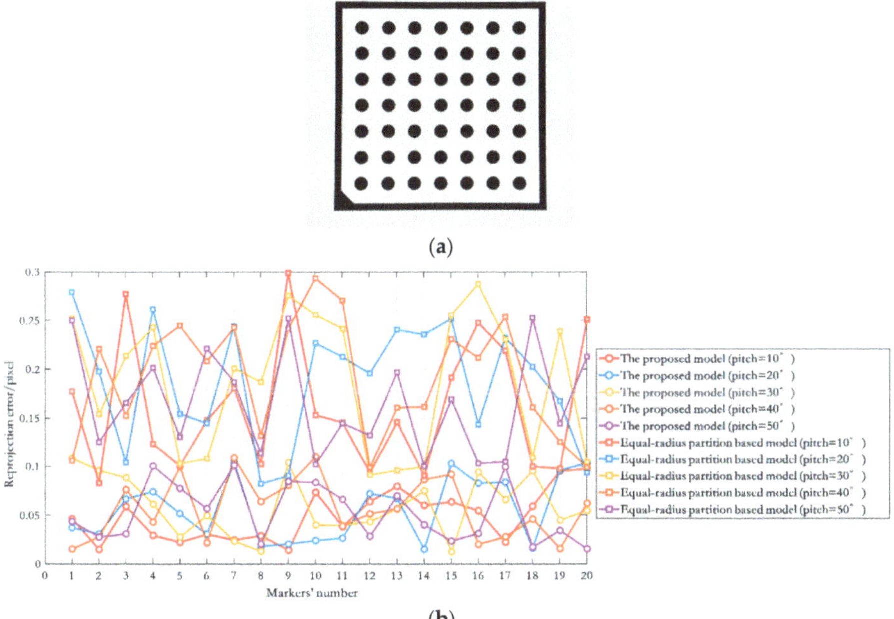

(a)

(b)

Figure 11. *Cont.*

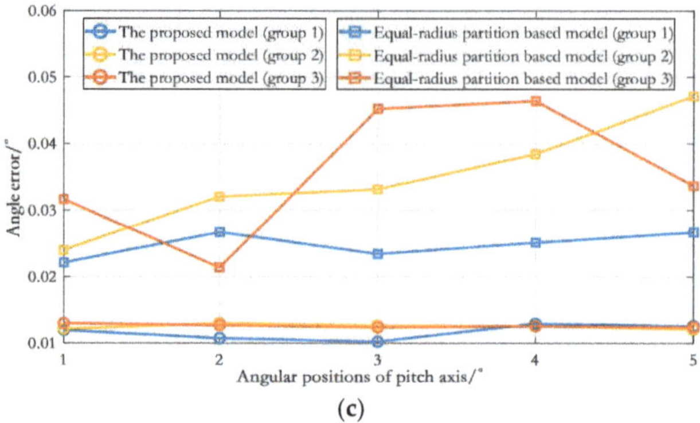

(**c**)

Figure 11. Camera calibration accuracy verification: (**a**) artifact; (**b**) re-projection error; (**c**) angular reconstruction error.

6. Conclusions

This paper has investigated the methods of modeling and calibration of lens distortions for close-range photogrammetry (e.g., short object distance and small focal length). Our work finds that the following:

(1) A focusing-state-independent DoF distortion model is constructed, and the distortion parameters at any object plane can be solved through the distortion on two defocus planes, which removes the human errors introduced by manual adjustment of the focus and zoom rings.

(2) A 2D-to-3D equal-increment partitioning method for lens distortion is proposed. After fusing with the DoF distortion model to form a DoF distortion partition model, the accuracy of lens distortion characterization is further improved.

(3) A two-step method is proposed to calibrate camera parameters, in which the DoF distortion is calculated independently by the plumb-line method, which eliminated the coupling effect among the parameters in the camera model.

(4) Experiments were performed to verify the accuracy of the 2D distortion partition model, DoF-dependent distortion partition model, and camera calibration. The results show that the maximum and average angular reconstruction errors by the proposed model were 0.013° and 0.011° respectively, which validates the accuracy and feasibility of the equal-increment partitioning based DoF distortion method.

The main limitation of the present study is that the number of partitions is not optimized to achieve higher calibration accuracy. Our future work will focus on this and extend our model to other optical systems with fisheye or catadioptric lenses.

Author Contributions: Conceptualization, W.L. and X.L.; formal analysis, X.L.; funding acquisition, W.L. and X.L.; investigation, X.L.; methodology, X.L.; project administration, W.L.; supervision, X.M. and X.Y. (Xiaokang Yin); writing—original draft, X.L.; writing—review & editing, X.Y. (Xin'an Yuan). All authors have read and agreed to the published version of the manuscript.

Funding: This research was funded by the Special National Key Research and Development Plan (No. 2016YFC0802303), National Natural Science Foundation of China (No. 52005513), and the Fundamental Research Funds for the Central Universities (No. 27RA2003015).

Conflicts of Interest: The authors declare no conflict of interest.

Abbreviations

(List of abbreviations and symbols present in this article)

Acronym or Symbols	Definition	Acronym or Symbols	Definition
DoF	depth of field	K_i^{s,s_n}	i-th radial distortion coefficient of the object plane at distance s_n when the lens is focused at the distance of s
FoV	field-of-view	K_i^{s,s_m}	i-th radial distortion coefficient of the object plane at distance s_m when the lens is focused at the distance of s
$p\left(u_{li} \quad v_{li} \right)$	undistorted coordinates	K_i^{s,s_k}	i-th radial distortion coefficient of the object plane at distance s_k when the lens is focused at the distance of s
$O_w X_w Y_w Z_w$	world coordinate system	$\delta P^{s,s_m}$	decentering distortion function in the defocused plane at the object distances of s_m when the lens is focused at distance s
ouv	image coordinate system	$\delta P^{s,s_k}$	decentering distortion function in the defocused plane at the object distances of s_k when the lens is focused at distance s
O_C	optical center	$\delta P^{s,s_n}$	decentering distortion function in the defocused plane at the object distances of s_n when the lens is focused at distance s
z'	scaling factor	P^{s,s_n}	i-th decentering distortion coefficient of the object plane at distance s_n when the lens is focused at the distance of s
K	intrinsic parameter matrix	P_i^{s,s_m}	i-th decentering distortion coefficient of the object plane at distance s_m when the lens is focused at the distance of s
CCD	Charge Coupled Device	P_i^{s,s_k}	i-th decentering distortion coefficient of the object plane at distance s_k when the lens is focused at the distance of s
CMOS	Complementary Metal-Oxide Semiconductor	δ_{max}	maximum value of image distortion
M	transformation matrix	r_{max}	maximum distortion radius of the image
VCS	vision coordinate system	I_l	length of the image
R	rotation matrix	I_h	height of the image
T	translation matrix	$\delta_{limited}$	minimum distortion value
$\left(\overline{u}_{li} \quad \overline{v}_{li} \right)$	distorted coordinates	$r_{limited}$	minimum value of the image distortion radius
$\delta_{u_{li}}$	distortion function of an image point in the u direction	n_p	number of partitions
$\delta_{v_{li}}$	the distortion function of an image point in the v direction	δ_{equ}	distortion increment
(u_0, v_0)	distortion center	ρ	partition radius
r	distortion radius	P_m	point in VCS
K_1	first-order coefficient of radial distortion	P_k	point in VCS
K_2	second-order coefficient of radial distortion	Π_m	m-th object plane in the VCS
P_1	first-order coefficient of decentering distortion	Π_k	k-th object plane in the VCS
P_2	second-order coefficient of decentering distortion	Π_n	n-th object plane in the VCS
δr_∞	radial distortion for a lens that is focused on plus infinity	$^g K_i^{s,s_m}$	i-th order radial distortion coefficient in the g-th partition of object plane Π_m

Acronym or Symbols	Definition	Acronym or Symbols	Definition
$\delta r_{-\infty}$	radial distortion for a lens that is focused on minus infinity	${}^{g}P_i^{s,s_m}$	i-th order decentering distortion coefficient in the g-th partition of object plane Π_m
m_s	vertical magnification in the focal plane at object distance s	${}^{g}K_i^{s,s_k}$	i-th order radial distortion coefficient in the g-th partition of object plane Π_k
δr_s	lens radial distortion in the focal plane	${}^{g}P_i^{s,s_k}$	i-th order decentering distortion coefficient in the g-th partition of object plane Π_k
δr_{s_m}	radial distortion in the focal plane when the lens is focused on the distance of s_m	${}^{g}K_i^{s,s_n}$	i-th order radial distortion coefficient in the g-th partition of object plane Π_n
δr_{s_k}	radial distortion in the focal plane when the lens is focused on the distance of s_k	${}^{g}P_i^{s,s_n}$	i-th order decentering distortion coefficient in the g-th partition of object plane Π_n
f	focal length	PnP	Perspective-n-Point
K_i^s	i-th radial distortion coefficient for focused object plane at distance s	θ	angle between the regression line and the u axis
$K_i^{s_m}$	i-th radial distortion coefficient when the lens is focused on the distance of s_m	D	average sum of squared distances from the points $\left(\ u_{li}\quad v_{li}\ \right)$ to all the lines
$K_i^{s_k}$	i-th radial distortion coefficient when the lens is focused on the distance of s_k	imR	undistorted image
δr_u	component of the decentering distortion in u direction	imD	distorted image
δr_v	component of the decentering distortion in v direction	RMS	root mean square
K^{s,s_p}	radial distortion coefficient in the defocused plane with the depth of s_p when the lens is focused at distance s	PSNR	Peak Signal-to-Noise Ratio
g	empirical coefficient	MSE	mean square error of the image before and after distortion correction
K^{s_p}	radial distortion coefficient in the focal plane at distance s_p	$\boldsymbol{R}_q$	rotation matrix in the q-th pose
K^s	radial distortion coefficient in the focal plane at distanced s	$\boldsymbol{T}_q$	translation matrix in the q-th pose
$\delta r^{s,s_n}$	radial distortion function of the object plane at distance s_n when the lens is focused at the distance of s	${}^{g}K_i$	i-th order radial distortion coefficient in the g-th partition of the q-th pose
$\delta r^{s,s_m}$	radial distortion function of the object plane at distance s_m when the lens is focused at the distance of s	${}^{g}P_j$	j-th order decentering distortion coefficient in the g-th partition of the q-th pose
$\delta r^{s,s_k}$	radial distortion function of the object plane at distance s_k when the lens is focused at the distance of s	LM	Levenberg-Marquardt

References

1. Fraser, C.S. Automatic camera calibration in close range photogrammetry. *Photogramm. Eng. Remote Sens.* **2013**, *79*, 381–388. [CrossRef]
2. Basu, A.; Licardie, S. Alternative models for fish–eye lenses. *Pattern Recognit. Lett.* **1995**, *16*, 433–441. [CrossRef]
3. Lee, H.; Han, D. Rectification of bowl-shape deformation of tidal flat DEM derived from UAV imaging. *Sensors* **2020**, *20*, 1602. [CrossRef] [PubMed]

4. Drap, P.; Lefèvre, J. An exact formula for calculating inverse radial lens distortions. *Sensors* **2016**, *16*, 807. [CrossRef] [PubMed]

5. Liu, M.; Sun, C.; Huang, S.; Zhang, Z. An accurate projector calibration method based on polynomial distortion representation. *Sensors* **2015**, *15*, 26567–26582. [CrossRef]

6. Devernay, F.; Faugeras, O. Straight lines have to be straight. *Mach. Vis. Appl.* **2001**, *13*, 14–24. [CrossRef]

7. Fitzgibbon, A.W. Simultaneous linear estimation of multiple view geometry and lens distortion. In Proceedings of the 2001 IEEE Computer Society Conference on Computer Vision and Pattern Recognition, Kauai, HI, USA, 8–14 December 2001; pp. 125–132.

8. Alemán-Flores, M.; Alvarez, L.; Gomez, L.; Santana-Cedrés, D. Automatic lens distortion correction using one-parameter division models. *Image Process. Line* **2014**, *4*, 327–343. [CrossRef]

9. Claus, D.; Fitzgibbon, A.W. A rational function lens distortion model for general cameras. In Proceedings of the 2005 IEEE Computer Society Conference on Computer Vision and Pattern Recognition, San Diego, CA, USA, 20–25 June 2005; pp. 213–219.

10. Huang, J.; Wang, Z.; Xue, Q.; Gao, J. Calibration of camera with rational function lens distortion model. *Chin. J. Lasers* **2014**, *41*, 0508001. [CrossRef]

11. Brown, D.C. Decentering distortion of lenses. *Photogramm. Eng.* **1966**, *32*, 444–462.

12. Brown, D.C. Close–range camera calibration. *Photogramm. Eng.* **1971**, *37*, 855–866.

13. Tsai, R.Y. A versatile camera calibration technique for high–accuracy 3D machine vision metrology using off–the–shelf TV cameras and lenses. *IEEE Rob. Autom Mag.* **2003**, *3*, 323–344. [CrossRef]

14. Weng, J.; Cohen, P.; Herniou, M. Camera calibration with distortion models and accuracy evaluation. *IEEE Trans. Pattern Anal. Mach. Intell.* **1992**, *14*, 965–980. [CrossRef]

15. Kakani, V.; Kim, H.; Kumbham, M.; Park, D.; Jin, C.B.; Nguyen, V.H. Feasible Self–Calibration of Larger Field–of–View (FOV) Camera Sensors for the Advanced Driver–Assistance System (ADAS). *Sensors* **2019**, *19*, 3369. [CrossRef] [PubMed]

16. Li, X.; Liu, W.; Pan, Y.; Liang, B.; Zhou, M.D.; Li, H.; Wang, F.J.; Jia, Z.Y. Monocular–vision–based contouring error detection and compensation for CNC machine tools. *Precis. Eng.* **2019**, *55*, 447–463. [CrossRef]

17. Zhang, Z. A flexible new technique for camera calibration. *IEEE Trans. Pattern Anal. Mach. Intell.* **2000**, *22*, 1330–1334. [CrossRef]

18. Liu, W.; Ma, X.; Li, X.; Pan, Y.; Wang, F.J.; Jia, Z.Y. A novel vision-based pose measurement method considering the refraction of light. *Sensors* **2018**, *18*, 4348. [CrossRef]

19. Yang, J.H.; Jia, Z.Y.; Liu, W.; Fan, C.N.; Xu, P.T.; Wang, F.J.; Liu, Y. Precision calibration method for binocular vision measurement systems based on arbitrary translations and 3D-connection information. *Meas. Sci. Technol.* **2016**, *27*, 105009. [CrossRef]

20. Zhao, Z.; Zhu, Y.; Li, Y.; Qiu, Z.; Luo, Y.; Xie, C.; Zhang, Z. Multi-camera-based universal measurement method for 6-DOF of rigid bodies in world coordinate system. *Sensors* **2020**, *20*, 5547. [CrossRef]

21. Prescott, B.; Mclean, G.F. Line–based correction of radial lens distortion. *Graph. Models Image Process.* **1997**, *59*, 39–47. [CrossRef]

22. Ahmed, M.; Farag, A. Nonmetric calibration of camera lens distortion: Differential methods and robust estimation. *IEEE Trans. Image Process.* **2005**, *14*, 1215–1230. [CrossRef]

23. Santana-Cedrés, D.; Gomez, L.; Alemán-Flores, M.; Salgado, A.; Esclarín, J.; Mazorra, L.; Alvarez, L. Estimation of the lens distortion model by minimizing a line reprojection error. *IEEE Sens. J.* **2017**, *17*, 2848–2855. [CrossRef]

24. Becker, S.C.; Bove, V.M., Jr. Semiautomatic 3D–model extraction from uncalibrated 2D–camera views. *Int. Soc. Opt. Photonics* **1995**, *2410*, 447–461.

25. Penna, M.A. Camera calibration: A quick and easy way to determine the scale factor. *IEEE Trans. Pattern Anal. Mach. Intell.* **1991**, *13*, 1240–1245. [CrossRef]

26. Magill, A.A. Variation in distortion with magnification. *J. Res. Natl. Bur. Stand.* **1955**, *45*, 148–149.

27. Fryer, J.G. Lens distortion for close–range photogrammetry. *Photogramm. Eng. Remote Sens.* **1986**, *52*, 51–58.

28. Treibitz, T.; Schechner, Y.Y.; Singh, H. Flat refractive geometry. *IEEE Trans. Pattern. Anal. Mach. Intell.* **2012**, *34*, 51–65. [CrossRef]

29. Fraser, C.S.; Shortis, M.R. Variation of distortion within the photographic field. *Photogramm. Eng. Remote Sens.* **1992**, *58*, 851–855.

30. Dold, J. Ein hybrides photogrammetrisches Industriemesssystem höchster Genauigkeit und seiner Überprüfung. Ph.D. Thesis, Schriftenreihe Studiengang Vermessungswesen, Heft 54. Universität der Bundeswehr, München, Germany, 1997.
31. Brakhage, P.; Notni, G.; Kowarschik, R. Image aberrations in optical three–dimensional measurement systems with fringe projection. *Appl. Opt.* **2004**, *43*, 3217–3223. [CrossRef]
32. Hanning, T. High precision camera calibration with a depth dependent distortion mapping. In Proceedings of the 8th IASTED International Conference on Visualization, Imaging, and Image Processing, Palma de Mallorca, Spain, 1–3 September 2008; pp. 304–309.
33. Alvarez, L.; Gómez, L.; Sendra, J.R. Accurate depth dependent lens distortion models: An application to planar view scenarios. *J. Math. Imaging Vis.* **2011**, *39*, 75–85. [CrossRef]
34. Sun, P.; Lu, N.; Dong, M. Modelling and calibration of depth–dependent distortion for large depth visual measurement cameras. *Opt. Express* **2017**, *25*, 9834–9847. [CrossRef]
35. Ricolfe-Viala, C.; Esparza, A. Depth-dependent high distortion lens calibration. *Sensors* **2020**, *20*, 3695. [CrossRef] [PubMed]
36. Li, X.; Liu, W.; Pan, Y.; Ma, J.; Wang, F. A knowledge–driven approach for 3D high temporal–spatial measurement of an arbitrary contouring error of CNC machine tools using monocular vision. *Sensors* **2019**, *19*, 744. [CrossRef] [PubMed]
37. Lowe, D.G. Object recognition from local scale–invariant features. In Proceedings of the IEEE Transactions on Pattern Analysis and Machine Intelligence, Kerkyra, Greece, 20–27 September 1999; pp. 51–65.
38. Zheng, Y.; Kuang, Y.; Sugimoto, S.; Astrom, K.; Okutomi, M. Revisiting the PnP problem: A fast, general and optimal solution. In Proceedings of the 2013 IEEE International Conference on Computer Vision, Sydney, NSW, Australia, 1–8 December 2013; pp. 2344–2351.

Publisher's Note: MDPI stays neutral with regard to jurisdictional claims in published maps and institutional affiliations.

 *sensors*

Article

Development of a Robust Multi-Scale Featured Local Binary Pattern for Improved Facial Expression Recognition

Suraiya Yasmin [1], Refat Khan Pathan [2], Munmun Biswas [2], Mayeen Uddin Khandaker [3,*] and Mohammad Rashed Iqbal Faruque [4]

1   Department of Computer Science and Engineering, International Islamic University Chittagong, Chittagong-4318, Bangladesh; suraiyabrishti@gmail.com
2   Department of Computer Science and Engineering, BGC Trust University Bangladesh, Chittagong-4381, Bangladesh; refatkhan93@gmail.com (R.K.P.); munmun@bgctub.ac.bd (M.B.)
3   Centre for Biomedical Physics, School of Healthcare and Medical Sciences, Sunway University, Bandar Sunway 47500, Selangor, Malaysia
4   Space Science Centre (ANGKASA), Institute of Climate Change (IPI), Universiti Kebangsaan Malaysia, UKM, Bangi 43600, Selangor, Malaysia; rashed@ukm.edu.my
*   Correspondence: mayeenk@sunway.edu.my

Received: 14 August 2020; Accepted: 14 September 2020; Published: 21 September 2020

Abstract: Compelling facial expression recognition (FER) processes have been utilized in very successful fields like computer vision, robotics, artificial intelligence, and dynamic texture recognition. However, the FER's critical problem with traditional local binary pattern (LBP) is the loss of neighboring pixels related to different scales that can affect the texture of facial images. To overcome such limitations, this study describes a new extended LBP method to extract feature vectors from images, detecting each image from facial expressions. The proposed method is based on the bitwise AND operation of two rotational kernels applied on LBP(8,1) and LBP(8,2) and utilizes two accessible datasets. Firstly, the facial parts are detected and the essential components of a face are observed, such as eyes, nose, and lips. The portion of the face is then cropped to reduce the dimensions and an unsharp masking kernel is applied to sharpen the image. The filtered images then go through the feature extraction method and wait for the classification process. Four machine learning classifiers were used to verify the proposed method. This study shows that the proposed multi-scale featured local binary pattern (MSFLBP), together with Support Vector Machine (SVM), outperformed the recent LBP-based state-of-the-art approaches resulting in an accuracy of 99.12% for the Extended Cohn–Kanade (CK+) dataset and 89.08% for the Karolinska Directed Emotional Faces (KDEF) dataset.

Keywords: facial expression recognition system; computer vision; multi-scale featured local binary pattern; unsharp masking; machine learning

---

1. Introduction

Facial expression recognition (FER) is a regular and incredible sign to decipher the state of human feelings and expectations, expressing human emotion without saying anything, as faces are considerably more than key to singular personalities. In a word, one can say that it is one of the most natural, current, and robust means for communicating people's intentions and emotions with others. As it is related to human emotion, which differs from one to another, researchers discovered many methods by both machine learning and deep learning techniques to obtain a critical understanding of this matter. Nowadays, things are becoming more mechanized through computer automation, where computer vision is playing a vital role in the automation process by training computers to interpret and understand the visual world. Thus, studies on FER show high demand in computer vision, which can be utilized in autonomy, neuro-advertising, scholastics, and altogether in security. Besides this, FER is one of the most challenging biometric recognition technologies due to its characteristics of nature, intuition, etc.

FER has two essential stages: feature extraction (geometric and appearance-based) and classification. While the geometrically-based feature extraction includes facial components like eye, mouth, nose, and eyebrow, the appearance-based one comprises the exact section of the face. On the other hand, the classification categorizes the expression, like a smile, sadness, anger, disgust, surprise, or fear. Researchers have worked with many neural networking concepts like Convolutional Neural Network (CNN), Recurrent Neural Network (RNN), and machine learning classifiers like Support Vector Machine (SVM), K Nearest Neighbour (KNN) to find, relatively, the most accurate FER technique. In connection to this, several researchers utilized the Neural Network based on different kinds of popular methods like CNN [1], CNN-RNN [2], 3DCNN-DAP [3,4], Weighted Mixture Deep Neural Network [5], CNN with attention mechanism (ACNN) where it empowers the model to move consideration from the impeded patches to other unhampered ones, just as distinct facial regions are dependent on patch-based ACNN (pACNN) and global-local based ACNN (gACNN) [6]. Although neural networks are easy to build with the latest programming languages like Python, R, and tools like Matlab and Weka, nevertheless, when it comes to the computational power, especially in facial image processing with many classes, it requires very high processing power with a high amount of random access memory (RAM) and a graphics processing unit (GPU). Additionally, suppose it is not a supercomputer. In that case, one needs hours to simply train a neural network model, which calculates too many features, where most of them are non-object-orientated, making the model prone to overfitting. However, since currently, the artificial intelligence (AI) receives a focal point to replicate or simulate human intelligence in machines, the incorporation of a multimodal concept (such as both machine learning and deep learning techniques) may produce a better FER compared to the typical models and sub-processes.

Machine learning classifiers like SVM, KNN, and Tree cannot extract features automatically from raw images like the Neural Network (NN). Moreover, many other classifiers such as Principal Component Analysis (PCA), Extreme Learning Machine (ELM), Conditional Random Fields (CRF), and so on can also be used to classify facial emotion. However, classifiers need a state-of-the-art descriptor to extract a feature-set from natural images to classify into different classes. A wide range of methods and innovations have been tested by many researchers to find the best way for the classification of human disclosure. Features for FER are generally extracted with appearance-based methods like local binary pattern (LBP), local derivative pattern (LDP), local geometric binary (GLBP), and geometric methods like the histogram of oriented gradients (HOG), salient facial patches, classifier for salient areas on the faces [7], local binary pattern from three orthogonal planes (LBP-TOP) [8], local texture coding operator [9], and differential geometry. For instance, with the appearance-based method LBP, Zhang et al. applied a new method named Multi-resolution Histograms of Local Variation Patterns (MHLVP) on Gabor wavelets [10] and obtained a very impressive outcome on the Facial Recognition Technology (FERET) dataset; however, the computational complexity and element measurement was excessive. One of LBP's universal drawbacks is relevant to its small 3×3 neighborhood, which cannot capture dominant features with large scale structures [11–16]. Zhao and Pietik Ainen extended the LBP operator to the spatiotemporal space, and they named it the volume local binary patterns model [17], which has been generally embraced in catching powerful features by rotating and concatenating different methods but worked with a single dataset, thus, the accuracy may fall for blurred images. Coming out from regular filtered images, features extraction with noise, and partial occlusions, a combined method of the histogram of oriented gradients (HOG) with the uniform-local ternary pattern (U-LTP) [18] is described, which gives a good filtering process as well. More discriminative features in higher-order derivative directions were captured by the LDP [19], which improved LBP.

However, it is mostly limited to the surrounding eight-pixel values by avoiding more significant dimensional relations.

Along with LBP, many geometric based methods were also used in FER. Images are partitioned into blocks and sub-blocks, and an active appearance model was used for revealing the essential facial portions and extracted by differential geometric features [20] which has more accuracy in FER than the static geometric features, also provides valuable geometric data with the time and sequence of facial expression images. For non-formed images, a method of cases that were out-of-plane head revolutions was taken care of using the turn inversion invariant histogram of oriented gradients [21], which has insufficient time complexity and improved the learning model of the cascade to collaborate with the classification technique. Tsai and Chang have applied the filter of Gabor, discrete cosine, change, and transformation of angular radial [22] to use HFs, consolidating with self-quotient image (SQI) channels for improving FER accuracy under different light source environments. Typically, there are some miss images in the examination, and it is essential to include a non-face class in outward appearance classifications that are not clarified there. The facial illustration is to infer a gathering of features from unique face images to viably speaking faces. It should limit the inside class varieties of articulations while amplifying between class contrasts. In general circumstances, the geometric method needs very well structured facial images. Practically, most of the time, it is not possible to capture well-textured images to perform geometric methods.

In addition to the many geometric and appearance-based methods, there are some more methods like the response method [23] that extracts features from directional texture and number patterns where performance is tested in constrained and unconstrained situations. Researchers have not been limited to static features only. There are some other methods for extracting dynamic and multilevel features [24], which have coordinated into an end-to-end network to participate flawlessly with one another. Moreover, to solve a small sample size (SSS) issue, using a novel method-directional multilinear independent component analysis (ICA) technique was demonstrated in [25], which prompts the dimensionality situation by encoding the input image or high dimensional data array as a general tensor. A different methodology for facial expression analysis is the use of the Human-Computer Interaction (HCI) context [26] disintegrated into smaller micro-decisions that are separately made by particular binary classifiers with higher accuracy of the general model. Besides the above-described methods, some methods are also used for the detection of real-time expressions such as embedded systems [27], Radon Barcodes [28], and many more. Classifiers acquire characteristic features from the above strategies as their sources as inputs. However, the classifier's execution relies on the nature of feature vectors. A summary of a few recent works in the field of FER is shown in Table 1.

Table 1. Key information on some similar recently studied methods on facial expression recognition (FER).

Year	Classifier	Features	Databases
2015 [1]	SVM	CNN	FER/SFEW
2017 [5]	WMDNN	LBP	CK+/JAFFE/CASIA
2017 [7]	PCA	LBP/HOG	CK+/JAFFE
2019 [8]	SVM	LBP-TOP	CASME II/SMIC
2019 [9]	ELM	CS-LGC	CK+/JAFFE
2005 [10]	KNN	MHLVP	FERET
2007 [17]	SVM	VLBP/LBP-TOP	DynTex/MIT/CK+
2017 [18]	HOG	Ri-HOG	CK+/MMI/AFEW
2018 [20]	SVM	Differential Geometric Features	CK+
2017 [21]	HOG	Ri-HOG	CK+/MMI/AFEW
2017 [22]	SVM	FERS	CKFI/FG-NET/JAFFE
2019 [28]	SVM	LBP/LTP/RBC	Infant COPE

In light of the information mentioned above, one can observe a non-negligible limitation, especially in appearance-based typical LBP methods. Therefore, this study proposes a feature extraction method based on a new extended LBP "Multi-Scale Featured Local Binary Pattern", which can be used not only in FER but also in various purposes to analyze an image. Since the automatic face expression recognition requires two significant angles: facial illustration and classifier style, this study utilizes four machine learning classifiers: SVM, KNN, Tree, and Discriminant Quadric Analysis. There are so many datasets, for example, Japanese Female Facial Expression (JAFFE), Chinese Academy of Sciences Institute of Automation (CASIA), Static Facial Expressions in the Wild (SFEW), Chinese Academy of Sciences Micro-expression-II (CASME), Spontaneous Micro-expression (SMIC), Acted Facial Expressions in the Wild (AFEW), and all are available in the literature. However, we used two well-known facial image datasets: Extended Cohn–Kanade Dataset (CK+) and Karolinska Directed Emotional Faces (KDEF) to verify our proposed method. Note that the Extended Cohn–Kanade Dataset (CK+) [29] is an extended version of Cohn–Kanade (CK) [30] and finds greater use in developing and evaluating facial expression analysis algorithms. It contains a better example of catching the sample space than the CK dataset, which includes 304 labeled videos with 5521 frames of test subjects from various ethnicities in varied age groups extending from 18 to 50.

On the other hand, the used KDEF dataset helps assess the emotional contents and appraise intensity and arousal scale. Moreover, it contains a legitimate arrangement of feeling the full facial images. More details about these datasets are shown in Table 2 and some sample faces are shown in Figure 1.

Table 2. Used datasets in the proposed method.

Dataset	No of Expressions Used	Image Size	No of Subject	Total Image
CK+	7	640×490	123	593 video sequence
KDEF	7	562×762	70	4900 Images

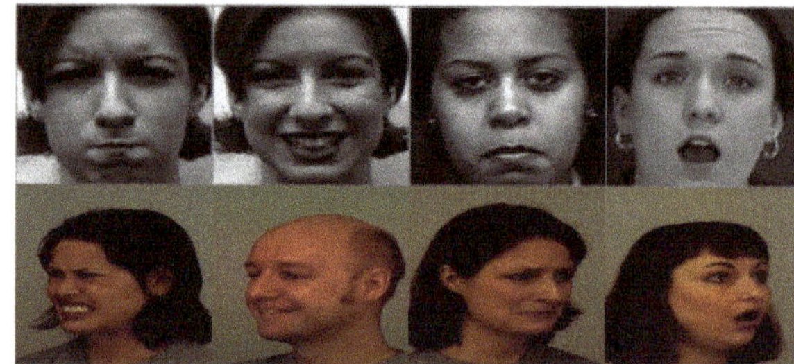

Figure 1. Sample face image from Extended Cohn–Kanade (CK+) and Karolinska Directed Emotional Faces (KDEF) datasets.

2. Contribution

Based on the available literature, we observed that if the images are not well textured and blurred, then the prediction value falls. Thus, we have proposed a new feature extraction process for images that makes the texture of an image more machine-readable and converts the sub-region to 58 Uniform LBP and gives a classifier friendly feature vector tested on four machine learning classifiers. In this research, we have implemented three different angles where all the members are told to attempt to inspire the feeling that should have been expressed and to make the expression sharp and clear. The main contribution in the global LBP method is the process of calculating bitwise AND for two neighboring pixel values to obtain the relation between them after applying two suggested kernel matrices. Here, we have justified this method by detecting facial expression from an image that greatly relies on the image texture.

This manuscript is arranged with the proposed method in Section 3, including Section 3.1 pre-processing, Section 3.2: feature extraction, and Section 3.3: normalization. The result analysis is discussed in Section 4, and the conclusion is in Section 5.

3. Proposed Method

3.1. Pre-Processing

As the colored image sensitively affects light impact, the images were converted into grayscale as it has various shades of dark in the center, so to convert the image into grayscale, we used Equation (1) where r is the pixel value of red, g is green, and b is blue.

$$gray = 0.3r + 0.59g + 0.11b \tag{1}$$

The grayscale image may have an environmental and useless background as well, which increases the computational complexity and misleading accuracy. From the CK+ and KDEF dataset of the raw image, it was observed that the images are size 640×490 and 562×762 pixels on average. Therefore, for better results and lower complexity, the facial part from the whole image was detected and the face was cropped by Haar cascade frontal face-based on the Viola-Jones detection algorithm, which precisely detects faces then crops and resizes them to 100×100 pixels. Each of the images was then compared with a 5×5 table cell and it was observed that key portions of models such as eyes, nose, and lips areas are in 3×3 table cells (60×60 pixels). Therefore, for avoiding the unnecessary parts, we have cropped this to 3×3 cells, shown in Figure 2.

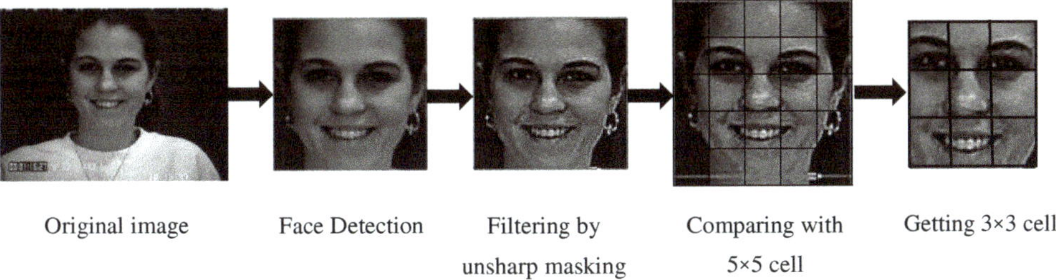

Figure 2. Pre-processing steps.

After detecting and cropping the images, the unsharp masking kernel [31] (shown in Figure 3) was used for sharpening the edges with Equation (2), which reduces some noises and gives a bright look. Grinding the images are essential for better understanding and communicating nearby grayscale change data by the contrast between each single points, and utilizes the weighted qualification in the eight directions as the local shade change data in the path, which is commotion and light-delicate and has no strength. The sharpening kernel was used in the side-by-side method where the Kernel moves in every one pixel.

$$S(x, y) = \sum_{i=-2}^{2} \sum_{j=-2}^{2} K(i, j) \times M(x - i, y - j) \tag{2}$$

where K is the Kernel in Figure 3, and M is the pixel values of the given image, and $S(x,y)$ is the central pixel value, which creates a sharpened image. The unsharp masking kernel was chosen in this study because it provides a good texture output in pixel values of different image datasets among many variants of kernels.

	-2	-1	0	1	2	Index
	1	4	6	4	1	-2
	4	16	24	16	4	-1
$(-1/256) \times$	6	24	-476	24	6	0
	4	16	24	16	4	1
	1	4	6	4	1	2

Figure 3. Unsharp masking kernel.

3.2. Feature Extraction

In this study, a method was developed for extracting features from an image to identify emotions. We depend not only on the shadow effect of the grayscale images but also on using a new kernel-based method to enhance the shadow effect to extract the features that are flexible and classifier friendly. We have proposed two kernels on the LBP of an image to be more precise about the shadow and light effect of the face parts, which mainly decides the face's emotional states. In this step, the pre-processed image was taken and applied to the serial process shown in Figure 4 to finally obtain the features using the algorithm indicated in Figure 5.

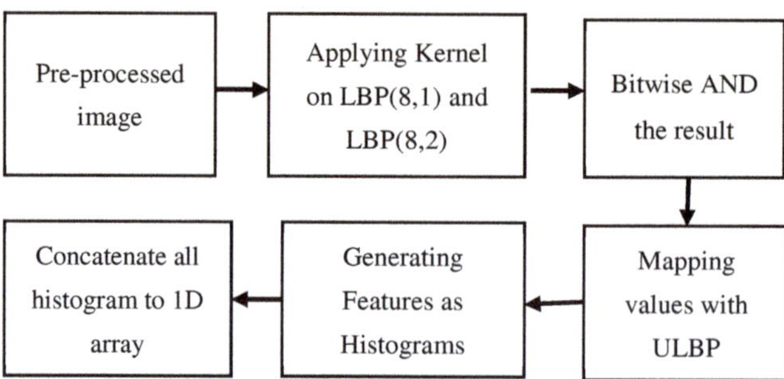

Figure 4. Feature extraction process.

Generally, LBP $_{(P, R)}$ is used in one radius on eight directional coordinates of the matrix value where P is the number of pixels to be considered and R is the radius from the central pixel. However, we used two LBP (LBP $_{(8, 1)}$ and LBP $_{(8, 2)}$) and applied two kernel matrix to calculate the central pixel of that cell. Considering the first stage of the image, we have divided it into sub-cells where 3 × 3 for LBP $_{(8, 1)}$ and 5 × 5 for LBP $_{(8, 2)}$ with two proposed kernels. A sample 3 × 3 image segment has been shown in Figure 6a and the model is shown in Figure 6b for the first Kernel, where each matrix is a 45° rotation, and the central matrix is the 3 × 3 cell of the pre-processed image. Considering that S_1 denotes the grey estimation of the pixel point in the 3 × 3 neighborhood of the pre-processed image, and the kernel value of pixel points in the area is K_1, the central pixel can be obtained by applying the first rotation kernel with Equation (3).

$$G(x,y) = \sum_{i=-1}^{1} \sum_{j=-1}^{1} K_1(i,j) \times S_1(x-i, y-j) \tag{3}$$

1. Let, d denotes the 8 rotation of kernel, i and j indicates the coordinate values of the image and kernel matrices.
2. **Input:** Processed image S, two kernel matrix K1 and K2, d denotes the 8 rotations.
3. **Initialization:** Initializing the variable Q_D = 0 and R_D = 0
4.
5. For d = 1 to 8
6.   For i = 1 to 3
7.   For j = 1 to 3
8.   Q = S(i,j) × K1(d,i,j)
9.   End For
10.   End For
11. C = Q > 0 ? 1 : 0
12. Q_D = Q_D + C × 2^d
13. For i = 1 to 5
14.   For j = 1 to 5
15.   R = S(i,j) × K2(d,i,j)
16.   End For
17. End For
18. C = R > 0 ? 1 : 0
19. R_D = R_D + C × 2^d
20. End For
21. Central_Pixel = Q_D AND R_D
22.
23. **Output:** Central pixel value of the given portion of image

Figure 5. Feature extraction algorithm.

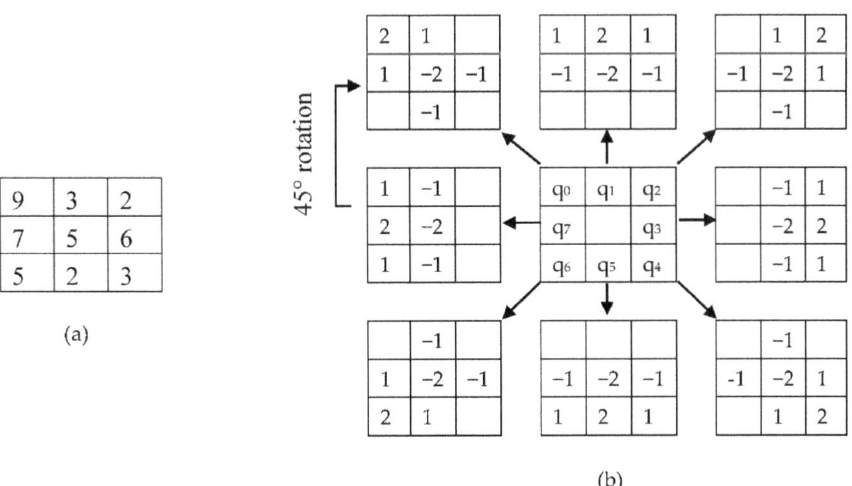

Figure 6. (**a**) Sample image segment of 3 × 3, (**b**) Description of Local Binary Pattern (LBP) $_{(8,1)}$: kernel value.

Here, K_1 is eight rotational kernels with 45° rotations each. Therefore, Equation (3) was applied eight times to obtain the value q_0 to q_7 in Figure 7, G (x, y) is the central pixel value, which will make the pixel matrix for 1st Kernel. After the calculation is shown in Figure 7, converting the positive value as one and the negative value as 0, we obtain the central decimal pixel value. By using the sample image segment in Figure 6a, we used Equation (3) to show the calculation to find the central pixel matrix values q_0 to q_7 (as shown in Figure 7). This same procedure has been followed with the 5×5 image segment and kernel are shown in Figure 8 to find the central pixel matrix of Figure 9.

$$q_0 = \{(7\times1) + (9\times2) + (3\times1)\} - \{(2\times1) + (5\times2) + (6\times1)\} = 10$$
$$q_1 = \{(9\times1) + (3\times2) + (2\times1)\} - \{(7\times1) + (5\times2) + (6\times1)\} = -6$$
$$q_2 = \{(3\times1) + (2\times2) + (6\times1)\} - \{(7\times1) + (5\times2) + (2\times1)\} = -6$$
$$q_3 = \{(2\times1) + (6\times2) + (3\times1)\} - \{(3\times1) + (5\times2) + (2\times1)\} = 2$$
$$q_4 = \{(6\times1) + (3\times2) + (2\times1)\} - \{(3\times1) + (5\times2) + (7\times1)\} = -6$$
$$q_5 = \{(5\times1) + (2\times2) + (3\times1)\} - \{(7\times1) + (5\times2) + (6\times1)\} = -11$$
$$q_6 = \{(6\times1) + (5\times2) + (2\times1)\} - \{(3\times1) + (5\times2) + (6\times1)\} = -1$$
$$q_7 = \{(9\times1) + (6\times2) + (5\times1)\} - \{(3\times1) + (5\times2) + (2\times1)\} = 11$$

q_0	q_1	q_2
q_7		q_3
q_6	q_5	q_4

10	−6	−6
11		2
−1	−11	−6

1	0	0
1		1
0	0	0

Binary: 10010001

Decimal: $128 + 0 + 0 + 16 + 0 + 0 + 0 + 1 = 145$

Figure 7. Calculation of LBP $_{(8, 1)}$.

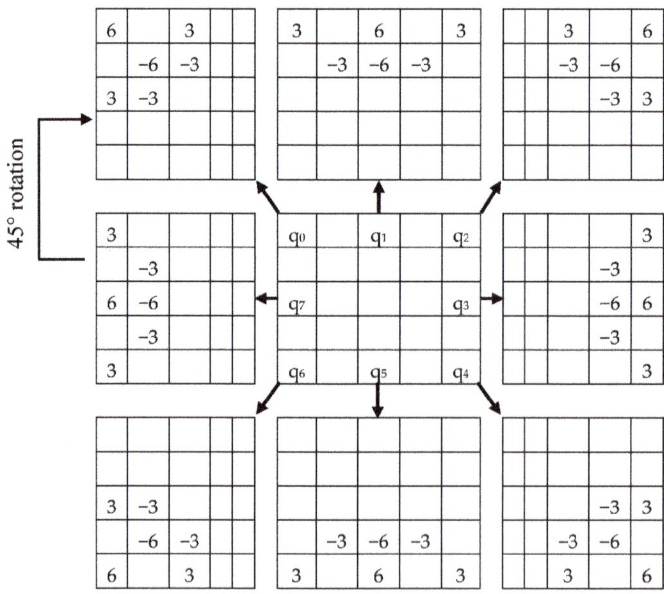

Figure 8. Description of LBP $_{(8, 2)}$: kernel value.

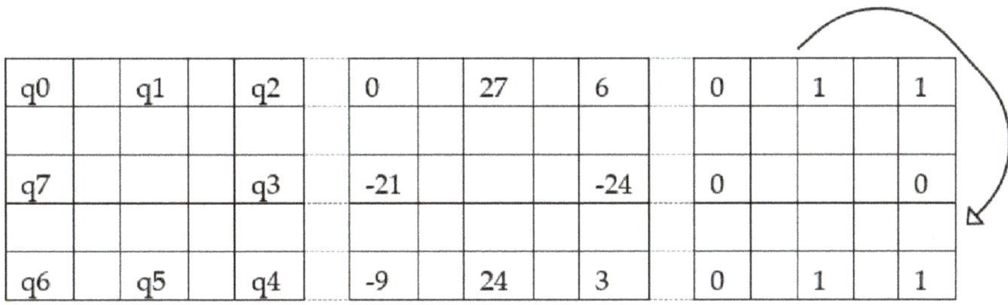

Binary: 01101100   Decimal: 0 + 64 + 32 + 0 + 8 + 4 + 0 + 0 = 108

Figure 9. Calculation of LBP $_{(8, 2)}$.

The model for the second Kernel is shown in Figure 8, where each matrix is a 45° rotation, and the central matrix is 5 × 5 cells of the pre-processed image. Again, accepting that S_2 denotes the grey estimation of the pixel point in the 5 × 5 neighborhood of the pre-processed image, and the kernel value of pixel points in the area is K_2, the value of the central pixel can be obtained by applying the second Kernel with Equation (4).

$$H(x,y) = \sum_{i=-2}^{2} \sum_{j=-2}^{2} K_2(i,j) \times S_2(x-i,y-j) \tag{4}$$

Similarly, kernel K_2 will have eight rotations with 45° each for obtaining q_0 to q_7 values in Figure 9. $H(x, y)$ is the central pixel which will make the pixel matrix for 2nd Kernel. Once again, converting the positive value as one and negative value as 0, we acquire the central decimal pixel value which is shown in Figure 9.

In the final stage, we have applied bitwise AND of $G(x, y)$, $H(x, y)$, where the binary output value of a model is determined to utilize Equation (5), which tells to the nearby change data between the center point and the 8-neighborhood pixels. It counts the number of spatial transitions from 0 to 1 or 1 to 0. In this stage, the equation will be as follows:

$$BM(x,y) = \left(\sum_{i=-1}^{1} \sum_{j=-1}^{1} K_1(i,j) \times S_1(x-i,y-j)\right) AND \left(\sum_{i=-2}^{2} \sum_{j=-2}^{2} K_2(i,j) \times S_2(x-i,y-j)\right) \tag{5}$$

Simplifying Equation (5) as:

$$BM(x,y) = G(x,y) \; AND \; H(x,y)$$

where $BM(x, y)$ is the binary matrix, the values of which are defined as 1 if $G(x, y) = H(x, y) = 1$ or 0 if any of $G(x, y)$ or $H(x, y)$ is 0.

We have used an assessment by applying a condition to find the output cell's central pixel in decimal in Equation (6).

$$MSLBP(x_c, y_c) = \sum_{n=0}^{7} BM(w_n) 2^n \tag{6}$$

where w_n corresponds to the neighboring binary value of the eight surrounding pixels of the binary matrix BM and $MSLBP(x_c, y_c)$ is the final central decimal pixel value.

After calculating the *MSLBP* matrix, we have divided the whole image into 6 × 6 = 36 cells and mapped each cell's value to the uniform local binary pattern (*ULBP*) by Equation (7). For *ULBP*, each cell pattern maps to 58-bin histograms. *ULBP* has unique 58 numbers where we will convert the

MSLBP pixel matrix to a one-dimensional array by mapping pixel values to *ULBP* values. A single-cell value of 255 will be converted to 58 by using *ULBP*.

$$FV = ULBP\big(MSLBP_{(x,y)}\big) \tag{7}$$

where *FV* is the feature vector, *ULBP* is the array of mapping values. *MSLBP* (x, y) is the pixel value of the image, which will be used as an index.

For one image, neighbor pixels are generally related; thus, the binary sequences of MSLBP (p, r) of the various radius can be seen as described. After ascertaining all values from left to right, we have obtained a binary pattern for every cell of an image. Taking all weighted values into account, we have found a decimal number in symmetric neighbor sets for various coordinates (x, y). The grey values of neighbors that are not the focal region for matrices can be evaluated by commitment. After that, we discovered one histogram for each cell, then we have concatenated all those histograms from each cell into a one-linear histogram shown in Figure 10. There will be a two-dimensional matrix for each image of seven classes where rows represent the image index, and the column represents the features. This long concatenated histogram is the initially featured vector with many noises and mismatched values within a class. We have normalized the histogram data to solve this kind of problem, which shows good accuracy in validation test cases compared with the original feature vectors.

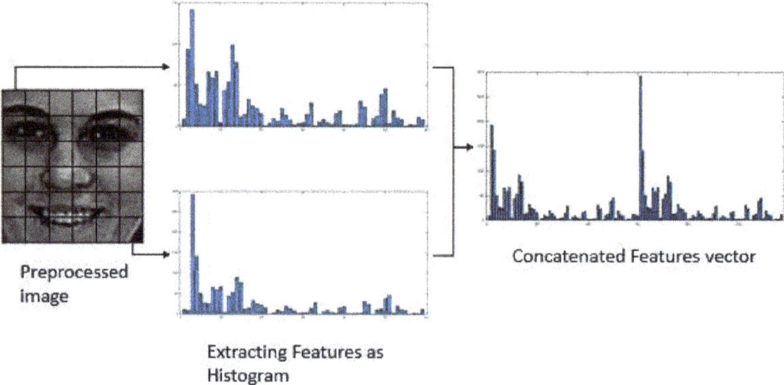

Figure 10. Converting process of selected geographical features of a histogram.

3.3. Normalization

Due to the so many images with different expressions and features, it is challenging to maintain continuity among the classes. Therefore, normalization of data becomes mandatory to handle within a range of values so that each class keeps some kind of consistency. We have used the Generalized Procrustes Analysis (GPA) [32] as normalization in our proposed method. It takes each level data individually and utilizes a measure of variance. The GPA generates a weighting factor by analyzing the differences in the scaling factor applied to respondent scale usages and individual scale usage. As a result, the distance between different classes' values was increased. Initially, we see the happy class's data situated on the scatter plot shown in Figure 11a (before normalization), then we can see that the images are getting closer to each other in Figure 11b (after normalization). In brief, the GPA takes all those features and reduces the fluctuation, and after using this, all related emotional state values have become at a closer level which causes the classification to act more precisely as the variance increases between different classes.

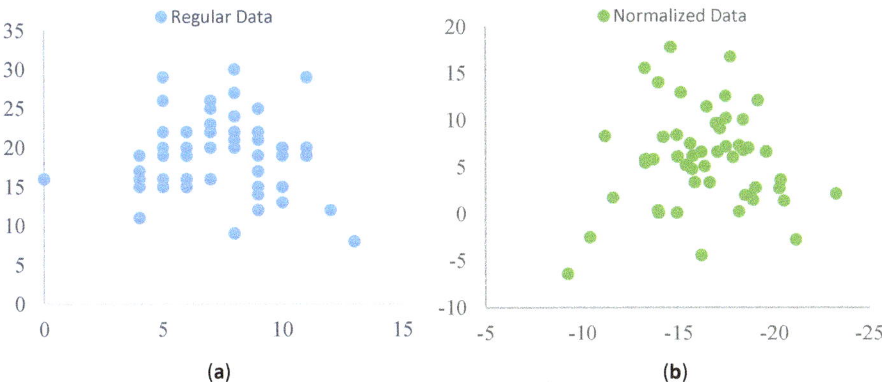

Figure 11. (**a**) Regular data, (**b**) Normalized data. Axis values are two feature values before (**a**) and after (**b**) normalization.

4. Results and Discussion

4.1. Performance Analysis of the Proposed Method

We have tested our proposed method on the CK+ and KDEF dataset. The given datasets are the most widely used for facial expression recognition, and this includes seven different facial expression labels or classes. We have used several machine-learning classifiers like K-nearest neighbors (KNN), Binary Tree, Quadric Discriminant Analysis (QA), and Support Vector Machine (SVM) shown in Figure 12. Among them, SVM gives the highest testing accuracy, which is shown in the confusion matrix for both dataset's test set following the 80-20 train-test split rule in Tables 3 and 4, respectively. From the CK+ dataset, almost 6000 images are used for training and 2000 for validation and testing, and for the KDEF dataset, almost 2900 images are used for training and 1000 for validation and testing. A total of 10 iterations of K-Fold cross-validation was used in all four classifiers. All values are shown in percentage (%).

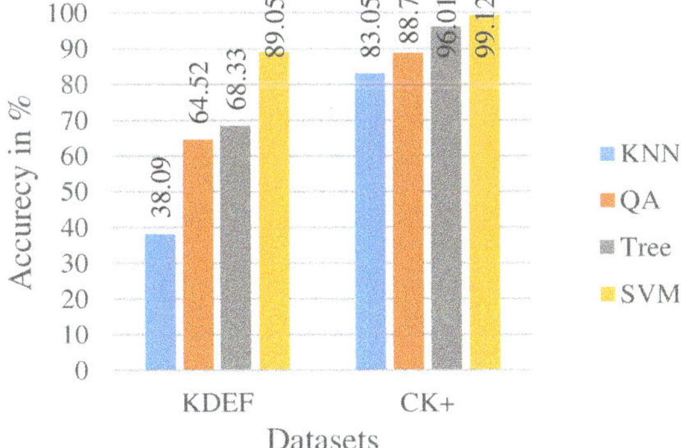

Figure 12. KDEF (KNN: 38.09, QA: 64.52, Tree: 68.33, SVM: 89.05), CK+ (KNN: 83.05, QA: 88.70, Tree: 96.01, SVM: 99.12).

The precision, recall, and F1 Score of the CK+ and KDEF dataset for SVM shows the outcome's excellent structure. For finding these values, we first have to analyze the confusion matrix. When the actual class is positive, and the predicted class is also positive, it is counted as True Positive (TP) value. When the actual class is negative, and the predicted class is too negative, it is counted as a True

Negative (TN) value. Along with these, if the actual class is positive but predicted as negative, it is counted as False Negative (FN). If the true class is negative but predicted as positive, it is counted as False Positive (FP).

Table 3. Confusion matrix of the CK+ dataset (SVM).

	Happy	Surprise	Sadness	Anger	Disgust	Fear	Neutral
Happy	100	0	0	0	0	0	0
Surprise	0	99.67	0	0	0	0.33	0
Sadness	2.3256	0	97.67	0	0	0	0
Anger	0	0	0	100	0	0	0
Disgust	0	0	0	1.78	98.22	0	0
Fear	0	0	0	0	0	100	0
Neutral	1.04	0	0.68	0	0	0	98.28

Table 4. Confusion matrix of the KDEF dataset (SVM).

	Happy	Surprise	Sadness	Anger	Disgust	Fear	Neutral
Happy	90.28	0	0	0	9.72	0	0
Surprise	0	98.28	0	0	1.04	0	0.64
Sadness	0	0	76.72	8.56	0	9.44	5.28
Anger	0	0	4.17	88.89	0	4.17	2.78
Disgust	2.78	0	0	0	97.22	0	0
Fear	0	0	9.72	6.94	0	83.33	0
Neutral	0	0	6.94	1.39	0	2.78	88.89

Precision: It is the ration of TP and the total positive predictions. High precision means less classification error.

$$Precision = TP/(TP + FP)$$

Recall: It is the ration of TP and the total true positive classes.

$$Recall = TP/(TP + FN)$$

F1 Score: F1 Score is sometimes more useful than accuracy. It is the weighted average of the values of Precision and Recall. F1 Score is important here because we have an uneven number of classes.

$$F1\ Score = 2*(Precision*Recall)/(Precision + Recall)$$

Table 5 shows the precision, recall, and F1 Score for datasets. We have presented the precision, recall, and F1 score comparatively in Figures 13 and 14 for CK+ and KDEF datasets for all the K-folding cross-validations. Values are shown for SVM classifier because it has the highest accuracy.

Table 5. Pre (Precision), Rec (Recall), F1 (F1 Score) shown for dataset CK+, and KDEF. Values are shown for the Support Vector Machine (SVM) classifier for seven classes.

Classes	CK+			KDEF		
	Pre	Rec	F1	Pre	Rec	F1
Happy	1	0.967	0.983	0.903	0.970	0.935
Surprise	0.996	1	0.998	0.983	1	0.992
Sadness	0.976	0.993	0.984	0.767	0.786	0.777
Anger	1	0.982	0.991	0.889	0.840	0.869
Disgust	0.982	1	0.991	0.972	0.900	0.935
Fear	1	0.996	0.998	0.833	0.836	0.835
Neutral	0.982	1	0.991	0.889	0.911	0.899

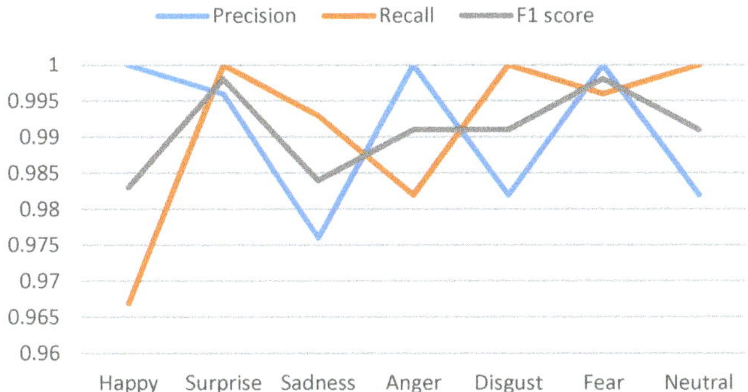

Figure 13. Dataset: CK+, Precision, Recall and F1 score shown for SVM.

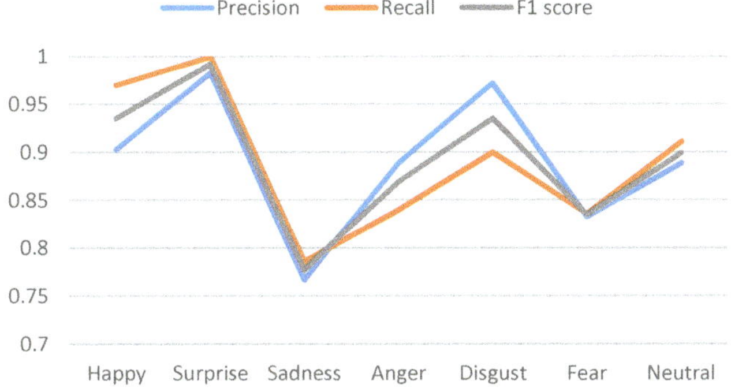

Figure 14. Dataset: KDEF, Precision, Recall, and F1 score is shown for SVM.

4.2. Analyses and Discussion of Results

Throughout this study, it is observed that classical LBP works with every pixel, which is contrasted and utilizes its eight surrounding 3×3 neighborhood by subtracting the center pixel value. Then, the resulting negative values are encoded with 0, otherwise 1. Finally, the encoded binary value is converted to decimal to obtain the center pixel value. The ongoing variety of LBP, for example, extended local binary patterns (ELBP) [15] operator not only performs the binary comparison of the center pixel and its neighbors but also encodes their exact grey-value differences (GDs) utilizing some extra binary units. In the completed modeling of the local binary pattern (CLBP) [16], it includes both the sign and the GDs between a given center pixel and its neighbors to improve the original LBP operator's discriminative intensity. The two strategies have utilized LBP $_{(8,1)}$ and compare the absolute value of GD with the given central pixel again to create an LBP-liked code. In Ref. [8], the authors first used the optical flow technique to obtain the Necessary Morphological Patches (NMPs) of micro-expressions; then, they calculated LBP-TOP operators by cascading them with optical flow histograms to make fusion features of dynamic patches. In local texture coding, the operator [9] enhances real-time system performance, utilizing four directional gradients on 5×5 grids for reducing sensitivity to noise. In Ref. [28], the authors present an observing framework using some features, such as LBP/LTP/red blood cell (RBC) for children, which utilizes an automatic pain detection system, and it could be accessed through wearable or mobile devices. A weighted fusion strategy [5] is proposed to completely utilize the features that were separated from various image channels with a partial Visual Geometry Group called the VGG16 network. Moreover, the method can develop consequently for extracting features of images on account of an absence of successful pre-prepared models dependent on LBP. The classical LBP and its varieties utilize pixel values of a different radius, but the relationships among them are missing. In this study, we have fulfilled the missing relational information among pixel values of varying radii. This study utilized an image into sub-cells where 3×3 for LBP $_{(8, 1)}$ and 5×5 for LBP $_{(8, 2)}$ with two proposed kernels with 45° rotations. After applying these kernels, bitwise AND operation occurred among the resulting matrices to establish the relation of different radii. Moreover, in pre-processing, we used the unsharp masking kernel to obtain a sharp image so that the intensity of pixel values can be more accurate. Compared with the neural network models, our method is a core algorithm to extract features where a neural network like CNN is a stack of automatic extraction of hidden layer features. Even though the latest neural network models are useful in the FER process, they still show unavoidable limitations. Different features like AAM/Arithmetic Unit system (AUs) [33] and Active Appearance Model (AAM)/Gabor [34] were used the CK+ dataset, and some other features like Gabor [35] and Facial Landmarks [36] used the KDEF datasets, all gaining different accuracies, which were much lower than our acquired accuracy. However, it can be expected that the addition of a neural network with our core algorithm to classify expressions might provide much higher efficiency on the other available standard FER datasets. Much readymade software, such as the Noldus network with Face-reader 8 [37] and Microsoft Emotion API [38], are available to obtain the facial expression easily from an image or live video. In Noldus face reader 8, besides FER, several things such as the detection of age, gender, ethnicity, facial hair, and glasses are performed. In doing so, a 3D model is created using the Active Appearance Method (AAM), and also an artificial neural network is used for training and classification. On the other hand, Microsoft Emotion API is a C# client-side library file, which is suitable for use as a third party API for detecting facial expressions in different projects under Microsoft Azure Cognitive Services. This API is licensed under the Massachusetts Institute of Technology (MIT), and the backend image processing model is developed and maintained by Microsoft. The primary comparison among Noldus Face-reader 8, Microsoft Emotion API and our work is incompatible as they are, in fact, software methods, and ours is a research method about MSFLBP. Moreover, only very little information is available on their methods, algorithms, and test results for building their FER models.

The outcome of SVM on the proposed MSFLBP method is shown in Table 6, compared with some of the most recent state-of-the-art methods. It demonstrates that the proposed feature extraction method outperforms the most recent state-of-the-art methods.

Table 6. Results of reviewed works for static image approaches (values are in %).

Year	Classifier	Features	Databases	Accuracy (%)
2017 [5]	WMDNN	LBP	CK+/JAFFE/CASIA	97.02
2019 [8]	SVM	LBP-TOP	CASME II/SMIC	73.51/70.02
2019 [9]	ELM	CS-LGC	CK+/JAFFE	98.33/95.24
2017 [18]	HOG	Ri-HOG	CK+/MMI/AFEW	93.8/72.4/56.8
2019 [28]	SVM	LBP/LTP/RBC	Infant COPE	89.43/95.12
2016 [33]	SVM	AAM/AUs	CK+	54.47
2016 [36]	KNN	Landmarks	KDEF/JAFFE	92.29
2017 [34]	SVM/CRF	AAM/Gabor	CK+	93.93
2020	SVM	The proposed method (MSFLBP)	CK+	99.12
			KDEF	89.08

5. Conclusions

The study demonstrates the recognition rate improvement based on the calculation time of facial expression recognition methods. In the classification performance, we have used two notable datasets, CK+ and KDEF, and analyzed, as a set of cell size and number of direction, containers for the seven fundamental universal expressions' exact characterization. We have used an unsharp masking kernel for sharpening the raw images. Then, we have applied two Kernel and bitwise AND to both binary matrices and converted the final binary matrix into a central decimal pixel value. After that, we have divided the output image into 64 cells and mapped each cell with ULBP mapping to obtain the features, like a histogram. By concatenating all cells' assigned values, we have finally obtained the feature vector, which was then trained and tested with four classifiers with 10 K-Fold cross-validations. Among them, SVM provides the best outcome. In this study, the traditional LBP method's limitations are overcome by applying bitwise AND on two rotational kernels by solving the pixel variance limitations. We have analyzed the neighboring pixel relation of traditional LBP and found two 3×3 and 5×5 kernels for obtaining the central pixel values, and after that, bitwise AND was applied to make the relation of the output central pixels of two kernels. Our described method can improve different texture recognition performance, utilize specific word applications with non-interrupting low-goals imaging, and also accomplish considerable accuracy. Several benefits of the described method include precise frequency extraction capability and less complexity, better efficiency in prediction, and fewer data storage. The addition of some more datasets from the different geographical regions can improve the real-time FER process. More combined methods like LBP-CNN can be used to identify augmented images.

Author Contributions: Conceptualization, M.B.; methodology, S.Y. and R.K.P.; software, S.Y. and R.K.P.; validation, M.B.; formal analysis, R.K.P.; S.Y.; and M.B.; investigation, S.Y. and R.K.P.; resources, M.B.; data curation, S.Y.; R.K.P.; and M.B.; writing—original draft preparation, S.Y.; R.K.P.; and M.B.; writing—review and editing, M.U.K.; visualization, M.U.K.; supervision, M.B.; project administration, M.R.I.F. and M.B.; funding acquisition, M.R.I.F. All authors have read and agreed to the published version of the manuscript.

Funding: This work was supported by the Research Universiti Grant, Universiti Kebangsaan Malaysia, Dana Impak Perdana (DIP), code: 2020-018.

Acknowledgments: The authors are appreciative of the Department of Computer Science and Engineering, BGC Trust University Bangladesh, and International Islamic University Chittagong, Bangladesh, for giving the workplaces to lead this research work. This work was supported by the Research Universiti Grant, Universiti Kebangsaan Malaysia, Dana Impak Perdana (DIP), code: 2020-018.

Conflicts of Interest: The authors declare no conflict of interest.

References

1. Yu, Z.; Zhang, C. Image based static facial expression recognition with multiple deep network learning. In Proceedings of the ICMI 2015-Proceedings of the 2015 ACM International Conference on Multimodal Interaction, Washington, DC, USA, 9–13 November 2015.
2. Kahou, S.E.; Michalski, V.; Konda, K.; Memisevic, R.; Pal, C. Recurrent neural networks for emotion recognition in video. In Proceedings of the ICMI 2015-Proceedings of the 2015 ACM International Conference on Multimodal Interaction, Washington, DC, USA, 9–13 November 2015.
3. Liu, M.; Li, S.; Shan, S.; Wang, R.; Chen, X. Deeply learning deformable facial action parts model for dynamic expression analysis. In *Asian Conference on Computer Vision, Proceedings of the Lecture Notes in Computer Science*; Springer International Publishing: Cham, Switzerland, 2015.
4. Shan, C.; Braspenning, R. Recognizing Facial Expressions Automatically from Video. In *Handbook of Ambient Intelligence and Smart Environments*; Nakashima, H., Aghajan, H., Augusto, J.C., Eds.; Springer International Publishing: Cham, Switzerland, 2010.
5. Yang, B.; Cao, J.; Ni, R.; Zhang, Y. Facial Expression Recognition Using Weighted Mixture Deep Neural Network Based on Double-Channel Facial Images. *IEEE Access* **2017**. [CrossRef]
6. Li, Y.; Zeng, J.; Shan, S.; Chen, X. Occlusion Aware Facial Expression Recognition Using CNN With Attention Mechanism. *IEEE Trans. Image Process.* **2018**. [CrossRef] [PubMed]
7. Liu, Y.; Li, Y.; Ma, X.; Song, R. Facial Expression Recognition with Fusion Features Extracted from Salient Facial Areas. *Sensors* **2017**, *17*, 712. [CrossRef]
8. Zhao, Y.; Xu, J. An Improved Micro-Expression Recognition Method Based on Necessary Morphological Patches. *Symmetry* **2019**, *11*, 497. [CrossRef]
9. Yang, J.; Wang, X.; Han, S.; Wang, J.; Park, D.S.; Wang, Y. Improved Real-Time Facial Expression Recognition Based on a Novel Balanced and Symmetric Local Gradient Coding. *Sensors* **2019**, *19*, 1899. [CrossRef]
10. Zhang, W.; Shan, S.; Zhang, H.; Gao, W.; Chen, X. Multi-resolution Histograms of Local Variation Patterns (MHLVP) for robust face recognition. In *International Conference on Audio-and Video-Based Biometric Person Authentication, Proceedings of the Lecture Notes in Computer Science*; Springer International Publishing: Cham, Switzerland, 2005.

11. Huang, D.; Shan, C.; Ardabilian, M.; Wang, Y.; Chen, L. Local binary patterns and its application to facial image analysis: A survey. *IEEE Trans. Syst. Man Cybern. Part C* **2011**, *41*, 765–781. [CrossRef]

12. Kumari, J.; Rajesh, R.; Pooja, K.M. Facial Expression Recognition: A Survey. *Procedia Comput. Sci.* **2015**, *58*, 486–491. [CrossRef]

13. Ahonen, T.; Hadid, A.; Pietikäinen, M. Face description with local binary patterns: Application to face recognition. *IEEE Trans. Pattern Anal. Mach. Intell.* **2006**. [CrossRef]

14. Canedo, D.; Neves, A.J.R. Facial Expression Recognition Using Computer Vision: A Systematic Review. *Appl. Sci.* **2019**, *9*, 4678. [CrossRef]

15. Huang, D.; Wang, Y.; Wang, Y. A robust method for near-infrared face recognition is based on extended local binary patterns. In *Proceedings of the Lecture Notes in Computer Science (Including Subseries Lecture Notes in Artificial Intelligence and Lecture Notes in Bioinformatics)*; Springer: Berlin/Heidelberg, Germany; pp. 437–446.

16. Guo, Z.; Zhang, L.; Zhang, D. A completed modeling of local binary pattern operator for texture classification. *IEEE Trans. Image Process.* **2010**. [CrossRef]

17. Zhao, G.; Pietikäinen, M. Dynamic texture recognition using local binary patterns with an application to facial expressions. *IEEE Trans. Pattern Anal. Mach. Intell.* **2007**. [CrossRef] [PubMed]

18. Sajjad, M.; Shah, A.; Jan, Z.; Shah, S.I.; Baik, S.W.; Mehmood, I. Facial appearance and texture feature-based robust facial expression recognition framework for sentiment knowledge discovery. *Cluster Comput.* **2017**. [CrossRef]

19. Zhang, B.; Gao, Y.; Zhao, S.; Liu, J. Local derivative pattern versus local binary pattern: Face recognition with high-order local pattern descriptor. *IEEE Trans. Image Process.* **2010**. [CrossRef]

20. Zangeneh, E.; Moradi, A. Facial expression recognition by using differential geometric features. *Imaging Sci. J.* **2018**. [CrossRef]

21. Chen, J.; Takiguchi, T.; Ariki, Y. Rotation-reversal invariant HOG cascade for facial expression recognition. *Signal Image Video Process.* **2017**. [CrossRef]

22. Tsai, H.H.; Chang, Y.C. Facial expression recognition using a combination of multiple facial features and a support vector machine. *Soft Comput.* **2018**. [CrossRef]

23. Alphonse, A.S.; Dharma, D. Novel directional patterns and a Generalized Supervised Dimension Reduction System (GSDRS) for facial emotion recognition. *Multimed. Tools Appl.* **2018**. [CrossRef]

24. Yu, Z.; Liu, G.; Liu, Q.; Deng, J. Spatio-temporal convolutional features with nested LSTM for facial expression recognition. *Neurocomputing* **2018**. [CrossRef]

25. Zhang, L.; Gao, Q.; Zhang, D. Directional independent component analysis with tensor representation. In Proceedings of the 26th IEEE Conference on Computer Vision and Pattern Recognition, CVPR, Anchorage, AK, USA, 23–28 June 2008.

26. Samara, A.; Galway, L.; Bond, R.; Wang, H. Affective state detection via facial expression analysis within a human-computer interaction context. *J. Ambient Intell. Humaniz. Comput.* **2019**. [CrossRef]

27. Turabzadeh, S.; Meng, H.; Swash, R.; Pleva, M.; Juhar, J. Facial Expression Emotion Detection for Real-Time Embedded Systems. *Technologies* **2018**, *6*, 17. [CrossRef]

28. Martínez, A.; Pujol, F.A.; Mora, H. Application of Texture Descriptors to Facial Emotion Recognition in Infants. *Appl. Sci.* **2020**, *10*, 1115. [CrossRef]

29. Lucey, P.; Cohn, J.F.; Kanade, T.; Saragih, J.; Ambadar, Z.; Matthews, I. The extended Cohn-Kanade dataset (CK+): A complete dataset for action unit and emotion-specified expression. In Proceedings of the 2010 IEEE Computer Society Conference on Computer Vision and Pattern Recognition-Workshops, CVPRW, San Francisco, CA, USA, 13–18 June 2010.

30. Kanade, T.; Cohn, J.F.; Tian, Y. Comprehensive database for facial expression analysis. In Proceedings of the Proceedings-4th IEEE International Conference on Automatic Face and Gesture Recognition, FG, Grenoble, France, 28–30 March 2000.

31. Kernel (Image Processing), n.d., para.2, Wikipedia. Available online: https://en.wikipedia.org/w/index.php?title=Kernel_(image_processing) (accessed on 30 August 2020).

32. Xiong, H.; Zhang, D.; Martyniuk, C.J.; Trudeau, V.L.; Xia, X. Using Generalized Procrustes Analysis (GPA) for normalization of cDNA microarray data. *BMC Bioinform.* **2008**. [CrossRef]

33. Sert, M.; Aksoy, N. Recognizing facial expressions of emotion using action unit-specific decision thresholds. In Proceedings of the 2nd Workshop on Advancements in Social Signal Processing for Multimodal Interaction-ASSP4MI '16, Tokyo, Japan, 16 November 2016. [CrossRef]

34. Liliana, D.Y.; Basaruddin, C.; Widyanto, M.R. Mix Emotion Recognition from Facial Expression using SVM-CRF Sequence Classifier. In Proceedings of the International Conference on Algorithms, Computing and Systems-ICACS '17, Jeju Island, Korea, 10–13 August 2017; pp. 27–31. [CrossRef]
35. Ruiz-Garcia, A.; Elshaw, M.; Altahhan, A.; Palade, V. A hybrid deep learning neural approach for emotion recognition from facial expressions for socially assistive robots. *Neural Comput. Appl.* **2018**, *29*, 359–373. [CrossRef]
36. Yaddaden, Y.; Bouzouane, A.; Adda, M.; Bouchard, B. A New Approach of Facial Expression Recognition for Ambient Assisted Living. In Proceedings of the 9th ACM International Conference on PErvasive Technologies Related to Assistive Environments-PETRA '16, Corfu, Greece, 29 June–1 July 2016. [CrossRef]
37. FaceReader 8, Technical Specifications. Noldus Information Technology. Available online: https://www.mindmetriks.com/uploads/4/4/6/0/44607631/technical_specs__facereader_8.0.pdf (accessed on 9 September 2020).
38. Face-An AI Service that Analyzes Faces in Images. Microsoft Azure. Available online: https://azure.microsoft.com/en-us/services/cognitive-services/face/ (accessed on 9 September 2020).

Article

Face and Body-Based Human Recognition by GAN-Based Blur Restoration

Ja Hyung Koo, Se Woon Cho, Na Rae Baek and Kang Ryoung Park *

Division of Electronics and Electrical Engineering, Dongguk University, 30 Pildong-ro 1-gil, Jung-gu, Seoul 04620, Korea; koo6190@dongguk.edu (J.H.K.); jsu319@dongguk.edu (S.W.C.); naris27@dongguk.edu (N.R.B.)
* Correspondence: parkgr@dongguk.edu; Tel.: +82-2-2260-3329; Fax: +82-2-2277-8735

Received: 21 July 2020; Accepted: 11 September 2020; Published: 14 September 2020

Abstract: The long-distance recognition methods in indoor environments are commonly divided into two categories, namely face recognition and face and body recognition. Cameras are typically installed on ceilings for face recognition. Hence, it is difficult to obtain a front image of an individual. Therefore, in many studies, the face and body information of an individual are combined. However, the distance between the camera and an individual is closer in indoor environments than that in outdoor environments. Therefore, face information is distorted due to motion blur. Several studies have examined deblurring of face images. However, there is a paucity of studies on deblurring of body images. To tackle the blur problem, a recognition method is proposed wherein the blur of body and face images is restored using a generative adversarial network (GAN), and the features of face and body obtained using a deep convolutional neural network (CNN) are used to fuse the matching score. The database developed by us, Dongguk face and body dataset version 2 (DFB-DB2) and ChokePoint dataset, which is an open dataset, were used in this study. The equal error rate (EER) of human recognition in DFB-DB2 and ChokePoint dataset was 7.694% and 5.069%, respectively. The proposed method exhibited better results than the state-of-art methods.

Keywords: multimodal human recognition; blur image restoration; DeblurGAN; CNN

1. Introduction

Currently, there are several methods of human recognition, including face, iris, fingerprint, finger-vein, and body. However, long-distance face recognition in indoor and outdoor environments is still limited. The human recognition methods can be largely divided into face, body, and iris. However, there are problems with face and iris recognition methods. In these methods, original images can be damaged due to motion blur or optical blur, which is generated when the images of human face or iris are obtained from a long distance. The human recognition performance is significantly degraded due to these types of damages. To solve this problem, the human body is typically used as for long-distance recognition in indoor and outdoor environments.

The data can still contain a blur when human body is used for recognition. However, the human body recognition is less affected than face or iris recognition. There are two methods for human body recognition: gait recognition of an individual and texture and shape-based body recognition, which is based on the still image of a human body. Gait recognition does not exhibit a blur problem. However, the time required for forming the dataset is long because continuous image acquisition is required. Thus, an experiment was conducted indoors for recognition using still images of a human body.

There are disadvantages to human body recognition in an indoor environment. The color of clothes significantly affects the recognition performance. Thus, the human body is divided into two parts to evaluate the recognition performance. In several studies, the body and face have been separated. However, blur restoration of the obtained data has never been performed before.

The method proposed in this study involves restoring the images of human body and face with a blur via a generative adversarial network (GAN). Subsequently, the features of body and face are extracted using a convolutional neural network (CNN) model. The final recognition performance is determined based on the weighted sum and weighted product, which is a score-level fusion approach, using the extracted features.

2. Related Work

Previous studies on long-distance human recognition can be divided into human recognition with or without blur restoration, and they can be further divided into single modality-based or multimodal-based methods.

2.1. Without Blur Restoration

Single modality-based methods include face recognition, body recognition based on texture, and body recognition based on gait. Several extant studies have been conducted on face recognition. Grgic et al. [1] obtained face data from three designated locations using five cameras. The recognition performance was determined based on principal component analysis (PCA) of the obtained face data. Banerjee et al. [2] used three types of datasets, namely FR_SURV, SCface, and ChokePoint, for the experiment. The recognition was performed through soft-margin learning for multiple feature-kernel combination (SML-MKFC) with domain adaptation (DA). The drawback of face recognition is that facial information is vulnerable to noise, such as blur. There are important features in a face, such as nasal bridge, eyebrow, and skin color, for recognizing an individual. The visibility of facial features is reduced when important features are combined with noise, such as a blur, thereby interfering with face recognition.

Most of the body recognition methods are gait-based, while others are texture and shape-based. For gait-based recognition, Zhou et al. [3] obtained data using two methods of original side-face image (OSFI) and gait energy image (GEI) fusion, as well as enhanced side-face image (ESFI) and GEI fusion. Furthermore, they proceeded with recognition based on PCA and multiple discriminant analysis (MDA). Gait-based recognition is less affected by noise, such a blur, because several images of an individual's gait are cropped based on the difference image of the background and object. The difference image is compressed into a single image. However, an extensive amount of time and data are required to obtain sufficient gait information. For texture and shape-based body recognition, Varior et al. [4] used the Siamese CNN (S-CNN) architecture. Nguyen et al. [5] obtained image features using AlexNet-CNN and then evaluated the recognition using PCA and support vector machine (SVM). Shi et al. [6] used the S-CNN architecture reported in an extant study [4]. However, they used five convolution blocks. Furthermore, a discriminative deep metric learning (DDML) was used in the study. This method is not significantly affected by a blur because the object's body information is included. However, the color of clothes worn by the object comprises of a large portion of the body information. Hence, the recognition performance is drastically reduced if the color of the clothes is similar to that of the object, which is being recognized.

Multimodal-based methods are categorized into two types, namely face and gait-based body recognition and face and texture and shape-based body recognition. For face and gait-based body recognition, Liu et al. [7] measured the performance using the dataset obtained by other researchers based on hidden Markov model (HMM) and Gabor features-based elastic bunch graph matching (EBGM). Hofmann et al. [8] used eigenface calculation for face recognition and α-GEI for gait recognition. This method exhibits the same advantages and disadvantages as gait-based body recognition. The common advantage is that it is less affected by a blur because a gait feature is used. The disadvantage is that it requires sufficiently high amount of data with continuous image motion for obtaining the gait image. In a previous study [9], human body and face were separately experimented in indoor environments for face and texture and shape-based body recognition. Visual geometry group (VGG) face net-16 for face and residual network (ResNet)-50 for body were used to obtain the features, and the final recognition performance was evaluated based on a score-level fusion approach using the obtained features. However, the problem with blur still persists when images are obtained in indoor environment. Therefore, in the study [9], only the images without a blur were used by determining the presence of a blur as per the threshold based on the method in the study [10].

2.2. With Blur Restoration

A blur is generated due to two main reasons. Motion blur is generated when an object moves, and optical blur is generated when a camera films the object. Thus, researchers improved the images using a deblur method and then proceeded with the evaluation of the recognition performance. Alaoui et al. [11] performed image blurring by applying point spread function (PSF) with the face recognition technology (FERET) database. The images were deblurred with fast total variation (TV)-l1 deconvolution, image features were obtained using PCA, and feature matching was performed with Euclidean distance. Hadid et al. [12] generated a blur using PSF and then proceeded with deblurring based on deblur local phase quantization (DeblurLPQ) and measured the recognition performance. Nishiyama et al. [13] used two types of datasets and generated an arbitrary blur using PSF with the FERET database and face recognition grand challenge (FRGC) 1.0. For blur restoration method, Wien filters or bilateral total variation (BTV) regularization was used. Mokhtari et al. [14] performed face restoration using two methods, namely centralized sparse representation (CSR) and adaptive sparse domain selection with adaptive regularization (ASDS-AR). Face recognition was performed using PCA, linear discriminant analysis (LDA), kernel principal component analysis (KPCA), and kernel Fisher analysis (KFA). Heflin et al. [15] used the FERET database wherein the face area was detected in the blurred image, motion blur and atmospheric blur were measured using a blur point spread function (PSF), and, finally, face deblurring was performed using a deconvolution filter, such as Wiener filter, to evaluate the recognition performance. Yasarla et al. [16] proposed uncertainty guided multi-stream semantic network (UMSN) and performed facial image deblurring. This method involves dividing the facial image region into four semantic networks and deblurring the blurred image and image divided into four regions via a base network (BN). Considering the aforementioned issues of previous researches, we propose a recognition method in which the blur on a body and face is restored using a GAN, and the features of body and face obtained using a deep CNN are used to fuse the matching score.

Although they are not the researches on long-distance human recognition, Peng et al. studied two challenges in clustering analysis, that is, how to cluster multi-view data and how to perform clustering without parameter selection on cluster size. For this purpose, they proposed a novel objective function to project raw data into one space where the projection embraces the cluster assignment consistency (CAC) and the geometric consistency (GC) [17]. In addition, Huang et al. proposed a novel multi-view clustering method called as multi-view spectral clustering network (MvSCN) which could be the first deep version of multi-view spectral clustering [18]. To deeply cluster multi-view data, MvSCN incorporates the local invariance within every single view and the consistency across different views into a novel objective function. They also enforced and reformulated an orthogonal constraint as a novel layer stacked on an embedding network.

Table 1 shows the summary of this study and previous studies on person recognition using surveillance camera environment.

Table 1. Summary of this study and previous studies on person recognition using surveillance camera environment.

Category			Method	Advantage	Disadvantage
Without blur restoration	Single modality	Face recognition	PCA [1]	Low performance degradation due to lighting changes.	Degraded recognition performance due to distortion of image features due to a blur.
			SML-MKFC with DA [2]		
		Texture and shape-based body recognition	S-CNN [4]	Low interference of a blur for recognition.	Different individuals wearing the same clothes are recognized as the same individual.
			CNN + PCA, SVM [5]		
			CNN + DDML [6]		
		Movement-based body recognition	PCA + MDA [3]	Less affected directly by a blur.	Requires an extensive time for acquiring data.
	Multimodal	Movement-based body and face recognition	HMM/Gabor features based EBGM [7]	Less affected by a blur because it is gait-based body recognition.	Sufficient data is required for continuous images. Distortion of face image due to noise, such as lighting changes or blur.
			Eigenface calculation and αGEI [8]		
		Texture and shape-based body and face recognition	VGG face net-16 and ResNet-50 [9]	Easy to acquire data because continuous images are not required.	Sufficient data is required for finetuning based on data characteristics.
With blur restoration	Single modality	Face recognition	Fast TV-l1 and Deconvolution+ PCA [11]	Improved recognition performance, as blurred face images are restored.	Most studies focused on comparing the deblurred facial image with the original image.
			DeblurLPQ [12]		
			Wien filters or BTV regularization [13]		
			CSR + ASDS-AR [14]		
			PSF + Wiener filter [15]		
			UMSN [16]		
	Multimodal	Texture and shape-based body and face recognition	Proposed method	Improved performance because body and face were separated for restoration.	Slow processing time due to restoration is performed twice for body and face.

3. Contribution of Our Research

Our research is novel in the following four ways in comparison to previous works:

- This is the first approach for multimodal human recognition by blur restoring the face and body images using GAN.
- Different from previous work [9], the presence of a blur was determined based on a focus score method in which blur restoration was applied via GAN for image in case that input image was determined as blur existence. The error was reduced, when compared to that without proposed focus score method and GAN.
- The structural complexity was reduced by separating the network for blur restoration and the CNN for human recognition. In addition, the processing speed is usually faster when one image of face and body is restored at simultaneously via GAN. However, our blur restoration proceeded separately through GAN because face images exhibit detailed information, and the generation of a blur exhibits different tendencies in face and body images.
- We make Dongguk face and body database version 2 (DFB-DB2), trained VGG face net-16 and ResNet-50, and GAN model for deblurring available by other researchers through [19] for fair comparisons.

4. Proposed Method

4.1. System Overview

Figure 1 shows the overall configuration of the system proposed in this study. A face image is obtained from the original image acquired in an indoor environment (step (1) in Figure 1). A body image is obtained from the original image excluding the face image (step (2) in Figure 1). The focus score of the face image is calculated (step (3) in Figure 1). An image exhibiting a focus score value of less than the threshold (step (4) in Figure 1) undergoes restoration using DeblurGAN (step (5) in Figure 1) and is combined with images exhibiting a focus score value that is greater than or equal to the threshold. The restoration of body image via DeblurGAN is conducted in the same manner. Image features of face and body are extracted by applying a CNN model to the image combined from the restored face and body images and the image with a focus score greater than or equal to the threshold (step (6) and (7) in Figure 1). The authentic/imposter matching distance is calculated using the feature vectors obtained above (step (8) and (9) in Figure 1). The score-level fusion is conducted using the matching distance (step (10) in Figure 1). The weighted sum and weighted product methods were for the score-level fusion in this study. The final recognition rate was measured using score-level fusion (step (11) in Figure 1).

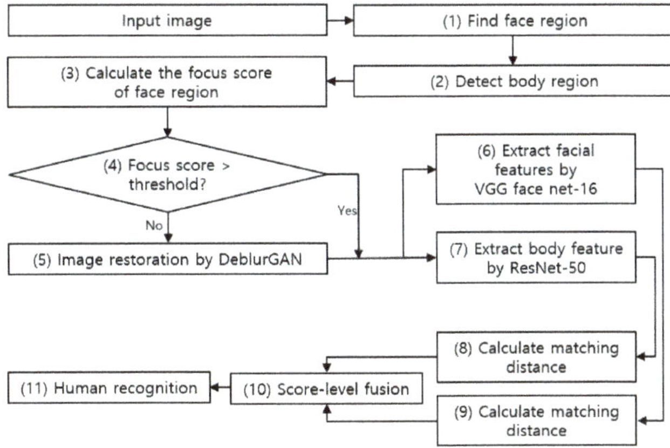

Figure 1. Overall procedure of proposed method.

4.2. Structure of GAN

A general description of a GAN is provided in this section. GAN consists of two networks, namely generator and discriminator. Generator aims to generate a fake image similar to a real image by considering Gaussian random noise as an input, whereas discriminator aims to find the fake image by discriminating the real image from the fake image generated by the generator. Therefore, a discriminator is trained to easily discriminate real and fake images, while a generator is trained to ensure that a fake image is close to the real image to the maximum possible extent. However, it is difficult to control the desired output for vanilla GAN because the input corresponds to Gaussian random noise.

First, cycle-consistent adversarial networks (CycleGAN) [20] were used. Unlike the existing GAN models, a CycleGAN does not distinguish between an input image and a target image. It uses a reference image as an input that is expected to be the result of input image and output image. There are two types of generators in CycleGAN, namely U-Net [21] architecture and residual blocks. The generator used in this study exhibits a residual block architecture [20]. One of the characteristics of a CycleGAN is the cycle-consistency loss. For example, if an input image X has generated an output Y through a generator, the output Y goes through the generator again to generate X′. The cycle-consistency loss refers to calculating the difference between X and X′.

Second, Pix2pix [22] was used. Pix2pix is a GAN applied with the concept of a conditional GAN (CGAN) mode. The generator of Pix2pix is similar to that of U-Net [21]. Unlike U-Net, skip-connection is applied between the encoder and decoder because a blur problem occurs due to the loss of image details when the size of the image is enlarged and then reduced. Furthermore, DeblurGAN [23] uses the input image and target image of a CGAN as an input. However, it exhibits a very different architecture. The architecture of the generator in DeblurGAN consists of two convolutional blocks, 9 residual blocks, and two transposed convolution blocks. Each convolution block contains instance normalization layer [24] and rectified linear units (ReLU) layer, as shown in Table 2. Instance normalization [24] is also referred as contrast normalization. ReLU layer serves as an activation function in residual blocks. The loss function of DeblurGAN uses adversarial loss and content loss. The total loss of the two loss functions can be calculated using Equation (1) as follows:

$$L_{total} = L_{Adv} + \lambda L_{Cont}. \tag{1}$$

First, adversarial loss (L_{Adv}) can be explained as follows. The adversarial loss discerns the blurred image restored via a generator by using a discriminator. In this case, the loss is considered as optimal when the difference between the loss discerned by the discriminator and the threshold value 1 is close to 0. Thus, L_{Adv} used in DeblurGAN is represented in Equation (2) as follows:

$$L_{Adv} = \sum_{k=1}^{N} -D_\theta(G_\theta(I_B)). \tag{2}$$

In Equation (2), N denotes the number of images, D_θ denotes the discriminator network, G_θ denotes the generator network, and I_B denotes a blurred image. As specified in DeblurGAN [23], Wasserstein GAN-gradient penalty (WGAN-GP) [25] was used for the adversarial loss. Next, L_{Cont} is explained in Equation (3).

$$L_{Cont} = \frac{1}{X_{n,m}Y_{n,m}} \sum_{k=1}^{X_{n,m}} \sum_{j=1}^{Y_{n,m}} (\varnothing_{n,m}(I_S)_{n,m} - \varnothing_{n,m}(G_\theta(I_B))_{n,m})^2. \tag{3}$$

With respect to content loss, either L1 or mean absolute error (MAE) loss or L2 or mean squared error (MSE) loss can be selected. However, perceptual loss was selected for the content loss of DeblurGAN. The perceptual loss of DeblurGAN can be distinguished by the difference between the restored image and target image obtained through conv3.3 features maps of VGG-19 pretrained with

ImageNet. In Equation (3), $X_{n,m}$ *and* $Y_{n,m}$ are the size of a feature map, and $\varnothing_{n,m}$ is the feature map obtained from the mth convolutional layer. Furthermore, I_S is the target image for restoring the blurred image [23]. Tables 2 and 3 summarize the architecture of the generator and discriminator in DeblurGAN. Figure 2a,b denote the architecture of a generator and discriminator in DeblurGAN, respectively.

Table 2. Generator of DeblurGAN. GAN = generative adversarial network.

	Layer Type	Size of Feature Map	Number of Filters	Size of Filters	Number of Strides	Number of Iterations
	Image input layer	256 (height) × 256 (width) × 3 (channel)				
	Convolution layer	256 × 256 × 64	64	7 × 7	1	
	Instance normalization layer					
	ReLU					
Convolution block 1	Convolution layer	128 × 128 × 128	128	3 × 3	2	
	Instance normalization layer					
	ReLU					
Convolution block 2	Convolution layer	64 × 64 × 256	256	3 × 3	2	
	Instance normalization layer					
	ReLU					
Resblocks	Convolution layer	64 × 64 × 256	256	3 × 3	1	9
	Instance normalization layer					
	ReLU					
	Convolution layer	64 × 64 × 256	256	3 × 3	1	
	Instance normalization layer					
Transposed blocks 1	Contraposed layer	128 × 128 × 128	128	4 × 4	2	2
	Instance normalization layer					
	ReLU					
Transposed blocks 2	Contraposed layer	256 × 256 × 64	64	4 × 4	2	
	Instance normalization layer					
	ReLU					
	Convolution layer (Output layer)	256 × 256 × 3	3	7 × 7	1	

Table 3. Discriminator of DeblurGAN (All convolution layers 1–5 * indicate that they have two paddings.).

Layer Type	Size of Feature Map	Number of Filters	Size of Filters	Number of Strides
Input image	256 × 256 × 3			
Target image	256 × 256 × 3			
Concatenator	256 × 256 × 6			
Convolution layer1 *	129 × 129 × 64	64	4 × 4 × 6	2
Convolution layer2 *	65 × 65 × 128	128	4 × 4 × 64	2
Convolution layer3 *	33 × 33 × 256	256	4 × 4 × 128	2
Convolution layer4 *	34 × 34 × 512	512	4 × 4 × 256	1
Convolution layer5 * (Output layer)	35 × 35 × 1	1	4 × 4 × 512	1

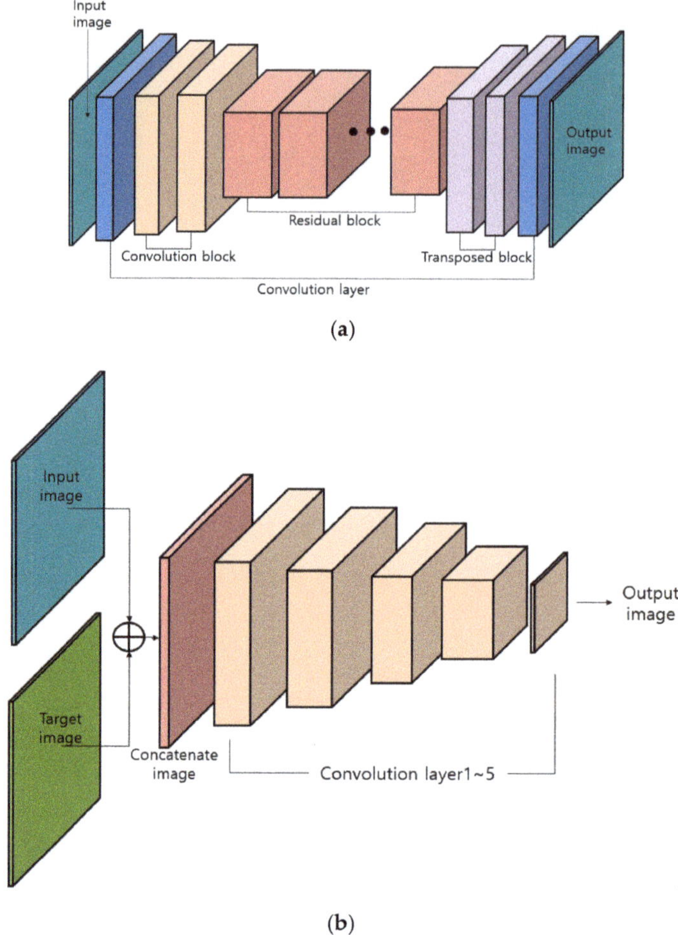

Figure 2. Architecture of DeblurGAN (**a**) Generator in DeblurGAN and (**b**) Discriminator in DeblurGAN.

4.3. Structure of Deep Learning (VGG Face Net-16 and ResNet-50)

The face and body images restored with DeblurGAN used VGG face net-16 and ResNet-50. In our previous research [9], we compared the recognition accuracies by VGG face net-16 and ResNet-50 with those by other CNN architectures on the custom-made Dongguk face and body database (DFB-DB1) whose acquisition environments including scenario and cameras were same to those of DFB-DB2 used in our research. According to the experimental results, VGG face net-16 and ResNet-50 outperform other CNN architectures, and we adopt these CNN models in our research. A pretrained model was used for two types of CNN models, which were fine-tuned based on the characteristics of the dataset used in this study.

The VGG face net-16, which was used for face images, consists of convolution filters and neural network. Specifically, it consists of 13 convolutional layers, five pooling layers, and three fully connected layers. The CNN pretrained model used in this study was trained with Labeled faces in the wild [26] and YouTube faces [27]. The size of the image restored with GAN corresponded to 256 × 256, and it was resized to 224 × 224 for using VGG face net-16 for fine-tuning. The resized image undergoes convolution calculation through the convolutional layer. The calculation is as follows:

output = (W − K + 2P)/S + 1. Here, W denotes the width and height of an input, K denotes the size of a convolutional layer filter, P denotes padding, and S denotes stride. For example, if a 224 × 224 image has convolution filter with K = 3, P = 0, and S = 1, then the output is (224 − 3 + 0)/1 + 1, i.e., 222.

There are many types of ResNet based on the number of convolutional layers. As the number of layers increase, the feature map of body images becomes smaller, and thereby causing a vanishing or exploding gradient problem. Thus, a shortcut is used for the ResNet architecture to avoid such a problem. In the shortcut, the input X goes through three convolutional layers and performs convolution calculation three times. If input X that has completed the convolution calculation is termed as F(x), then the shortcut is the sum of the features, or F(x) + X, which is then used as an input for the next convolutional layer. To reduce the convolution calculation time, 1 × 1, 3 × 3, and 1 × 1 convolutional layers were used as opposed to two 3 × 3 convolutional layers. This is termed as the bottleneck architecture wherein 1 × 1 in the front reduces the dimension of the input image, while the 1 × 1 in the back enlarges the dimensions.

5. Experimental Results and Analysis

5.1. Experiments for Database and Environment

Two types of cameras were used in this study to acquire the DFB-DB2. The cameras were Logitech BCC950 [28] and Logitech C920 [29]. The cameras were also used for Dongguk face and body dataset version 1 (DFB-DB1). There was no difference in the scenario used for DFB-DB2 and DFB-DB1 in the study [9]. Furthermore, the DFB-DB1 only consists of images above the threshold based on the method of an extant study [10]. However, the DFB-DB2 used in this study included images below the threshold that were restored with DeblurGAN. Figure 3 shows the scenario of the images with respect to DFB-DB2. In the figure, (a) shows the images acquired via the Logitech BCC950 camera, whereas (b) shows those acquired via the Logitech C920 camera.

Table 4 summarizes the details of face and body images of two databases, namely DFB-DB2 and ChokePoint dataset [30], used in this study. Two-fold cross validation was applied to both databases and each dataset was divided into sub-dataset 1 and 2. For example, if sub-dataset 1 is used for training, then sub-dataset 2 is used for testing. Furthermore, if sub-dataset 2 is used for training, then sub-dataset 1 is used for testing to evaluate the performance.

Table 4. Total images of DFB-DB2 and ChokePoint dataset.

		DFB-DB2			Chokepoint Dataset		
		Number of Classes in Each Fold	Number of Augmented Images (For Training)	Number of Images (For Testing)	Number of Classes in Each Fold	Number of Augmented Images (For Training)	Number of Images (For Testing)
Face	Sub-Dataset1	11	200,134	827	14	519,050	10,381
	Sub-Dataset2	11	239,338	989	14	513,450	10,269
Body	Sub-Dataset1	11	200,134	827	14	519,050	10,381
	Sub-Dataset2	11	239,338	989	14	513,450	10,269

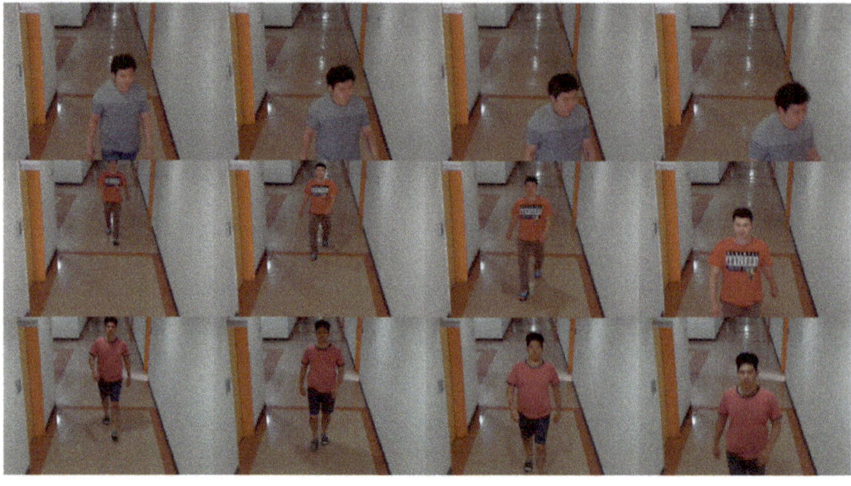

(a)

(b)

Figure 3. Representative Dongguk face and body dataset version 2 (DFB-DB2) images captured by (**a**) Logitech BCC950 camera and (**b**) Logitech C920 camera.

The ChokePoint dataset is provided at no cost by National ICT Australia Ltd. (NICTA) and consists of Portal 1 and 2. Portal 1 contains 25 individuals (19 males and 6 females), and Portal 2 contains 29 individuals (23 males and 6 females). A total of three cameras were used from six locations to constitute the dataset. The dataset of the study [9] was maintained. Furthermore, the images considered exhibit a blur, based on the threshold value in an extant study [10], were restored with DeblurGAN and included for evaluating the recognition performance. Figure 4 shows the examples of the ChokePoint dataset.

Figure 4. Example images for ChokePoint dataset.

5.2. Training DeblurGAN and CNN Models

5.2.1. DeblurGAN Model Training Process and Results

Blur image and clear image were distinguished for training DeblurGAN based on the focus score threshold value [9]. The values below the threshold were set as test images for DeblurGAN; the focused image exhibiting a value greater than or equal to the threshold was used as a reference image. Pytorch version of DeblurGAN [31] was used for the program. All the images for training and testing DeblurGAN were resized to 256 × 256. The learning rate was 0.0001, and the batch size was 1 for training DeblurGAN.

5.2.2. CNN Model Training Process and Results

After performing image deblurring with DeblurGAN, face images were trained with VGG face net-16 [32] and body images were trained with ResNet-50 [33]. The number of data points for training each deep CNN model was insufficient, thus the number of data points was increased via data augmentation for training.

As shown in Table 4, data augmentation was performed only in the training data, whereas the original non-augmented data were used as test data. The number of test data points for the DFB-DB2 is less than that of the ChokePoint dataset, which is an open dataset, and therefore center image crop was performed during augmentation. The cropped image was applied with image translation and cropping for five pixels in top, bottom, left, and right directions. Furthermore, the image was horizontally flipped (mirroring). The training data that was processed accordingly included 440,000 augmented images from sub-datasets 1 and 2. For the ChokePoint dataset, after performing center image crop, image translation and cropping were applied for two pixels in top, bottom, left, and right directions. Furthermore, horizontal flipping was applied to obtain images that were magnified by 50 times. The sub-datasets 1 and 2 in Table 4 include a total of 1.03 million augmented images. Figure 5 shows the data augmentation method used in this study.

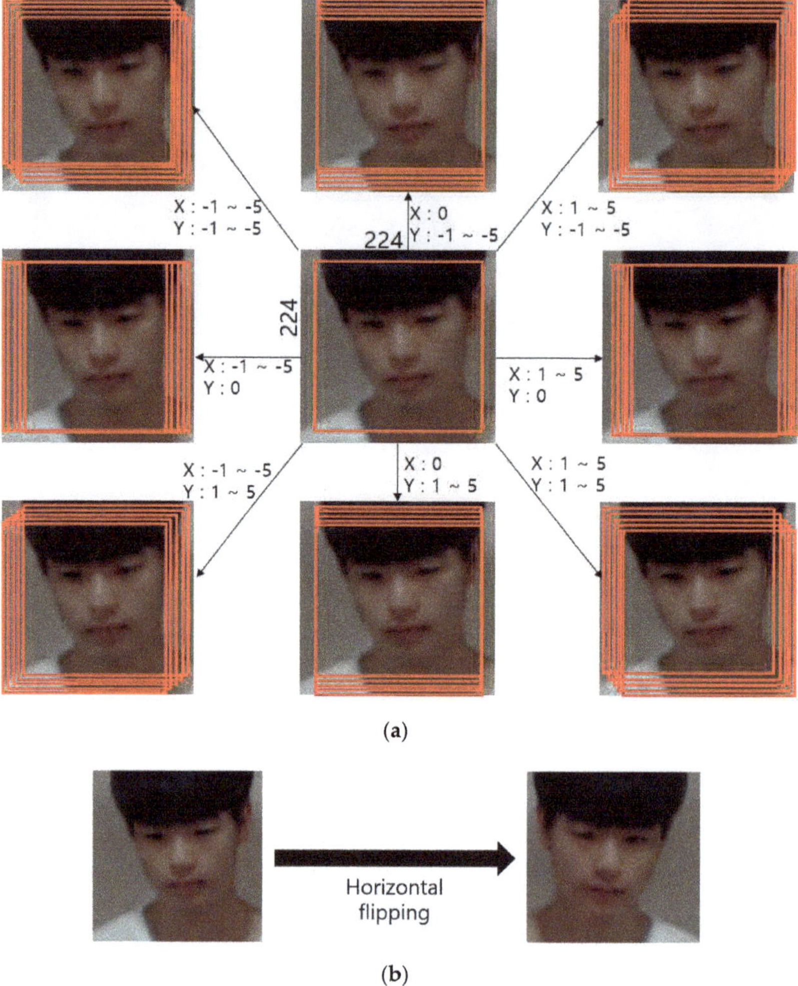

Figure 5. Data augmentation method involving (**a**) image translation and cropping and (**b**) horizontal flipping.

Given that VGG face net-16 is pretrained with Oxford face database, it was appropriately fine-tuned for the characteristics of the images in DFB-DB2. Furthermore, ResNet-50 also uses the pretrained model, and thus was appropriately fine-tuned for the characteristics of the image database used in this study. The learning rate was 0.0001, and the batch size was 20 for the training of VGG face net-16 and 15 for the training of ResNet-50.

Figure 6 illustrates the plots of the loss-accuracy of the training CNN model for trained face and body images. The specifications of the computer used for the experiment are as follows: CPU Intel(R) Core(TM) i7-6700 CPU @ 3.40 GHz, 16 GB RAM, NVIDIA GeForce GTX 1070 graphic card, and CUDA version 8.0.

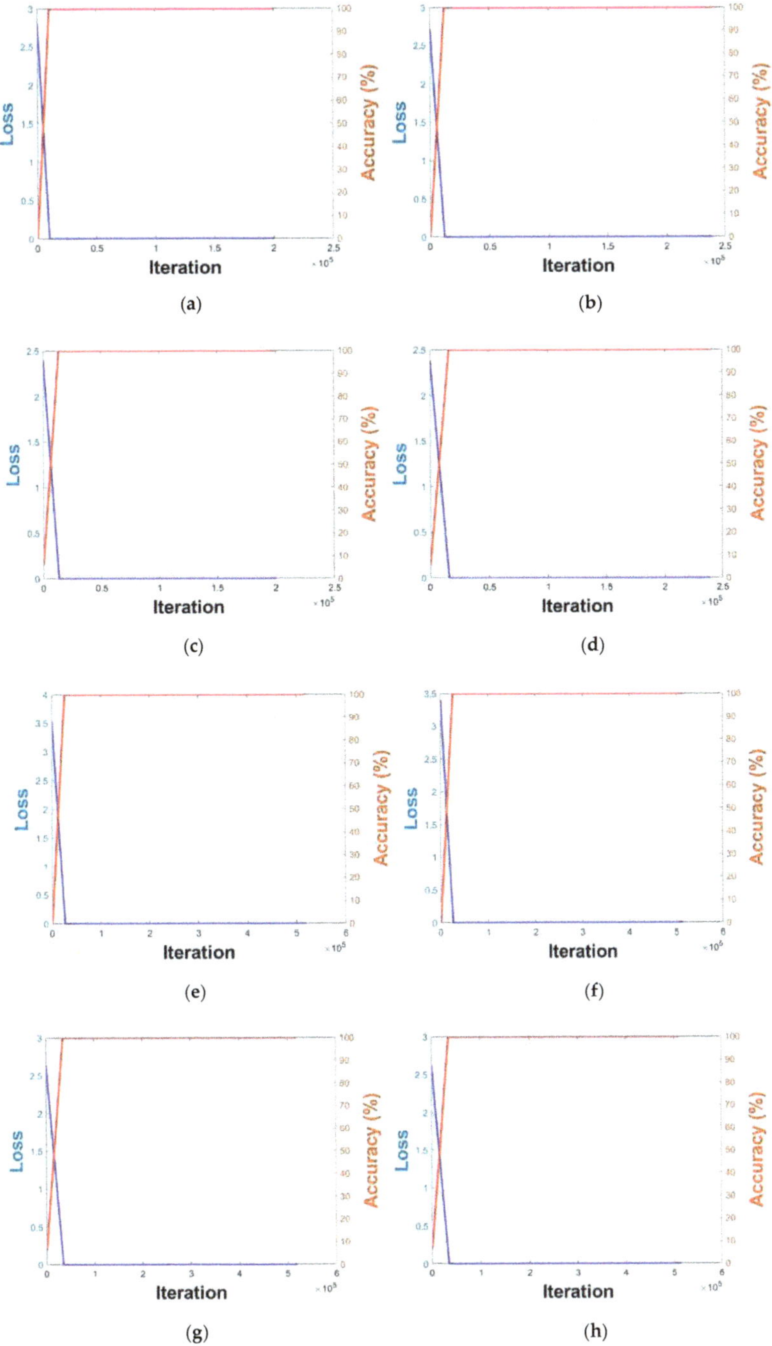

Figure 6. Plots depicting training loss and accuracy of DFB-DB2 ((**a**)–(**d**)) and ChokePoint dataset ((**e**)–(**h**)). Visual geometry group (VGG) face net-16 was used in the case of (**a**,**e**), the 1st fold was used for (**b**,**f**), the 2nd fold ResNet-50 was used in the case of (**c**), 1st fold in the case of (**g**), and the 2nd fold in the case of (**d**,**h**).

5.3. Testing Results from DeblurGAN and CNN Model

For comparing the original image and deblurred image during the deblurring process, signal-to-noise ratio (SNR) [34], peak signal-to-noise ratio (PSNR) [35], and structural similarity (SSIM) [36] can be used. However, the aforementioned methods, such as SNR, PSNR, and SSIM, cannot be compared with the proposed method because the blur or noise in the blurring images used in this study was naturally generated during the acquisition of the data as opposed to artificial generation of blur or noise in the original image.

5.3.1. Testing with CNN Model for DFB-DB2

Two-fold cross validation was performed to test the training CNN model. For a face image, 4096 features were obtained from the 7th fully connected layer of VGG face net-16. For a body image, 2048 features were obtained from the average pooling layer of ResNet-50. Given the features obtained from the CNN model, the image feature geometric center was calculated by using the Euclidean distance to determine the gallery image. The authentic and imposter distance was calculated by finding the normalized Euclidean distance between the gallery image and other probe images. The distance was used to calculate the equal error rate (EER).

Ablation Study

The performance of DFB-DB2 was compared with or without DeblurGAN. Here, "without DeblurGAN" means that both the procedures of focus score checking and DeblurGAN were not operated, whereas "with DeblurGAN" represents that both the procedures of focus score checking and DeblurGAN were adopted. The same DFB-DB2 and ChokePoint dataset were used for the experiment, while VGG face net-16 and ResNet-50 were used for the CNN model. The values in Tables 5 and 6 show that the recognition performance was improved after using DeblurGAN because there was a reduction in the number of changes in pixels between the original image and image generated after using DeblurGAN.

Table 5. Comparison of equal error rate (EER) for face recognition and body recognition on DFB-DB2 without or with DeblurGAN (unit: %).

Method		Face	Body
Without DeblurGAN (without focus score checking)	1st fold	45.29	33.68
	2nd fold	41.85	28.7
	Average	43.52	31.19
With DeblurGAN (with focus score checking)	1st fold	12.44	21
	2nd fold	7.94	16.48
	Average	10.19	18.74

Table 6. Comparison of EER for score-level fusion on DFB-DB2 without or with DeblurGAN (unit: %).

Method		Score-Level Fusion	
		Weighted Sum	Weighted Product
Without DeblurGAN (without focus score checking)	1st fold	32.44	32.57
	2nd fold	27.94	27.93
	Average	30.19	30.25
With DeblurGAN (with focus score checking)	1st fold	9.835	9.901
	2nd fold	5.552	5.557
	Average	7.694	7.729

As shown in Figure 7, the performance of 'with DeblurGAN (Face)' and 'with DeblurGAN (Body)' was improved. Face and body refer to face images and body images, respectively. Based on the score-level fusion approach, the weighted sum method exhibited a better performance than the weighted product method.

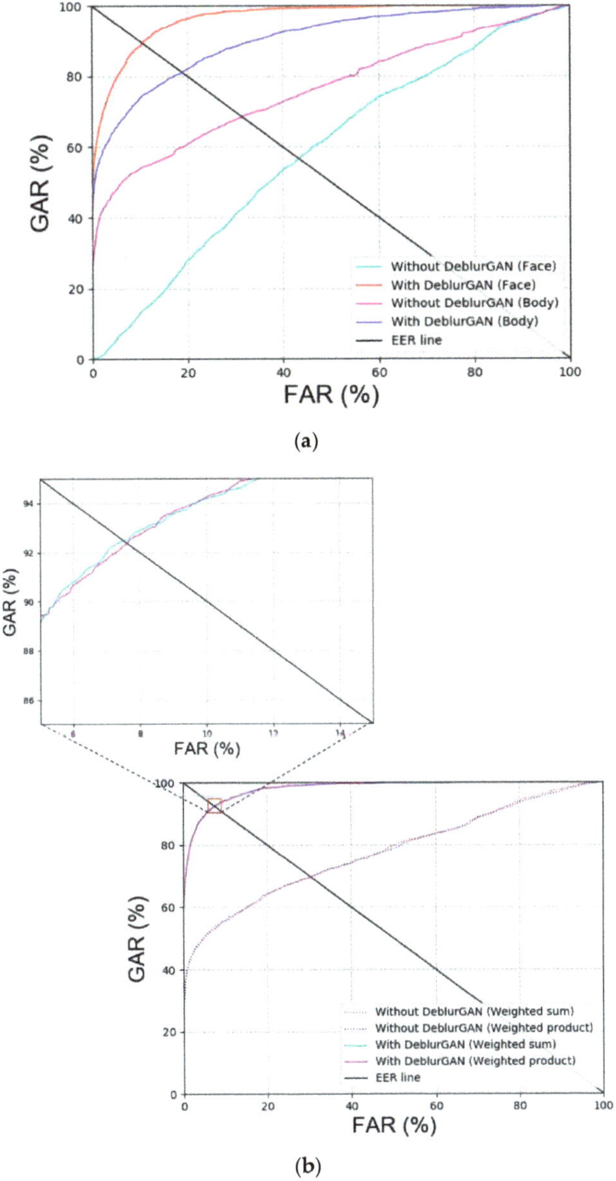

Figure 7. Receiver operating characteristic (ROC) curves with or without DeblurGAN performance of DFB-DB2 (**a**) face and body recognition result and (**b**) score-level fusion result.

Comparison between Previous Method and Proposed Methods

First, blur restoration is performed using other GAN methods besides DeblurGAN, which was proposed in this study for comparison. Specifically, CycleGAN [20], Pix2pix [22], attention-guided GAN (AGGAN) [37,38], and DeblurGAN version 2 (DeblurGANv2) [39] were used for GAN models. Table 7 and Figure 8 show the comparison results of GAN for DFB-DB2, and our method outperforms the state-of-the-art methods. As shown in Table 7, the recognition performance of CycleGAN, which restored the body image in DFB-DB2, was outstanding because DeblurGAN is a CGAN type method wherein the input image and target image are paired. However, when the target image is composed in this study, only the image that is similar to the input image is used for restoration. Therefore, the background, texture of clothes, and the individual's gait can be different, and this makes the restoration more difficult.

Table 7. Comparisons of EER for recognition by proposed method with those by other GAN-based methods in DFB-DB2 (unit: %).

Method		Average
Face	Proposed method	10.19
	CycleGAN [20]	13.255
	Pix2pix [22]	13.315
	AGGAN [37,38]	23.01
	DeblurGANv2 [39]	15.64
Body	Proposed method	18.74
	CycleGAN [20]	15.745
	Pix2pix [22]	23.195
	AGGAN [37,38]	22.52
	DeblurGANv2 [39]	22.71
Weighted sum	Proposed method	7.694
	CycleGAN [20]	8.29
	Pix2pix [22]	11.49
	AGGAN [37,38]	14.649
	DeblurGANv2 [39]	11.801
Weighted product	Proposed method	7.729
	CycleGAN [20]	8.41
	Pix2pix [22]	11.5605
	AGGAN [37,38]	14.342
	DeblurGANv2 [39]	11.869

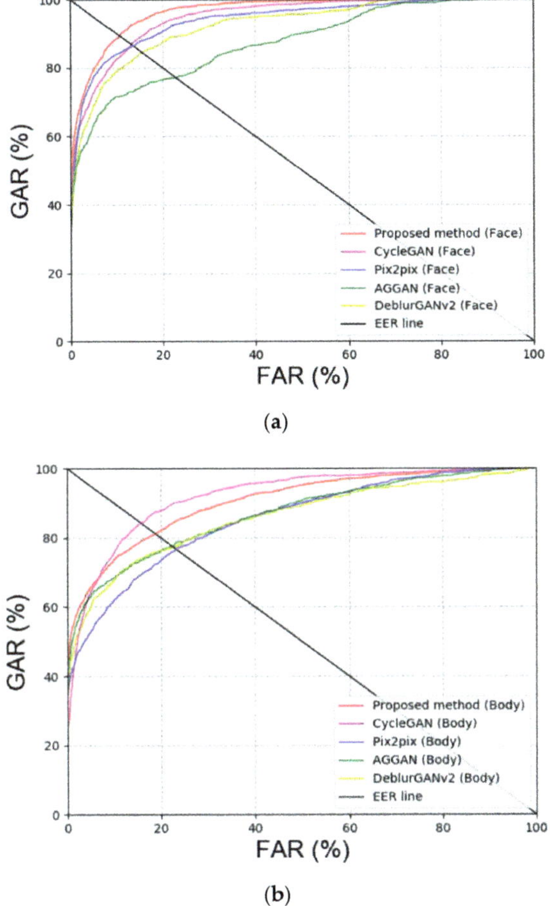

(**a**)

(**b**)

Figure 8. *Cont.*

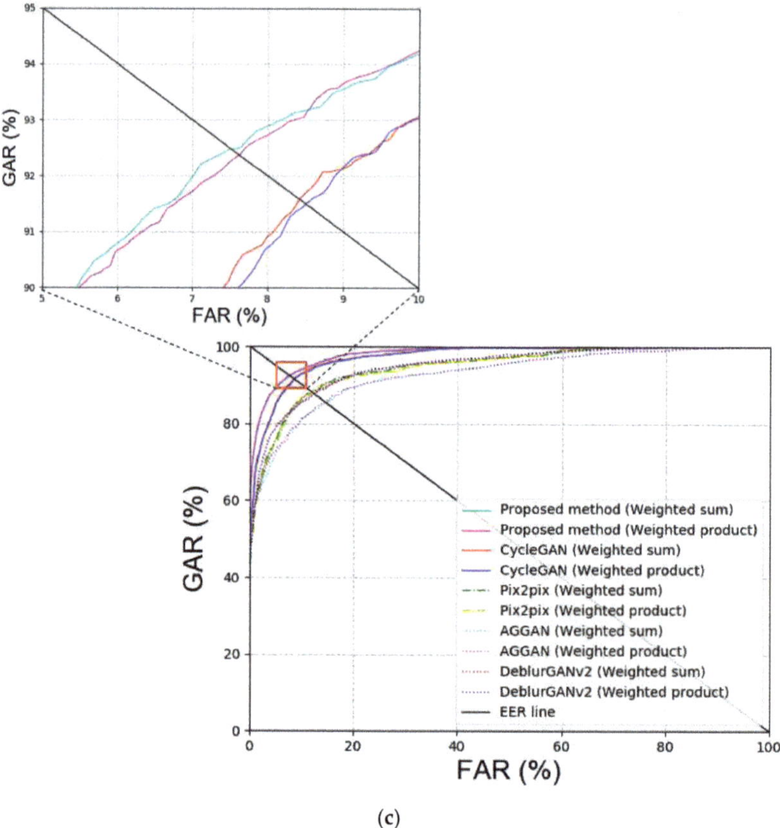

(c)

Figure 8. ROC curves via proposed and other GAN methods in DFB-DB2 (**a**,**b**) face and body image recognition results and (**c**) score-level fusion result.

Second, the experiment was conducted to compare face and face and body recognition. The experiment to compare face recognition was conducted with VGG face net-16 [40] and ResNet-50 [41,42]. Multi-level local binary pattern (MLBP) + PCA [43,44], histogram of gradient (HOG) [45], local maximal occurrence (LOMO) [46] and ensemble of localized features (ELF) [47] were used for the experiment to compare face and face and body recognition. Table 8 summarizes the comparison results of face recognition, and Table 9 summarizes the comparison results of face and face and body recognition. Figure 9 shows the receiver operating characteristic (ROC) curve of the results in Tables 8 and 9.

Table 8. Comparison of EER for the results of the proposed method and previous face recognition methods (unit: %).

Method	1st Fold	2nd Fold	Average
Proposed method	9.835	5.552	7.694
VGG face net-16 [40]	17.44	16.71	17.075
ResNet-50 [41,42]	13.96	14.06	14.01

Table 9. Comparison of EER for the results of the proposed and previous face and body recognition methods (unit: %).

Method	1st Fold	2nd Fold	Average
Proposed method	9.835	5.552	7.694
MLBP + PCA [43,44]	29.38	27.84	28.61
HOG [45]	38.09	44.14	41.12
LOMO [46]	21.98	23.4	22.69
ELF [47]	20.91	23.87	22.39

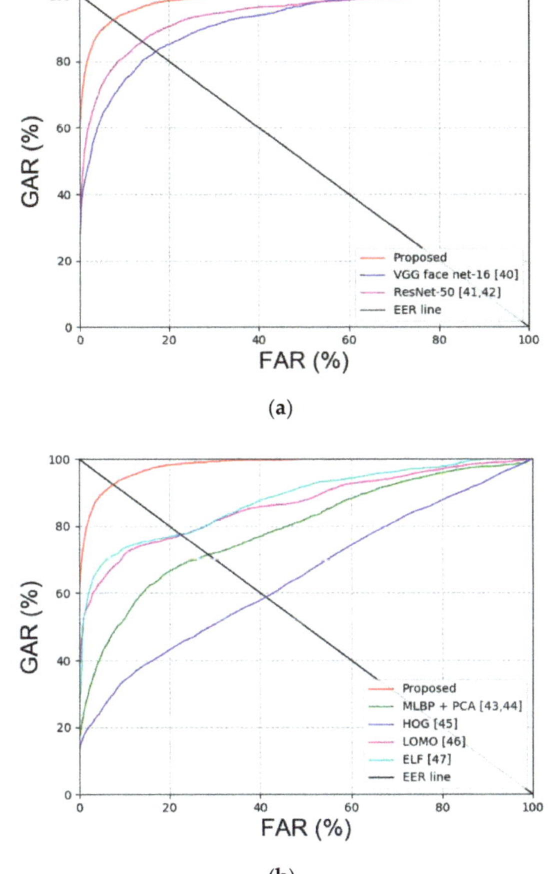

Figure 9. ROC curves obtained via comparing proposed and the state-of-art-methods. (**a**) Face image recognition results and (**b**) face and body image recognition results.

Third, the accuracy of recognition was evaluated via the cumulative match characteristic (CMC) curve. Figure 10 shows the comparison results of the proposed method and methods in Tables 8 and 9. The horizontal axis corresponds to the rank, and the vertical axis corresponds to the genetic acceptance rate (GAR) accuracy for each rank. Table 4 shows that the DFB-DB2 consists of 11 individuals, as shown in Figure 10.

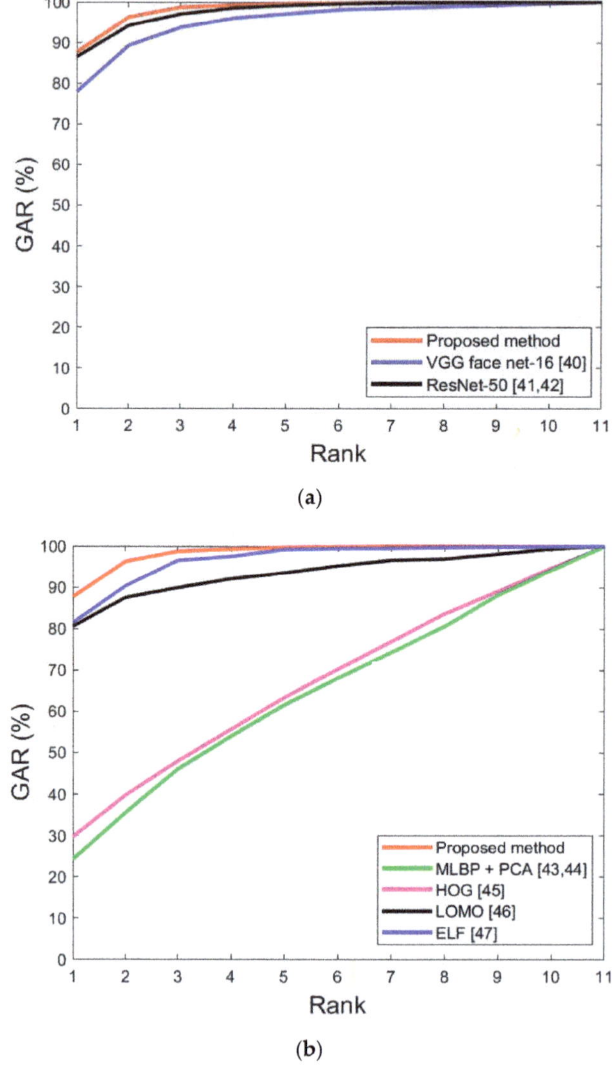

Figure 10. Cumulative match characteristic (CMC) curves of the proposed and previous methods on DFB-DB2. (**a**) Face image recognition results by the proposed and previous methods and (**b**) face and body image recognition results via the proposed and previous methods.

Figure 11 shows the difference in the performance by measuring the Cohen's d-value and t-test results of face recognition and face and body recognition and comparisons with the proposed method. With respect to face recognition, the difference in the Cohen's d-value between the proposed method and ResNet-50 [41,42] was 2.95. This significantly exceeds the effect size of 0.8 and is thus high. The p-value of the t-test is approximately 0.098, which differs from the proposed method by 99.902%. With respect to face and body recognition, the Cohen's d-value and t-test results were measured for the ELF [47] that exhibited the second-best performance when compared to that of the proposed method with a Cohen's d-value of 5.65. This exhibited a large effect size, and the t-test exhibited a difference of 99.97%.

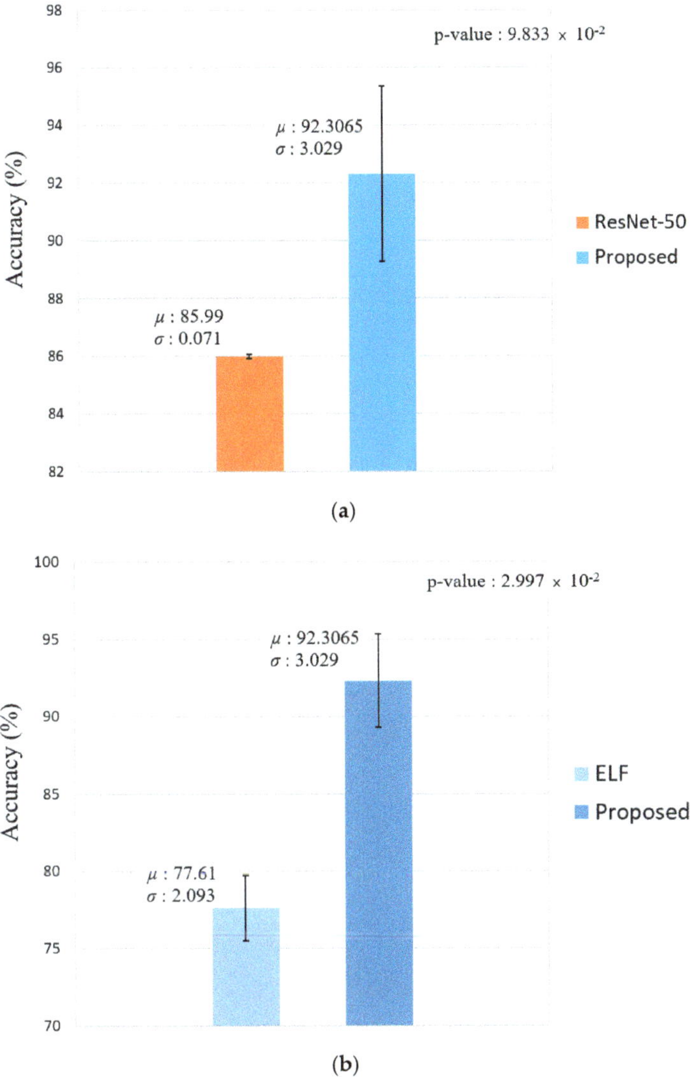

Figure 11. T-test performance of our proposed method and the second-best model in terms of average accuracy. (**a**) Comparison of the proposed method and ResNet-50 and (**b**) comparison of the proposed method and ensemble of localized features (ELF).

The false acceptance ratio (FAR), false rejection ratio (FRR), and correct case of the previous experimental results are analyzed in the plots. Figure 12 illustrates different cases, in which the image on the left corresponds to the enrolled image, and the image on the right corresponds to the probe image. The portion in the red box of the image on the right is restored via DeblurGAN.

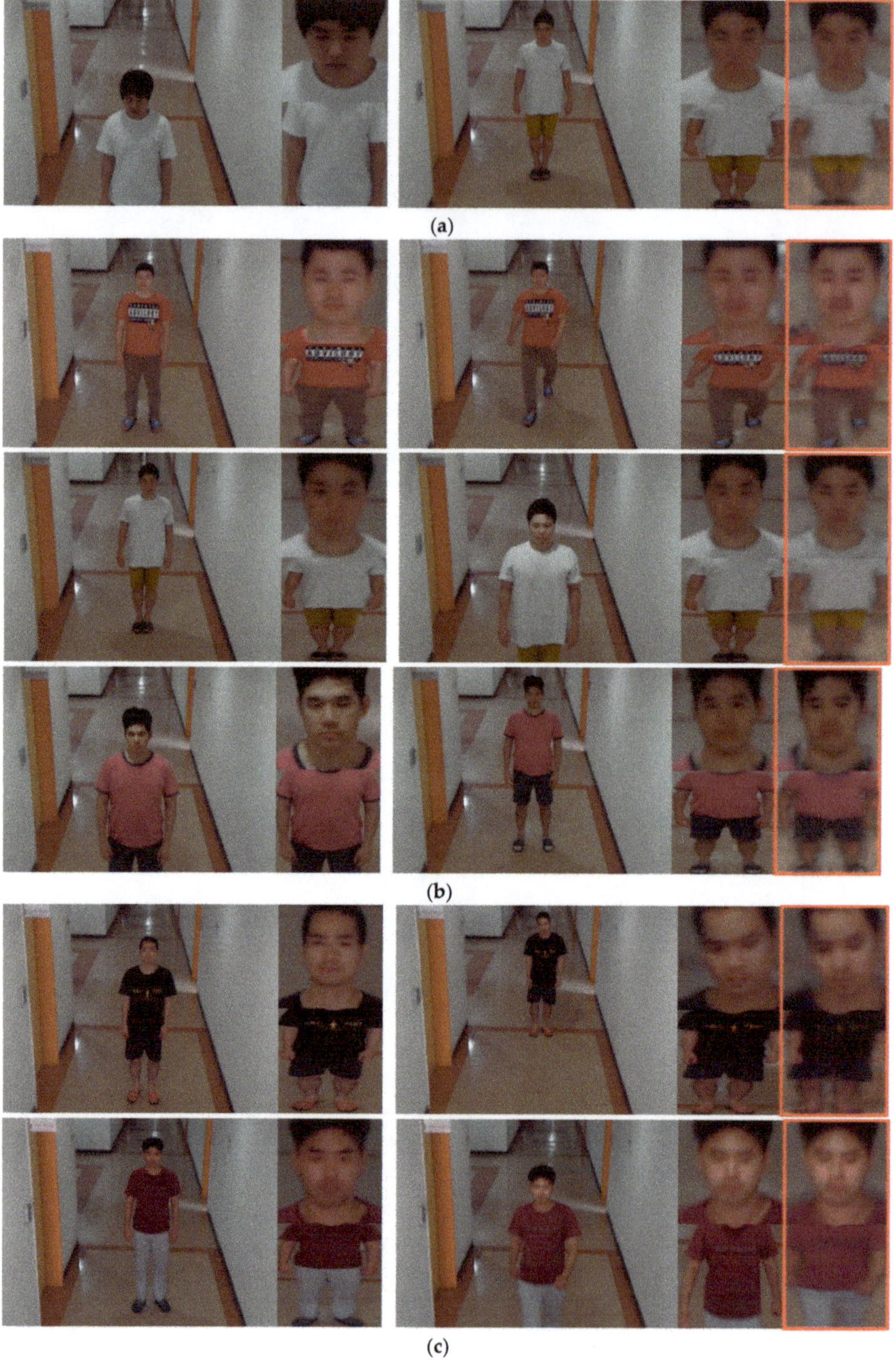

Figure 12. Cases of false acceptance (FA), false rejection (FR), and correction recognition (**a**)–(**c**) in DFB-DB2. (**a**) FA cases, (**b**) FR cases, and (**c**) cases of correct recognition.

5.3.2. Class Activation Map

Subsequently, we analyzed the class activation feature map of VGG face net-16 and ResNet-50 that were used for the DFB-DB2 to evaluate the recognition performance for face and body images. Figure 13 shows the class activation feature map from a specific layer using Grad-CAM method [48]. Furthermore, the important features shown through the distribution. Figure 13a,d,g,j correspond to the input face and body images of the CNN model, and Figure 13b,c,e,f,h,i,k,l show the class activation feature map results of face and body images.

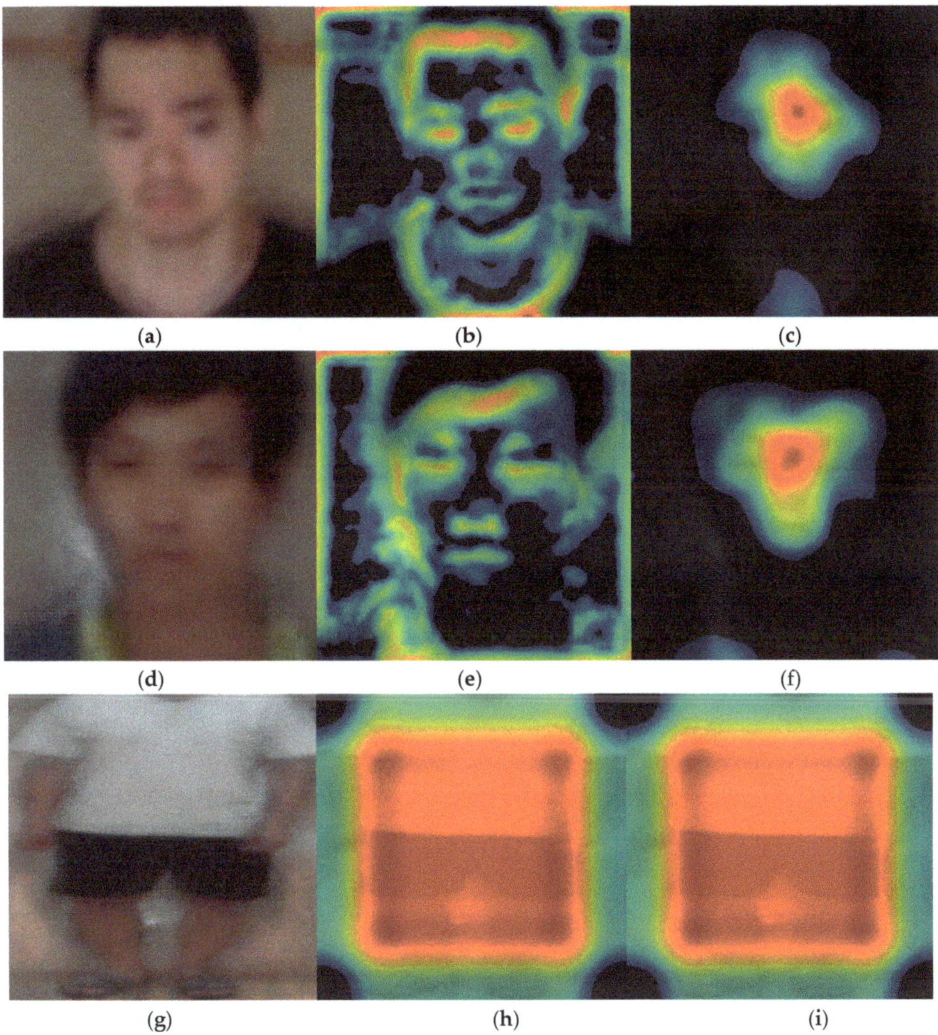

Figure 13. *Cont.*

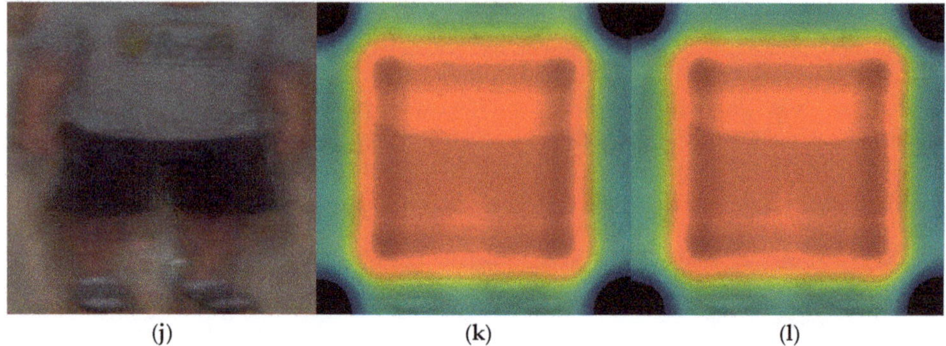

(j)   (k)   (l)

Figure 13. Results on class activation feature map on DFB-DB2. (**a**,**d**) Input face images, (**b**,**c**,**e**,**f**) results via VGG face net-16 in rectified linear units (ReLU) layer, (**b**,**e**) images from 7th ReLU layer, (**c**,**f**) images from 13th ReLU layer, (**g**,**j**) input body images, (**h**,**i**,**k**,**l**) results via ResNet-50 in batch normalized layer, (**h**,**i**) images from last batch normalized layer on conv5 2nd block, (**k**,**l**) images from last batch normalized layer on conv5 3rd block.

Specifically, when the input (a) is processed through VGG face net-16, (b) corresponds to the class activation feature map of the 7th ReLU layer, and (c) corresponds to the class activation feature map of the 13th ReLU layer. The image in (c) shows the distribution focused around the face area where the red color represents the main feature, while the blue color represents less important features. The black color indicates that no features were detected. When the process goes from (b) to (c), the features are more focused around the face region. Additionally, body images were extracted from the batch normalized layer. In contrast to the face image results, the main features were observed around the body region because the trained part of the ResNet-50 model considers information with respect to the individual's body and clothes as important features.

5.3.3. Testing with CNN Model for ChokePoint Dataset

Ablation Study

The images restored with DeblurGAN and images with a score exceeding the threshold value were combined in the experiment, as proposed in the study. Based on the results in Tables 10 and 11, the weighted sum method, among the score-level fusion methods, exhibited better results. Figure 14 shows the results in Tables 10 and 11 in the form of plots. As shown in the plots in Figure 14, the recognition performance improves when DeblurGAN is applied. Furthermore, the weighted product method exhibited better results among score-level fusion methods.

Table 10. Comparison of EER for face recognition and body recognition on ChokePoint dataset without or with DeblurGAN (unit: %).

Method		Face	Body
Without DeblurGAN (without focus score checking)	1st fold	11.76	18.50
	2nd fold	8.09	18.46
	Average	9.925	18.48
With DeblurGAN (with focus score checking)	1st fold	7.05	15.97
	2nd fold	6.39	15.2
	Average	6.72	15.585

Table 11. Comparison of EER for score-level fusion on ChokePoint dataset without or with DeblurGAN (unit: %).

Method		Score-Level Fusion	
		Weighted Sum	**Weighted Product**
Without DeblurGAN (without focus score checking)	1st fold	9.84	9.79
	2nd fold	6.55	6.43
	Average	8.195	8.11
With DeblurGAN (with focus score checking)	1st fold	5.162	5.163
	2nd fold	4.99	4.975
	Average	5.076	5.069

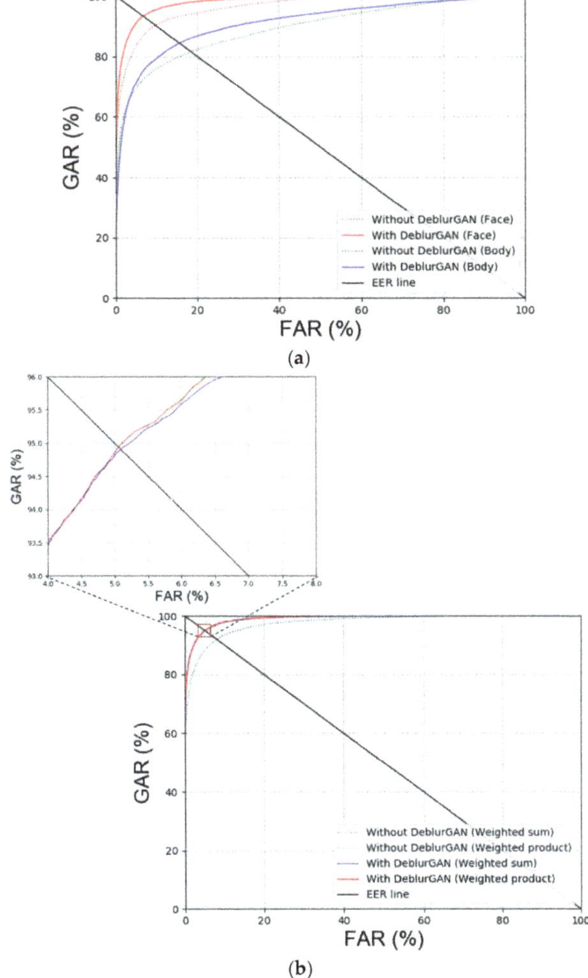

Figure 14. ROC curves with and without DeblurGAN performance on ChokePoint dataset. (**a**) Face and body image recognition results and (**b**) score-level fusion result.

Comparison between Previous Methods and Proposed Method

With respect to the GAN models for blur image restoration, the performance of CycleGAN and DeblurGAN was compared. Table 12 and Figure 15 show the results and plots, respectively. The results indicated that DeblurGAN exhibited better recognition performance than CycleGAN.

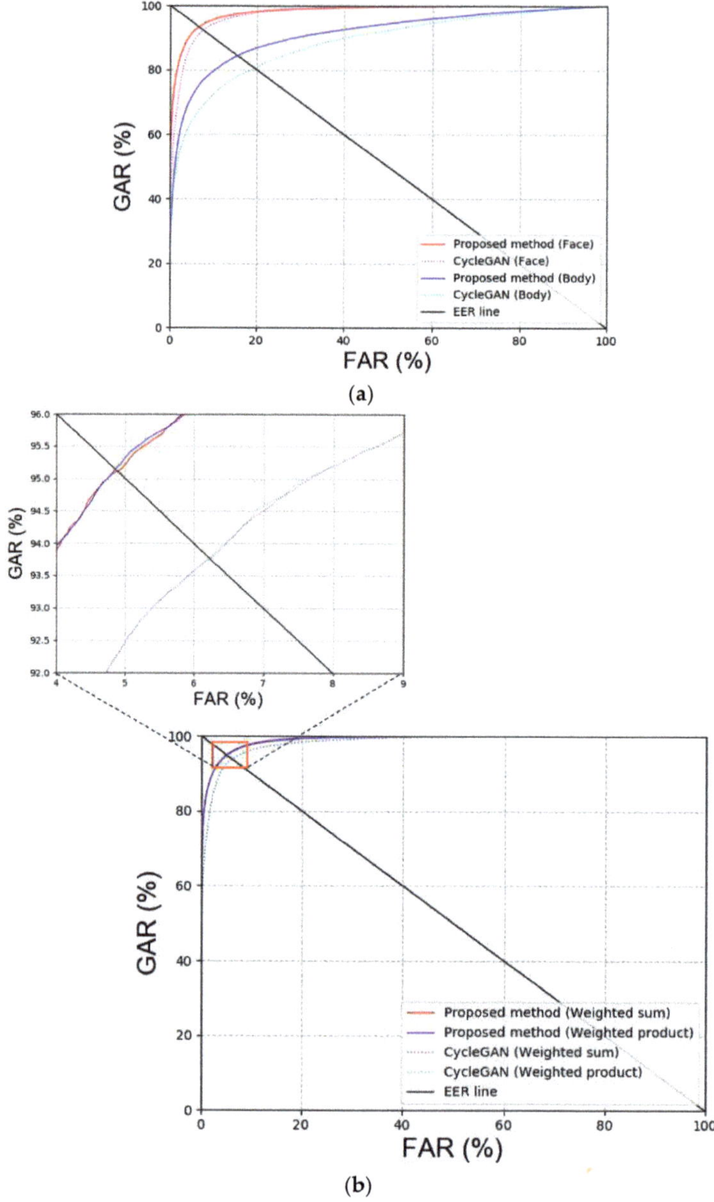

Figure 15. ROC curves for the proposed and CycleGAN in ChokePoint dataset. (**a**) Face and body recognition result and (**b**) score-level fusion result.

Table 12. Comparisons of EER for recognition by proposed method with that by CycleGAN (unit: %).

	Method	1st Fold	2nd Fold	Average
Face	Proposed method	7.05	6.39	6.72
	CycleGAN [20]	9.05	6.43	7.74
Body	Proposed method	15.97	15.2	15.585
	CycleGAN [20]	18.41	20.41	19.41
Weighted sum	Proposed method	5.162	4.99	5.076
	CycleGAN [20]	7.05	5.331	6.1905
Weighted product	Proposed method	5.163	4.975	5.069
	CycleGAN [20]	7.023	5.362	6.1925

Second, the existing face recognition and face and face and body recognition methods were compared with the proposed method. Tables 13 and 14 show the experimental results, and Figure 16 illustrates the results in the plots.

Table 13. Comparison of EER for recognition results via the proposed method and previous face recognition methods (unit: %).

Method	1st Fold	2nd Fold	Average
Proposed method	5.163	4.975	5.069
VGG face net-16 [40]	49.29	48.25	48.77
ResNet-50 [41,42]	12.93	16.97	14.95

Table 14. Comparison of EER for recognition results via the proposed and previous face and body recognition methods (unit: %).

Method	1st Fold	2nd Fold	Average
Proposed method	5.163	4.975	5.069
MLBP + PCA [43,44]	37.75	42.38	40.07
HOG [45]	41.84	41.47	41.66
LOMO [46]	29.7	21.57	25.635
ELF [47]	31.93	23.94	27.935

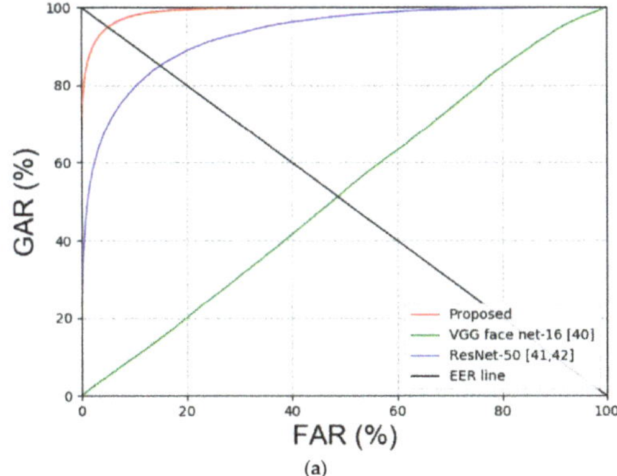

Figure 16. *Cont.*

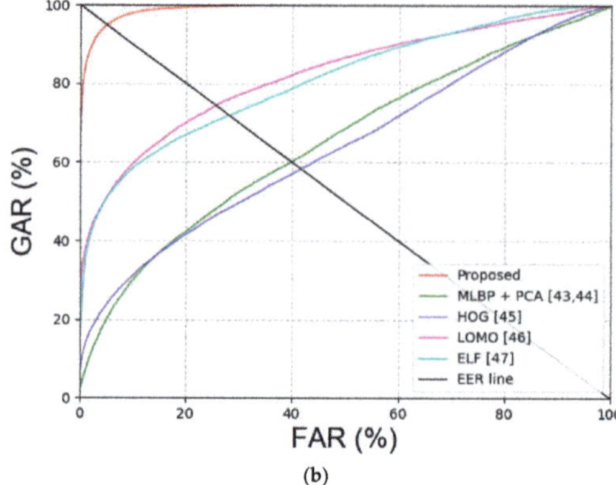

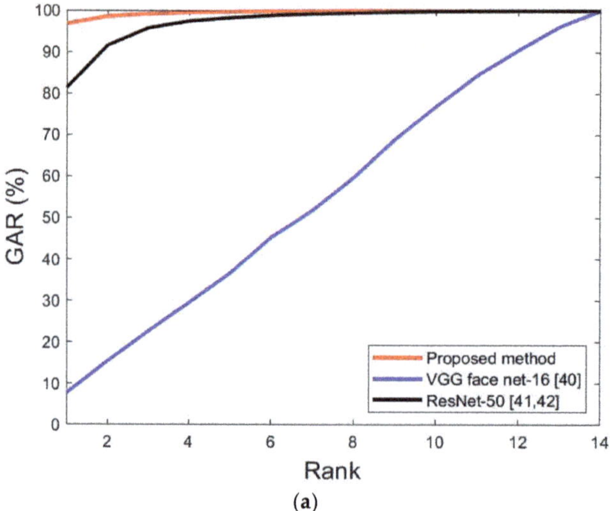

Figure 16. ROC curves for proposed and the state-of-art-method on ChokePoint dataset. (**a**) Face image results and (**b**) face and body image results.

Figure 17 shows the comparison of the CMC curve of the proposed method and previous methods for face and face and body recognition. As shown in Figure 17a,b, the performance of the proposed method exceeded that of other methods.

Figure 17. *Cont.*

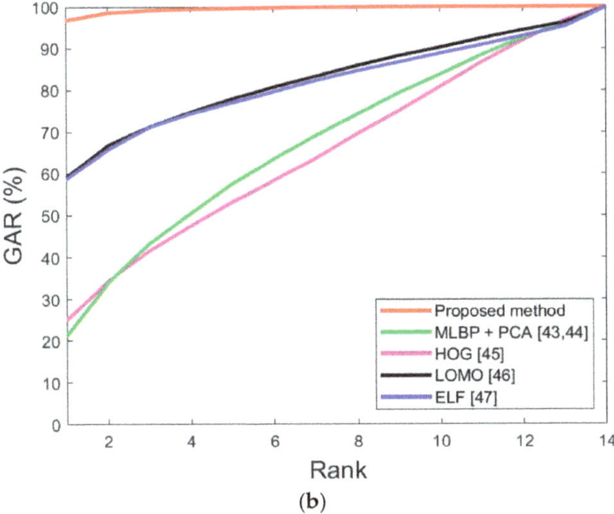

Figure 17. CMC curves for the proposed and previous method on ChokePoint dataset. (**a**) Face image recognition results via the proposed and previous methods; (**b**) face and body image recognition results for the proposed and previous methods.

The results of the proposed method using the ChokePoint dataset are shown for the cases of FAR, FRR, and correct recognition in Figure 18.

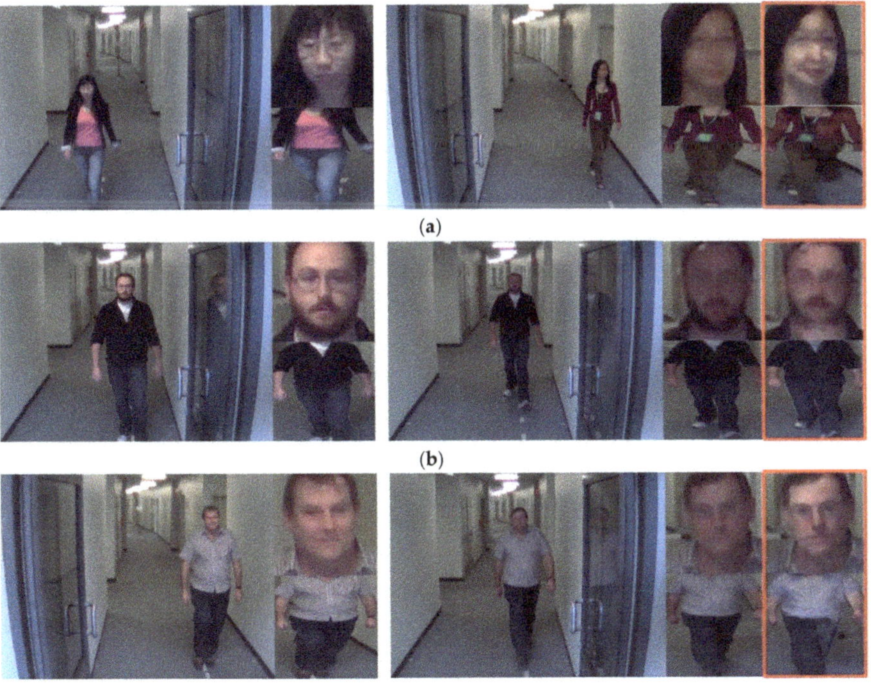

Figure 18. *Cont.*

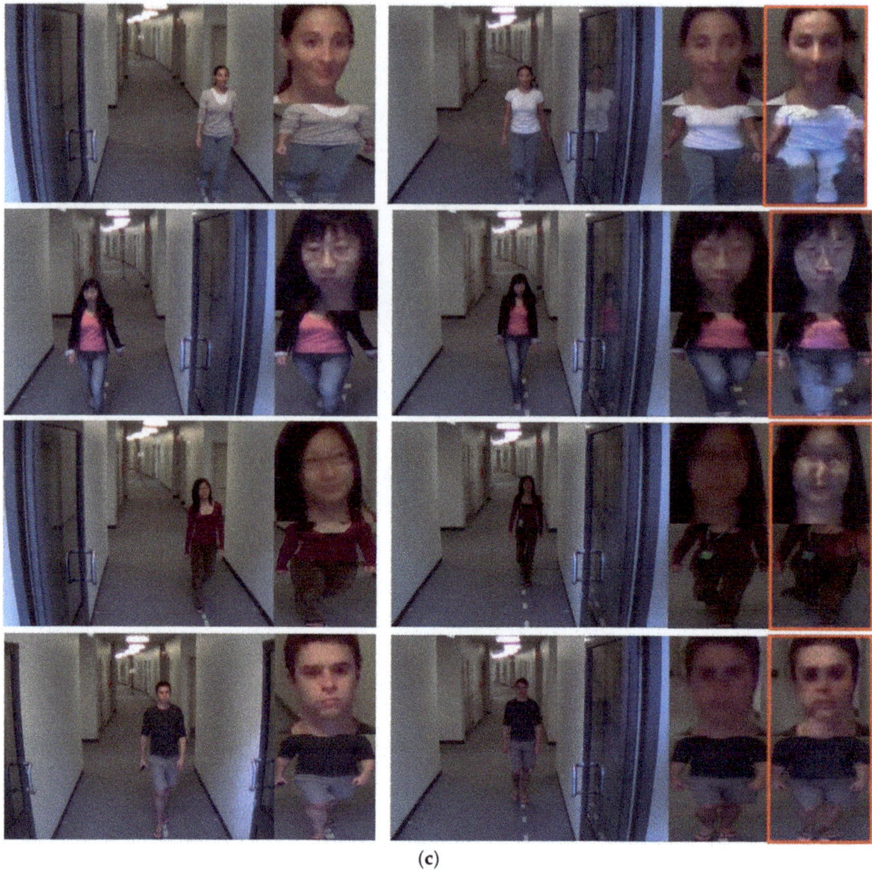

Figure 18. Cases corresponding to false acceptance (FA), false rejection (FR), and correction recognition (a)–(c) Cases from ChokePoint dataset (**a**) FA cases, (**b**) FR cases, and (**c**) cases of correct recognition.

Figure 19 shows the difference in the performance by measuring the Cohen's d-value and t-test results of face recognition and face and body recognition and comparison with the proposed method. With respect to face recognition, the Cohen's d-value between the proposed method and ResNet-50 [41,42] is 4.89, and this significantly exceeds the effect size of 0.8, thus its being high. The p-value of the t-test is approximately 0.03941, which differs from the proposed method by 99.961%. With respect to face and body recognition, Cohen's d-value and t-test results were measured for the ELF [47] that exhibited the second-best performance when compared to that of the proposed method. The Cohen's d-value is 5.06, thereby exhibiting a large effect size, and the t-test exhibited a difference of 99.963%.

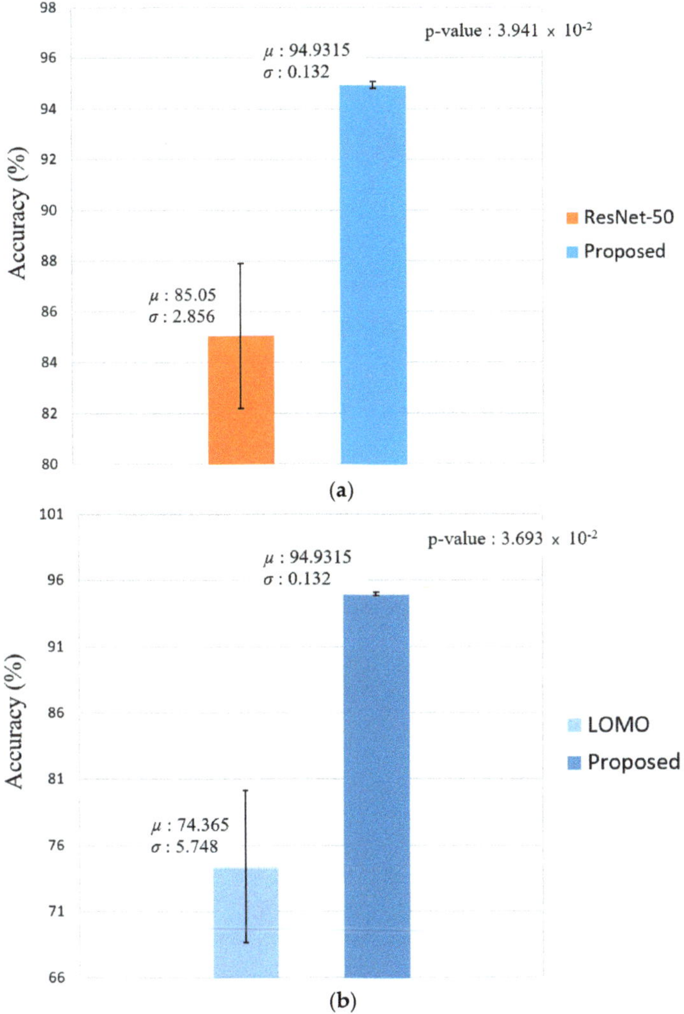

Figure 19. T-test performance of the proposed method and second-best model in terms of average accuracy. (**a**) Comparison of the proposed method and ResNet-50 and (**b**) comparison of the proposed method and local maximal occurrence (LOMO).

5.3.4. Class Activation Map

In the subsequent experiment, the class activation feature map of the ChokePoint dataset was examined. Figure 20 shows the class activation feature map results. The face image signifies the class activation feature map obtained from the ReLU layer of VGG face net-16. Figure 20h,i,k,l of Figure 20b,c,e,f body image represent the class activation feature map of the image that passed through the batch normalized layer. In the result of the images, the red color represents the main feature, and the blue color represents less important features. Similar results to the experiment using DFB-DB2 are obtained in Figure 20.

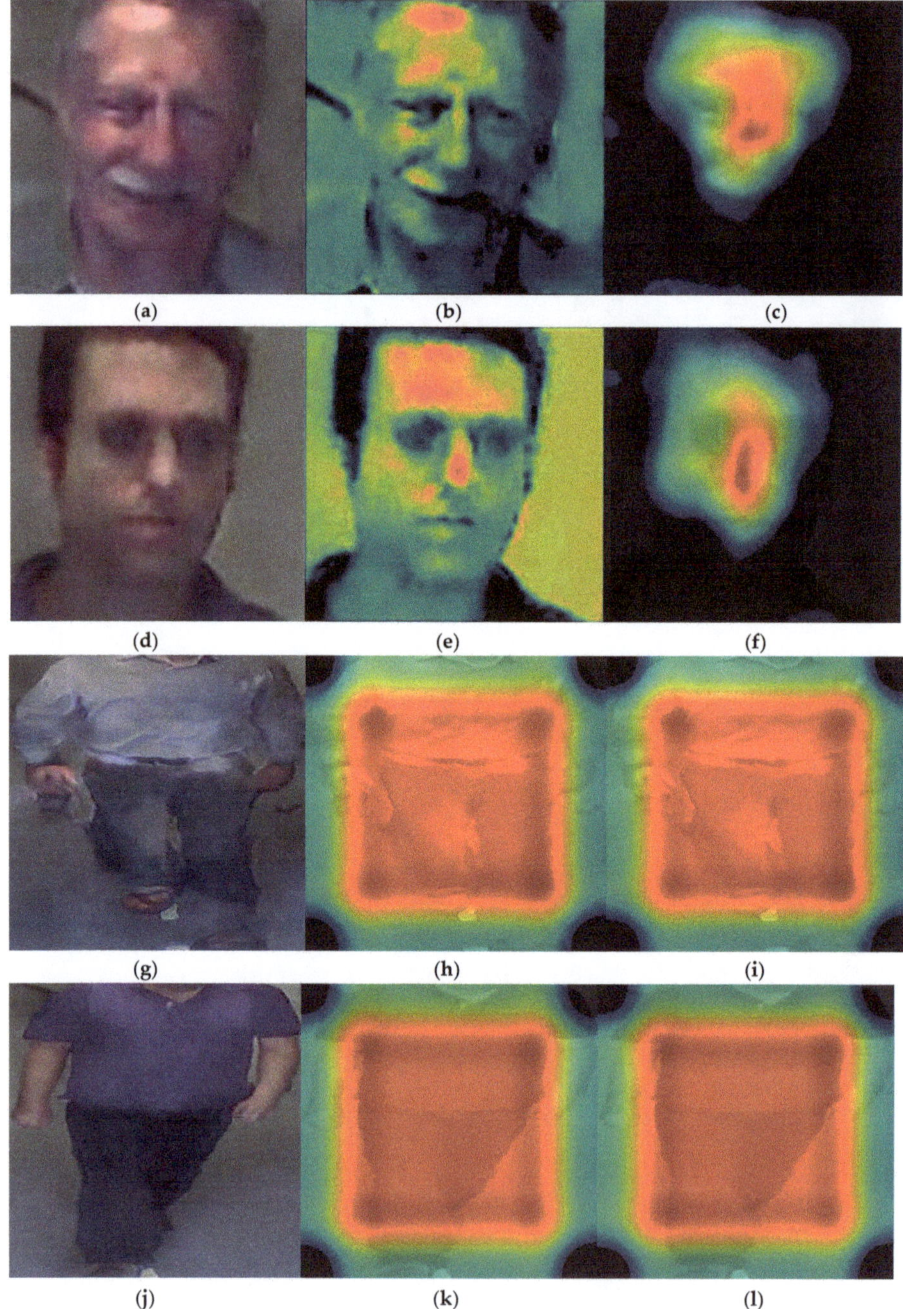

Figure 20. Result on class activation feature map on ChokePoint dataset. (**a,d**) Input face images, (**b,c,e,f**) results for the VGG face net-16 in ReLU layer, (**b,e**) images from 7th ReLU layer, (**c,f**) images from 13th ReLU layer, (**g,j**) input body images, (**h,i,k,l**) results for the ResNet-50 in batch normalized layer, (**h,i**) images from the last batch normalized layer on conv5 2nd block, and (**k,l**) images from the last batch normalized layer on conv5 3rd block.

5.3.5. Comparisons of Processing Time on Jetson TX2 and Desktop Computer

In the next experiment, the computing speed of the proposed method was compared using Jetson TX2 board [49] as shown in Figure 21 and a desktop computer including NVIDIA GeForce GTX 1070 graphic processing unit (GPU) card. Jetson TX2 board is an embedded system equipped with NVIDIA Pascal™ GPU architecture with 256 NVIDIA CUDA cores, 8 GB 128-bit LPDDR4 memory, and dual-core NVIDIA Denver 2 64-Bit CPU. The power consumption is less than 7.5 watts. The proposed method was ported with Keras [50] and TensorFlow [51] in Ubuntu 16.04 OS. The versions of the installed framework and library include Python 3.5 and TensorFlow 1.12; NVIDIA CUDA® toolkit [52] and NVIDIA CUDA® deep neural network library (CUDNN) [53] versions are 9.0 and 7.3, respectively.

Figure 21. Jetson TX2 embedded system.

As shown in Tables 15 and 16, our method requires the time cost of a total of 75.72 ms and 481.7 ms on desktop computer and Jetson TX2 embedded system, respectively, which means that our method can be operates at the speed of 13.2 frames/s (1000/75.72) and 2.08 frames/s (1000/481.7) on desktop computer and Jetson TX2 embedded system, respectively. The Jetson TX2 embedded system has less computing resource and GPU of lower speed compared to those in the desktop computer. Therefore, the processing speed on Jetson TX2 is slower than that on desktop computer. However, more advanced and cheaper GPU card and embedded GPU system have been fast commercialized, and our method can be operated at faster speed on those systems.

Table 15. Comparison of processing time on Jetson TX2 and desktop computer by DeblurGAN (unit: ms).

Platform	DeblurGAN
Desktop computer	32
Jetson TX2	349

Table 16. Comparison of processing time on Jetson TX2 and desktop computer by VGG face net-16 and ResNet-50 (unit: ms).

Platform	VGG Face Net-16	ResNet-50	Total
Desktop computer	24.8	18.92	43.72
Jetson TX2	91.9	40.8	132.7

6. Conclusions

There were lots of works that use GAN for deblur [38,39,54–56]. However, most previous works aimed at the visibility enhancement of general scene images, whereas the main purpose of our research is to enhance the recognition accuracy of face and body images. In the previous works, GAN tried to generate the image of high visibility and distinctiveness although limited amount of noise is additionally included in the generated image. However, GAN in our research tries to generate the face and body images with which the higher recognition accuracies can be obtained. It means that the maximization of intra-class consistency (from matching between same people) and inter-class variation (from matching between different people) in the generated image is more important than the visibility enhancement in our GAN. Therefore, we compared the recognition accuracies of face and body images by our GAN with those by other GANs, as shown in Tables 7 and 12 and Figures 8 and 15, instead of the metrics showing the image visibility, such as peak signal-to-noise ratio (PSNR) and structural similarity (SSIM), like previous works [38,39,54–56]. Consequently, it is not appropriate to use our method to handle the natural images.

The study proposed a deep CNN-based recognition method involving a score-level fusion approach for face and body images in which a GAN is applied to restore the blur problem that is generated when body recognition data is obtained in indoor environments from a long distance. Previous studies focused on minimizing a blur if discovered in face images although deblurring is typically omitted for body images because detailed information is considered as absent in body images when compared to the face images. However, the blur problem in body images affects recognition performance. To solve the problem, face images and body images were separated, and a blur was then restored using a GAN model in the study. Higher processing time is obtained if restoration is performed independently for face and body images using a GAN model. However, better restoration of distinctive features of face and body is observed. For impartial comparison experiments, the GAN model was used for restoration, VGG face net-16 and ResNet-50 were used for training in the study, and the DFB-DB2 built by the researchers was disclosed.

In future work, we would research about the advanced GAN model which can process the face and body images simultaneously. For that, we also consider the scheme of pre-classification of input image into face and body image, as well as adopting different loss functions according to input image. In addition, we would study the combined structure of GAN and recognition CNN models for the reduction of training time, and the measures to increase the processing speed of an embedded system would be explored via examining a lighter GAN for deblurring. Furthermore, our deblur-based recognition method would be applied to various biometric systems, including iris and finger-vein to evaluate recognition performance.

Author Contributions: J.H.K. and K.R.P. designed the face and body-based recognition system based on two CNNs and GAN model. S.W.C. and N.R.B. helped to experiments and analyzed results, and collecting databases. All authors have read and agreed to the published version of the manuscript.

Funding: This research was supported by the Ministry of Education and Ministry of Science and ICT, Korea.

Acknowledgments: This research was supported by the Basic Science Research Program through the National Research Foundation of Korea (NRF) funded by the Ministry of Education (NRF-2018R1D1A1B07041921), in part by the Ministry of Science and ICT (MSIT), Korea, under the ITRC (Information Technology Research Center) support program (IITP-2020-2020-0-01789) supervised by the IITP (Institute for Information & communications Technology Promotion), and in part by the Bio and Medical Technology Development Program of the NRF funded by the Korean government, MSIT (NRF-2016M3A9E1915855).

Conflicts of Interest: The authors declare no conflict of interest.

References

1. Grgic, M.; Delac, K.; Grgic, S. SCface–surveillance cameras face database. *Multimed. Tools Appl.* **2011**, *51*, 863–879. [CrossRef]
2. Banerjee, S.; Das, S. Domain adaptation with soft-Margin multiple feature-kernel learning beats deep learning for surveillance face recognition. *arXiv* **2016**, arXiv:1610.01374v2.
3. Zhou, X.; Bhanu, B. Integrating face and gait for human recognition at a distance in video. *IEEE Trans. Syst. Man Cybern. Part B-Cybern.* **2007**, *37*, 1119–1137. [CrossRef] [PubMed]
4. Varior, R.R.; Haloi, M.; Wang, G. Gated siamese convolutional neural network architecture for human re-identification. In Proceedings of the 14th European Conference on Computer Vision, Amsterdam, The Netherlands, 8–16 October 2016; pp. 791–808.
5. Nguyen, D.T.; Hong, H.G.; Kim, K.W.; Park, K.R. Person recognition system based on a combination of body images from visible light and thermal cameras. *Sensors* **2017**, *17*, 605. [CrossRef] [PubMed]
6. Shi, H.; Yang, Y.; Zhu, X.; Liao, S.; Lei, Z.; Zheng, W.; Li, S.Z. Embedding deep metric for individual re-identification: A study against large variations. In Proceedings of the 14th European Conference on Computer Vision, Amsterdam, The Netherlands, 8–16 October 2016; pp. 732–748.
7. Liu, Z.; Sarkar, S. Outdoor recognition at a distance by fusing gait and face. *Image Vision. Comput.* **2007**, *25*, 817–832. [CrossRef]
8. Hofmann, M.; Schmidt, S.M.; Rajagopalan, A.N.; Rigoll, G. Combined face and gait recognition using Alpha Matte preprocessing. In Proceedings of the 5th IAPR International Conference on Biometrics, New Delhi, India, 29 March–1 April 2012; pp. 390–395.
9. Koo, J.H.; Cho, S.W.; Baek, N.R.; Kim, M.C.; Park, K.R. CNN-based multimodal human recognition in surveillance environments. *Sensors* **2018**, *18*, 3040. [CrossRef] [PubMed]
10. Kang, B.J.; Park, K.R. A robust eyelash detection based on iris focus assessment. *Pattern Recognit. Lett.* **2007**, *28*, 1630–1639. [CrossRef]
11. Alaoui, F.; Ghlaifan Abdo Saleh, A.; Dembele, V.; Nassim, A. Application of blind deblurring algorithm for face biometric. *Int. J. Comput. Appl.* **2014**, *105*, 20–24.
12. Hadid, A.; Nishiyama, M.; Sato, Y. Recognition of blurred faces via facial deblurring combined with blur-tolerant descriptors. In Proceedings of the 20th IAPR International Conference on Pattern Recognition, Istanbul, Turkey, 23–26 August 2010; pp. 1160–1163.
13. Nishiyama, M.; Takeshima, H.; Shotton, J.; Kozakaya, T.; Kozakaya, O. Facial deblur inference to improve recognition of blurred faces. In Proceedings of the IEEE Conference on Computer Vision and Pattern Recognition, Miami Beach, FL, USA, 22–24 June 2009; pp. 1115–1122.
14. Mokhtari, H.; Belaidi, I.; Said, A. Performance comparison of face recognition algorithms based on face image retrieval. *Res. J. Recent Sci.* **2013**, *2*, 65–73.
15. Heflin, B.; Parks, B.; Scheirer, W.; Boult, T. Single image deblurring for a real-time face recognition system. In Proceedings of the IECON 2010-36th Annual Conference on IEEE Industrial Electronic Society, Glendale, AZ, USA, 7–10 November 2010; pp. 1185–1192.
16. Yasarla, R.; Perazzi, F.; Patel, V.M. Deblurring face images using uncertainty guided multi-stream semantic networks. *arXiv* **2020**, arXiv:1907.13106v2. [CrossRef]
17. Peng, X.; Huang, Z.; Lv, J.; Zhu, H.; Zhou, J.T. Comic: Multi-view clustering without parameter selection. In Proceedings of the 36th International Conference on Machine Learning, Long Beach, CA, USA, 10–15 June 2019; pp. 5092–5101.
18. Huang, Z.; Zhou, J.T.; Peng, X.; Zhang, C.; Zhu, H.; Lv, J. Multi-view spectral clustering network. In Proceedings of the 28th International Joint Conference on Artificial Intelligence, Macao, China, 10–16 August 2019; pp. 2563–2569.
19. Dongguk Face and Body Database Version 2 (DFB-DB2) and CNN Models for Deblur and Face & Body Recognition. Available online: http://dm.dgu.edu/link.html (accessed on 11 March 2020).
20. Zhu, J.-Y.; Park, T.S.; Isola, P.; Efros, A.A. Unpaired image-to-image translation using cycle-consistent adversarial networks. In Proceedings of the IEEE International Conference on Computer Vision, Venice, Italy, 22–29 October 2017; pp. 2242–2251.

21. Ronneberger, O.; Fischer, P.; Brox, T. U-net: Convolutional networks for biomedical image segmentation. In Proceedings of the International Conference on Medical Image Computing Computer-Assisted Intervention, Munich, Germany, 5–9 October 2015; pp. 234–241.

22. Isola, P.; Zhu, J.-Y.; Zhou, T.; Efros, A.A. Image-to-image translation with conditional adversarial networks. In Proceedings of the IEEE Conference on Computer Vision and Pattern Recognition, Honolulu, HI, USA, 21–26 July 2017; pp. 5967–5976.

23. Kupyn, O.; Budzan, V.; Mykhailych, M.; Mishkin, D.; Matas, J. DeblurGAN: Blind motion deblurring using conditional adversarial networks. In Proceedings of the IEEE Conference on Computer Vision and Pattern Recognition, Salt Lake City, UT, USA, 18–22 June 2018; pp. 8183–8192.

24. Ulyanov, D.; Vedaldi, A.; Lempitsky, V. Instance normalization: The missing ingredient for fast stylization. *arXiv* **2017**, arXiv:1607.08022v3.

25. Gulrajani, I.; Ahmed, F.; Arjovsky, M.; Dumoulin, V.; Courville, A. Improved training of wasserstein GANs. In Proceedings of the Neural Information Processing System, Long Beach, CA, USA, 4–9 December 2017; pp. 5767–5777.

26. Huang, G.B.; Ramesh, M.; Berg, T.; Learned-miller, E. Labeled faces in the wild: A database for studying face recognition in unconstrained environments. In Proceedings of the Workshop on Faces in 'Real-Life' Images: Detection, Alignment, and Recognition, Marseille, France, 17 October 2008; pp. 1–11.

27. Wolf, L.; Hassner, T.; Maoz, I. Face recognition in unconstrained videos with matched background similarity. In Proceedings of the IEEE Conference on Computer Vision and Pattern Recognition, Colorado Springs, CO, USA, 20–25 June 2011; pp. 529–534.

28. Logitech BCC950 Camera. Available online: https://www.logitech.com/en-roeu/product/conferencecam-bcc950?crid=1689 (accessed on 20 November 2019).

29. Logitech C920 Camera. Available online: https://www.logitech.com/en-us/product/hd-pro-webcam-c920?crid=34 (accessed on 23 November 2019).

30. ChokePoint Dataset. Available online: http://arma.sourceforge.net/chokepoint/ (accessed on 26 September 2019).

31. DeblurGAN. Available online: https://github.com/KupynOrest/DeblurGAN/ (accessed on 23 November 2019).

32. Simonyan, K.; Zisserman, A. Very deep convolutional networks for large-scale image recognition. In Proceedings of the 3rd International Conference on Learning Representations, San Diego, CA, USA, 7–9 May 2015; pp. 1–14.

33. He, K.; Zhang, X.; Ren, S.; Sun, J. Deep residual learning for image recognition. In Proceedings of the IEEE Conference on Computer Vision and Pattern Recognition, Las Vegas, NV, USA, 26 June–1 July 2016; pp. 770–778.

34. Stathaki, T. *Image Fusion: Algorithms and Applications*; Academic Press: Cambridge, MA, USA, 2008.

35. Salomon, D. *Data Compression: The Complete Reference*, 4th ed.; Springer: New York, NY, USA, 2006.

36. Wang, Z.; Bovik, A.C.; Sheikh, H.R.; Simoncelli, E.P. Image quality assessment: From error visibility to structural similarity. *IEEE Trans. Image Process.* **2004**, *13*, 600–612. [CrossRef]

37. Mejjati, Y.A.; Richardt, C.; Tompkin, J.; Cosker, D.; Kim, K.I. Unsupervised attention-guided image-to-image translation. In Proceedings of the 32th Annual Conference on Neural Information Processing Systems, Montreal, QC, Canada, 3–8 December 2018; pp. 3693–3703.

38. Qi, Q.; Guo, J.; Jin, W. Attention network for non-uniform deblurring. *IEEE Access* **2020**, *8*, 100044–100057. [CrossRef]

39. Kupyn, O.; Martyniuk, T.; Wu, J.; Wang, Z. DeblurGAN-v2: Deblurring (orders-of-magnitude) faster and better. In Proceedings of the IEEE/CVF International Conference on Computer Vision, Seoul, Korea, 27 October–2 November 2019; pp. 1–10.

40. Parkhi, O.M.; Vedaldi, A.; Zisserman, A. Deep face recognition. In Proceedings of the British Machine Vision Conference, Swansea, UK, 7–10 September 2015; pp. 1–12.

41. Gruber, I.; Hlaváč, M.; Železný, M.; Karpov, A. Facing face recognition with ResNet: Round one. In Proceedings of the International Conference on Interaction Collaborative Robotics, Hatfield, UK, 12–16 September 2017; pp. 67–74.

42. Martínez-Díaz, Y.; Méndez-Vázquez, H.; López-Avila, L.; Chang, L.; Enrique Sucar, L.; Tistarelli, M. Toward more realistic face recognition evaluation protocols for the youtube faces database. In Proceedings of the IEEE Conference on Computer Vision and Pattern Recognition Workshops, Salt Lake City, UT, USA, 18–22 June 2018; pp. 526–534.

43. Khamis, S.; Kuo, C.-H.; Singh, V.K.; Shet, V.D.; Davis, L.S. Joint learning for attribute-consistent individual re-identification. In Proceedings of the 13th European Conference on Computer Vision Workshops, Zurich, Switzerland, 6–12 September 2014; pp. 134–146.
44. Köstinger, M.; Hirzer, M.; Wohlhart, P.; Roth, P.M.; Bischof, H. Large scale metric learning from equivalence constraints. In Proceedings of the IEEE Conference on Computer Vision and Pattern Recognition, Providence, RI, USA, 16–21 June 2012; pp. 2288–2295.
45. Li, W.; Wang, X. Locally aligned feature transforms across views. In Proceedings of the IEEE Conference on Computer Vision and Pattern Recognition, Portland, OR, USA, 23–28 June 2013; pp. 3594–3601.
46. Liao, S.; Hu, Y.; Zhu, X.; Li, S.Z. Person re-identification by local maximal occurrence representation and metric learning. In Proceedings of the IEEE Conference on Computer Vision and Pattern Recognition, Boston, MA, USA, 7–12 June 2015; pp. 2197–2206.
47. Gray, D.; Tao, H. Viewpoint invariant pedestrian recognition with an ensemble of localized features. In Proceedings of the 10th European Conference on Computer Vision, Marseille, France, 12–18 October 2008; pp. 262–275.
48. Selvaraju, R.R.; Cogswell, M.; Das, A.; Vedantam, R.; Parikh, D.; Batra, D. Grad-cam: Visual explanations from deep networks via gradient-based localization. In Proceedings of the IEEE International Conference on Computer Vision, Venice, Italy, 22–29 October 2017; pp. 618–626.
49. Jetson TX2 Module. Available online: https://www.nvidia.com/en-us/autonomous-machines/embedded-systems-dev-kits-modules/ (accessed on 12 December 2019).
50. Keras: The Python Deep Learning Library. Available online: https://keras.io/ (accessed on 11 March 2020).
51. Tensorflow: The Python Deep Learning Library. Available online: https://www.tensorflow.org/ (accessed on 19 July 2019).
52. CUDA. Available online: https://developer.nvidia.com/cuda-90-download-archive (accessed on 11 March 2020).
53. CUDNN. Available online: https://developer.nvidia.com/cudnn (accessed on 11 March 2020).
54. Lenka, M.K. Blind deblurring using GANs. *arXiv* **2017**, arXiv:1907.11880v1.
55. Zhang, S.; Zhen, A.; Stevenson, R.L. GAN based image deblurring using dark channel prior. In Proceedings of the IS&T International Symposium on Electronic Imaging, San Francisco, CA, USA, 13–17 January 2019; pp. 1–5.
56. Zhang, X.; Lv, Y.; Li, Y.; Liu, Y.; Luo, P. A modified image processing method for deblurring based on GAN networks. In Proceedings of the 5th International Conference on Big Data and Information Analytics, Kunming, China, 8–10 July 2019; pp. 29–34.

 *sensors*

Article

LdsConv: Learned Depthwise Separable Convolutions by Group Pruning

Wenxiang Lin [1], **Yan Ding** [1,*], **Hua-Liang Wei** [2], **Xinglin Pan** [3] **and Yutong Zhang** [1]

[1] Key Laboratory of Dynamics and Control of Flight Vehicle, Ministry of Education,
     School of Aerospace Engineering, Beijing Institute of Technology, Beijing 100081, China;
     wenxianglineut@163.com (W.L.); zhang1123034978@163.com (Y.Z.)
[2] Department of Automatic Control and Systems Engineering, University of Sheffield, Sheffield S1 3JD, UK;
     w.hualiang@sheffield.ac.uk
[3] Department of SMILE Lab, School of Computer Science and Engineering, University of Electronic Science
     and Technology of China, Chengdu 610031, China; kp600168@gmail.com
* Correspondence: dingyan@bit.edu.cn

Received: 20 July 2020; Accepted: 30 July 2020; Published: 4 August 2020

Abstract: Standard convolutional filters usually capture unnecessary overlap of features resulting in a waste of computational cost. In this paper, we aim to solve this problem by proposing a novel Learned Depthwise Separable Convolution (LdsConv) operation that is smart but has a strong capacity for learning. It integrates the pruning technique into the design of convolutional filters, formulated as a generic convolutional unit that can be used as a direct replacement of convolutions without any adjustments of the architecture. To show the effectiveness of the proposed method, experiments are carried out using the state-of-the-art convolutional neural networks (CNNs), including ResNet, DenseNet, SE-ResNet and MobileNet, respectively. The results show that by simply replacing the original convolution with LdsConv in these CNNs, it can achieve a significantly improved accuracy while reducing computational cost. For the case of ResNet50, the FLOPs can be reduced by 40.9%, meanwhile the accuracy on the associated ImageNet increases.

Keywords: convolutional neural network; convolutional filter; classification

---

1. Introduction

Convolutional neural networks (CNNs) have shown remarkable achievements in various vision tasks [1–8]. Most of the achievements benefit from the innovative design of network architectures [9–14], with applications in a variety of areas including phishing detection (see, e.g., [15]). Recent designs usually use the convolutional filter as the basic unit and achieve good training results through special network architectures. However, the manual design of the network architecture has been gradually replaced by architecture searching [16–22] with the rapid development of the computation ability of the hardware. Compared with architecture searching, which often requires strong computing power and expensive time cost, the model compression method and other new convolutional filter design techniques [23–25] provide an economic choice to improve the efficiency of CNNs.

At present, the commonly used convolutions are Groupwise Convolution [2], Depthwise Convolution [26] and Pointwise Convolution [27]. Pointwise Convolution is able to adjust the dimension of the channels or feature maps. It is widely used in the design of architectures. Groupwise Convolution can reduce the connection density and computation cost of convolutional filters, while Depthwise Convolution is the extreme version of Groupwise Convolution which sets the number of groups to be the same as the number of input channels. However, if we simply replace the standard convolution with Depthwise or Groupwise Convolution without special adjustment of the architecture, the resulting model may not work well. Therefore, some new convolutional filters have been proposed recently. HetConv [23] proposes the heterogeneous kernel-based convolution. OctConv [24] designs a convolutional filter that can extract multi-scale information from features. These convolutional filters have the ability to improve the performance of model by simply replacing the standard convolutions without any adjustment of the baseline. The present study proposes a similar but different plug and play convolutional unit. Our proposed LdsConv pays more attention on the learning ability of the model and aims to transform a standard convolutional filter into a learned depthwise separable convolutional filter.

Model compression is considered as another reliable and economic method to improve the efficiency of the convolutional neural network, which can be roughly divided into three categories: (a) Connection pruning [28,29]; (b) Filter pruning [30–36]; and (c) Quantization [28,37–39]. These methods can effectively reduce the computation of the convolutional neural network, but this is always achieved at the price of sacrificing the accuracy. Sometimes, special hardware support is also required for compression methods.

Instead of directly pruning the whole model, we choose to integrate the pruning technique into the design of convolutional filters. In this way, the model can automatically learn to know which input features are most valuable for each single output, so that it enables to extract better features with fewer filters. To achieve this objective, we design a new type of convolutional filter—Learned Depthwise Separable Convolution (LdsConv), which can be directly plugged into existing standard architecture to reduce floating point of operations (FLOPs) and meanwhile improve the accuracy.

To integrate the pruning methods, we develop the two-stage training framework to divide the training task into picking and combining. In the first stage, the LdsConv picks out the most valuable input features and applies more filters to them by pruning technique. In the second stage, the additional pointwise convolution combines the output of the first stage and produces the output features. The idea of division of labour and progressive working has been reflected in computer vision. For example, the two-stages detection framework [40] divides the task into region proposed stage and classification as well as location stage. Cascade RCNN [41] further refines the second stage into three parts and each part is based on the front one. Similarly, we adopt this idea in the convolutional operation and thus divide the training task into picking up useful filters and mixing up the results of picking up. The relationship between two stages is progressive and inseparable. The two-stage training process simplifies the training task for each stage and finally improves the efficiency of the model.

Our experiments show that by replacing the standard/depthwise convolution with the LdsConv in CNNs, it can improve the accuracy and reduce computational costs in the following models: ResNet [1], DenseNet [42], MobileNet [9], and SE-ResNet [43].

Our main contributions are three-fold:

1. We integrate the weight pruning method into the depthwise separable convolutional filter and develop the two-stage training framework.
2. We design an efficient convolution filter named Learned Depthwise Separable Convolution, which can be directly inserted into the existing CNNs. It can not only reduce and computational cost, but also improve the accuracy of the model.
3. We validate the effectiveness of the proposed LdsConv through extensive ablation studies. To facilitate further studies, our source code, as well as experiment results, will be available at https://github.com/Eutenacity/LdsConv.

2. Related Work

2.1. High Efficiency Convolutional Filter

Ever since the pioneering work on Alexnet [2] and VGG [3], researchers have studied how to improve the efficiency of CNNs from various perspectives. However, much less work has been devoted to developing innovative convolutional filters. Among those proposed convolutional filters, the most popular ones are Groupwise Convolution [2], Depthwise Convolution [26] and Pointwise Convolution [27]. They are widely used in the design of efficient CNNs. ResNet [1,44] uses Pointwise Convolution to build bottleneck layers that allow the network to go deeper without increasing too many parameters. For example, ResNeXt [45] and ShuffleNet [12] use Groupwise Convolution to reduce redundancy in internal connections. Xception [10] and Mobilenet [9] use Depthwise Convolution to further reduce the connection density. SENet [43] and CBAM [46] design a module that can automatically weigh the output of convolutional filters at the cost of a small number of parameters. Hetconv [23] uses convolutional filters with heterogeneous kernels to replace the standard convolutional filters. OctConv [24] reduces the spatial redundancy in CNNs by designing special convolutional filters with multi-scale input features. The Multi-Kernel Depthwise Convolution proposed in [47] can better extract information with multiple kernel sizes and effectively utilize the computational efficiency of Depthwise Convolution. The fully learnable group convolution (FLGC) proposed in [48] can be integrated into a deep neural network and automatically learn the group structure in the training stage in a fully end-to-end manner; its can achieve high computational efficiency. In [49], a new dynamic grouping convolution (DGConv) was proposed, which is able to learn the number of groups in an end-to-end manner; it has been proven to have several advantages. The training-free method, called network decoupling (ND), proposed in [50] is interesting; it achieves high computational efficiency and accuracy performance via pre-trained CNN models which are transferred to the MobileNet-like depthwise separable convolution structure. Compared to these methods, the proposed LdsConv chooses to incorporate weight pruning technique into the design of convolutional filters and further develops the two-stage training framework to simplify the training task for each stage.

2.2. Model Compression

Model compression is another popular method to improve the efficiency of the convolutional neural network. Refs. [28,29] remove redundancy in the model by pruning connection. Refs. [28,37–39] compress the calculation amount of the model via quantization. Refs. [30–36] prune filters that have a minimal contribution in the model. After removing these filters, the model is usually fine-tuned to maintain its performance. Among these methods, filter pruning methods generally do not require special hardware and software, but they need a pre-trained model which may use a computationally expensive training to obtain.

The proposed LdsConv inserts the weight pruning process into the training. Therefore, the LdsConv embedded model is able to be trained from scratch without a pre-trained model. Different from [51] which only integrates the pruning and fine-tuning process with training, LdsConv further develops the two-stage training framework dividing the training task into picking and combing. Moreover, LdsConv conducts the group pruning by replacing the original convolution with the groupwise convolution before training and use an additional balanced loss function to make the pruning procedure more smooth. Additionally, LdsConv adds an additional pointwise convolution at the end of the pruning, to integrate the pruning results and build a regular depthwise separable convolution, allowing for efficient computation in practice at test time.

3. Method

In this section, we first introduce Depthwise Separable Convolution and LdsConv. Then we describe the details about the utilization of LdsConv. We also discuss implementation details and show how to replace Depthwise Separable Convolution with LdsConv.

3.1. Depthwise Separable Convolution

Consider a standard convolution that takes an $R \times D_h \times D_w$ feature as an input and produces an $O \times D_h \times D_w$ feature as an output, where R, O, D_h and D_w denote the numbers of input channels, output channels, and the height and the width of the feature. Usually a standard convolution applies R filters to every input channel for each output. Thus, a standard convolution has the weight matrix with the size of $R \times O \times H \times W$ where H and W denote the height and the width of the filter. To reduce the computational cost, the depthwise separable convolution splits the standard convolution into two: a depthwise convolution for filtering, that only applies a single filter to the corresponding input channel for the output one, and a pointwise convolution for combing the outputs of the depthwise convolution and producing final output channels. The depthwise convolution is parameterized by the kernel of the size $R \times 1 \times H \times W$ and the pointwise convolution is of the size $R \times O \times 1 \times 1$.

3.2. Learned Depthwise Separable Convolution

Considering the strength of the depthwise separable convolution, it is highly desirable to design a more complex architecture to enhance the capability of the convolution so that the neural network can decide on which feature should be applied. In doing so, we need a novel convolution architecture, named Learned Depthwise Convolution (LdsConv). As shown in Figure 1, the training process is divided into picking stages and the combining stage. Moreover, the training task is also divided into picking and combining. In picking stages, we focus on removing little influence filters repeatedly to pick out valuable input features. In the combining stage, similarly to Depthwise Separable Convolution, an additional 1×1 convolution is applied to combine features.

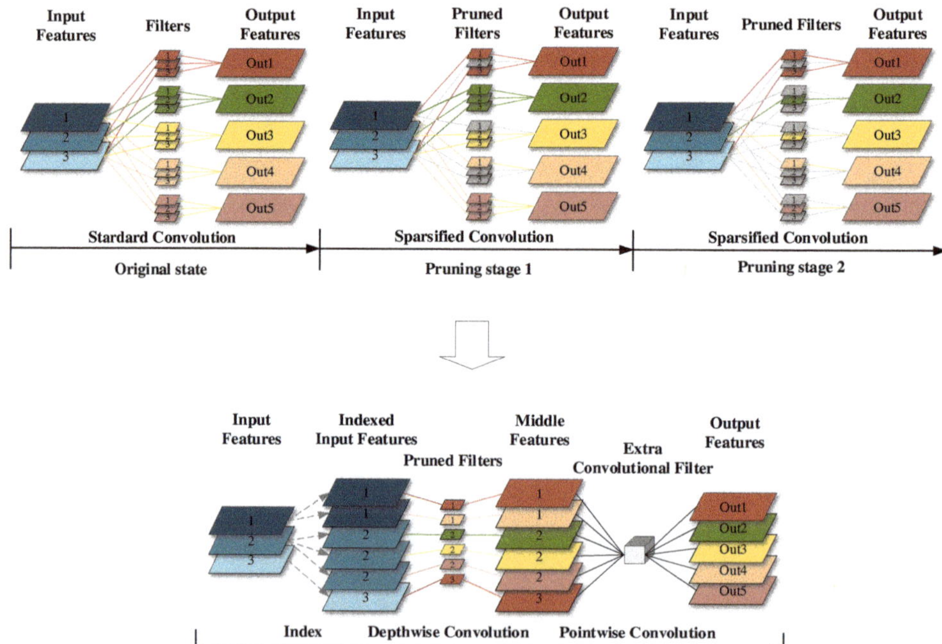

Figure 1. The illustration of the LdsConv with a input channel of $R = 3$, an output channel of $O = 5$, a group cardinality of $N_O = 5$, a group number of $G = 1$, a pruning factor of $k = 2$ and a stage factor of $s = 2$. At the end of picking stages, we remove filters with the number of $(N_O - k)R$. After the picking stages, an additional 1×1 standard convolution is added into the convolutional module to form a standard depthwise separable convolution.

3.2.1. Group Pruning

Initially, we adopt a group convolution which divides a standard convolution of size $R \times O \times H \times W$ into G groups of 4D tensors F^g with the size of $N_R \times N_O \times H \times W$ to initialize our architecture. For convenience of description, define $N_R = \frac{R}{G}$ and $N_O = \frac{O}{G}$. Given the fact that the size of convolution layers is widely different which needs different G for the division operation, in the experiment we set a unify hyper-parameter N_O, named group cardinality, to represent our model and analyze its influence on the accuracy. Group pruning aims to relieve the effect of the pruning to the accuracy by making pruning results more uniform.

3.2.2. Pruning Criterion

During the training process, we gradually screen out less important filters for each group. The importance of the filters is evaluated by the L_1-norm of its weight F^{gij} that corresponds to the weight of the i-th input for the j-th output within group g. In other words, we remove filters with the L_1-norm.

3.2.3. Pruning Factor

It is important to consider and determine how many filters should be removed before the combining stage. Formally, we set a hyper-parameter k with a range from 1 to 4 to represent that the number of remaining filters is $k \times R$. In Section 4, discussions and analysis on how to choose k is presented, which both has a good balance of parameter and accuracy and fits all around dataset and network scale.

3.2.4. Stage Factor

In contrast to methods that prune weights in pre-trained models, our weight pruning process is plugged into the training procedure. Thus, we define the stage factor to determine the times of pruning. For a group filter weight F^g with size of $N_R \times N_O \times H \times W$, the number of filters that need to be pruned can be calculated by the equation $N_d = N_R N_O - k N_R$. Thus, the total number of pruned filters is $G N_d = R N_O - k R$. Then, at the end of each picking stage, we prune $G N_d / s$ filters.

3.2.5. Balance Loss Function

To reduce the negative impact on the accuracy induced by pruning, we deliberately set the number of remaining filters of each input feature to be even avoid the case that most of remained filters extract information from only a small number of input features. As we know, it is hard to optimize the number of filters as they are non-differentiable. We thus define the coefficient of M to ensure that filters belong to input features with a bigger number of possible remained filters would be penalized more strongly.

In each training iteration in picking stages, we first find the filters that have the highest probability to remain. Then, we check their input features to get the number of probably remaining filters of each input feature. Finally, we restrain these filters belong to input features having a big number of probably remaining filters. To this end, we use the following regularizer for a group filter weight F^g during training:

$$L_{bal} = \sum_{j=1}^{N_O} \sum_{i=1}^{N_R} M_i \left(\sum_{l=1}^{HW} \left| w_{l,i,j} \right| \right)^2 \tag{1}$$

where M_i denotes the coefficient for filters belong to the i-th input feature and $w_{l,i,j}$ denotes every parameter in F^{gij}. By adjusting the coefficient of M_i, the input feature having higher number of probably remaining filters will force its filters to be penalized more strongly. The equation for M_i is defined as:

$$M_i = max(e^{(N_i^R - \lambda k)/\gamma} - 1, 0) \tag{2}$$

where N_i^R denotes the number of probably remaining filters belonging to the i-th input feature. We introduce a parameter λ to define the threshold over which the filter belonging to the i-th input feature will receive the penalty since the average value of N_i^R is k. Furthermore, γ is set to adjust the penalty level. In this paper, we set $\lambda = 1.5$ and $\gamma = 10$ in all experiments empirically.

3.2.6. Additional Pointwise Convolution

At the end of picking stages, we convert the sparsified model into a network with regular modules that can be efficiently deployed on devices without special hardware and software support. For this reason, we add additional pointwise convolutions to each LdsConv to build Depthwise Separable Convolution (see Figure 1). This operation also highly broadens the expression ability of LdsConv filters and lead the training task to combining the output of picking stages and producing the final output features. The weight of the additional pointwise convolution has the size of $kR \times O \times 1 \times 1$ related to the number of input channel R and output channel O of the original convolution and the pruning factor k. The initial value of the weight is set by the index information of the remaining filters. Figure 2 shows the initial value of the example in Figure 1. We set the value of the position in the weight matrix to 1 only when the middle feature extract by the remaining filter matches the output feature. The color in Figure 1 represents this matching relationship. This kind of initial value can narrow the negative effect of the newly additional pointwise convolution added in the training process.

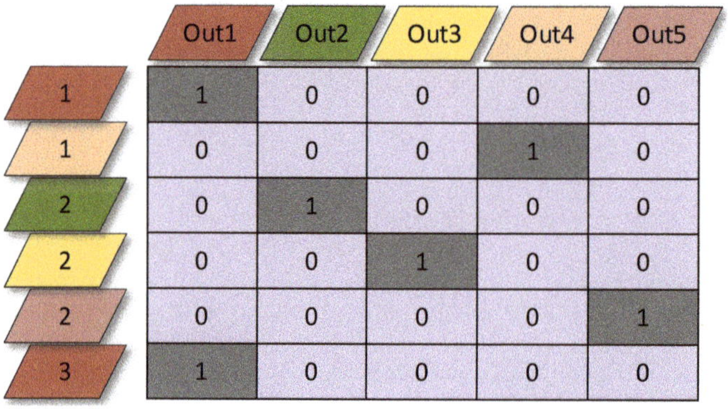

	Out1	Out2	Out3	Out4	Out5
1	1	0	0	0	0
1	0	0	0	1	0
2	0	1	0	0	0
2	0	0	1	0	0
2	0	0	0	0	1
3	1	0	0	0	0

Figure 2. Initial value assignment of the example shown in Figure 1. The left set of parallelograms represents the middle features. The numbers in these parallelograms mean the index in the input features. The upper set of parallelograms represents the output features. The same color between the left parallelogram and the top one means that they are matched in picking stages. The value in the matrix means the initial value of the additional convolution.

3.2.7. Learning Rate

We adopt the cosine shape learning rate schedule during training, which smoothly changes the learning rate, and usually improves the accuracy [18,52,53]. Figure 3 demonstrates the learning rate as a function of training epoch, and the corresponding training loss of a ResNet50 using LdsConv filters on the ImageNet dataset [54]. Before we enter the combining stage, we add additional pointwise convolution and reset the learning rate to reduce the negative effect of the learning rate to the newly added weights. Thus, the abrupt increase occurs in the loss at epoch 45. However, the plot shows that the loss gradually recovers from this accident.

3.3. The Implementation of LdsConv

In addition to the use of LdsConv, we briefly describe how to replace standard convolutional filters and depthwise separable convolutional filters with LdsConv filters.

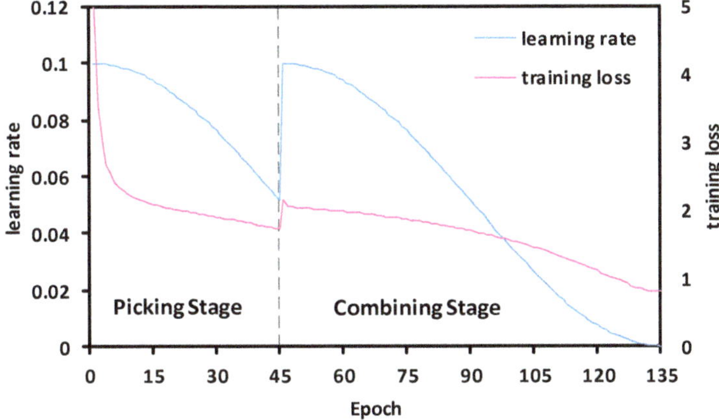

Figure 3. The cosine shape learning rate and a typical training loss curve trained on ImageNet. The vertical gray bar in the figure marks the end of picking stage and the begin of combing stage.

3.3.1. Standard Convolution

When we try to replace a standard convolution with our proposed LdsConv, the most important hyper-parameter is the group cardinality N_O. In general, we suggest setting N_O to the value from 8 to 32. But if the number of the channels of the original convolution is too small to divide, we need to set N_O to the same value as the number of output channels ensuring the group to be 1. For other hyper-parameters, we can simply use the recommended value given by Section 4. In addition to the fact that we replace the standard convolution with the group one first, a 1×1 convolution should exist to mix all channels information after the group convolution. In Figure 4, we demonstrate the replacement in the ResNet.

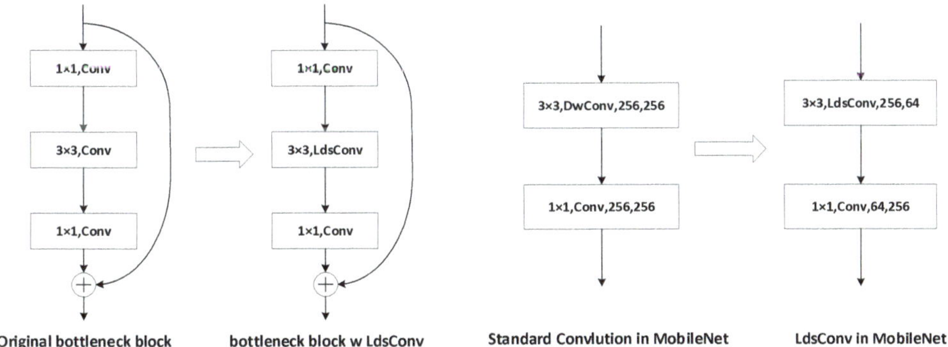

Figure 4. The replacement of original convolutional filters with LdsConv. Left: The replacement in ResNet. We directly replace the 3×3 convolution with our proposed LdsConv in the bottleneck block. Right: The replacement in MobileNet. We replace the original 3×3 convolution and reduce the number of output channel of the LdsConv and input channel of the sequent 1×1 convolution.

3.3.2. Depthwise Separable Convolution

In general, a pointwise convolution exists in each depthwise separable convolution. So, we do not need to worry about the problem mentioned above. In other words, we can simply replace the depthwise convolution with our proposed LdsConv. However, parameters and FLOPs may increase if we do not make any adjustments. Therefore, we suggest adding an additional convolution before or after the LdsConv to reduce the number of input or output channels of the LdsConv. The right part of Figure 4 shows our implementation of LdsConv filters in MobileNet.

4. Experiment

In this section, we validate the effectiveness and efficiency of the proposed LdsConv. We first present ablation studies for image classification on Cifar [55]. Then, we perform a set of experiments on ImageNet [54] to check the performance of the proposed LdsConv.

4.1. Ablation Study on Cifar

We conduct a series of ablation studies to find the best situation to implement LdsConv filters and then check its robustness in different models.

4.1.1. Training Details

We use stochastic gradient descent (SGD) algorithm to train all the models. Specifically, we adopt Nesterov momentum with a momentum weight of 0.9 without dampening, and use a weight decay of $1e^{-4}$. Unless otherwise specified, the size of the training batch is set to be 64 and the number of total training epochs is 300, in which the picking stages take 150 epochs and the combining stage has 150 epochs. For the convenience of network accuracy comparison, we all use the standard cosine learning rate change strategy without reset which starts from 0.1 and gradually reduces to 0. It is worth mentioning that special modification on learning rate dose not affect too much. Therefore, we remove the reset described in Section 3.2.7 for the convenience.

4.1.2. Implement on DenseNet-BC-100

We do experiment with DenseNet-BC-100 architecture having a growth rate of 12 [42] on the CIFAR-100 dataset.When we implement our proposed LdsConv, we simply replace the 3×3 convolutional filters in dense blocks with the LdsConv filters. Specifically, we set the group cardinality N_O to the same as the number of output channels since the number is too small to divide. Then we start experiments on the effect of pruning factor k and stage factor s for the LdsConv.

4.1.3. Effect of Stage Factor

The first part of Table 1 compares DenseNet-BC-100 models having LdsConv filters with different stage factors. In particular, we set the pruning factor k to 2. The result shows that $s = 4$ seems to be the best value. While reaching the peak at 4, the accuracy drops down for higher stage factors. We attribute this change to the decreasing of gap epochs between pruning which is calculated by the equation $E_G = E_P/s$ where E_P denotes training epochs of picking stages. To expel its effect, we conduct two more experiments with $s = 6$ and $s = 8$ and set E_G to be the same value as the one when $s = 4$ in the second part of Table 1. In other words, the picking stages of these two experiments take 225 and 300 epochs, respectively. The result shows that the accuracy can increase a lot without the decreasing of gap epochs E_G. By taking into account the training time, we suggest to set the stage factor to 4 in the ordinary course of events.

4.1.4. Effect of Pruning Factor

We do experiment with several pruning factors k, which vary from 1 to 4. In addition, we set the stage factor s to 4 which means all models have the same times of pruning. The results presented in the third part of Table 1 demonstrate that parameters of the model raise while the accuracy rise ups and downs with the increasing of the pruning factor. The risk of overfitting and the decreasing of pruning proportion battle with each other resulting in this change. In particular, it suggests that setting the pruning factor k to 2 is a good choice which balances both the accuracy and the number of parameters. We can also reduce the pruning factor k to 1 or even integrate the additional pointwise convolution with the sequent convolution to reach a higher reduction to weights.

Table 1. The table shows the ablation study results in different setups on CIFAR-100. '*' refers to the LdsConv using the balance loss. '#' refers to the model trained with gap epochs $E_G = E_P/4$.

Model	Accuracy (%)	GFLOPs	Params (M)
Lds-DenseNet-BC-100 ($s = 2$)	76.9	0.23	0.64
Lds-DenseNet-BC-100 ($s = 4$)	77.3	0.23	0.64
Lds-DenseNet-BC-100 ($s = 6$)	76.3	0.23	0.64
Lds-DenseNet-BC-100 ($s = 8$)	76.6	0.23	0.64
Lds-DenseNet-BC-100 # ($s = 6$)	77.4	0.23	0.64
Lds-DenseNet-BC-100 # ($s = 8$)	77.9	0.23	0.64
Lds-DenseNet-BC-100 ($k = 1$)	76.3	0.21	0.6
Lds-DenseNet-BC-100 ($k = 2$)	77.3	0.23	0.64
Lds-DenseNet-BC-100 ($k = 3$)	77.3	0.25	0.71
Lds-DenseNet-BC-100 ($k = 4$)	76.8	0.28	0.74
Lds-DenseNet-BC-100 * ($k = 1$)	76.8	0.21	0.6
Lds-DenseNet-BC-100 * ($k = 2$)	77.7	0.23	0.64
Lds-DenseNet-BC-100 * ($k = 3$)	77.6	0.25	0.71
Lds-DenseNet-BC-100 * ($k = 4$)	77.1	0.28	0.74
Lds-ResNet50 * ($N_O = 4$)	80.1	2.87	14.97
Lds-ResNet50 * ($N_O = 8$)	80.9	2.87	14.97
Lds-ResNet50 * ($N_O = 16$)	79.8	2.87	14.97
Lds-ResNet50 * ($N_O = 32$)	79.8	2.87	14.97
Dw-DenseNet-BC-100	74.6	0.21	0.6
Lds-DenseNet-BC-100 ($k = 2$) w/o AC	76.2	0.3	0.79

4.1.5. Effect of Balance Loss Function

To check the effectiveness of our balance loss function, we apply it to the models with varied pruning factors. The fourth part of Table 1 shows that the accuracy is improved by adding the balance loss regularization.

4.1.6. Effect of Group Cardinality

To evaluate the effect of the group cardinality N_O, we experiment with ResNet50 [1] which is designed to train on ImageNet and thus has large number of channels. We remove the first three downsampling operations and retain only the last two ones since images in cifar have smaller resolution. The fifth part of Table 1 compares ResNet50 models using LdsConv filters with varied group cardinality. Specifically, we set the group cardinality N_O to 4,8,16 and 32. The stage factor s is set to 4 and the pruning factor k is set to 2 for all models. The result shows that the accuracy first rises up and then goes down. When $N_O = 8$, the model reaches its best accuracy. While reaching the accuracy peak at 8, the accuracy drops down for lower N_O indicating over-group can also have negative effects. We own the negative effects to the shrink in expression ability when the convolution is grouped.

4.1.7. Effect of Two-Stage Training Framework

To verify the function of each stage, we first explore the norm value of the picking results and then evaluate the effect of the additional convolution. The three panels of Figure 5a illustrates the weights of the last 3×3 convolution for orignal DenseNet-BC-100, Dw-DenseNet-BC-100 and Lds-DenseNet-BC-100. We replace the 3×3 standard convolutions in dense blocks with depthwise separable convolutions in Dw-DenseNet-BC-100 which can be regarded as the typical one-stage training form of LdsConv. Each block in the figure represents the L1 norm (normalized by the maximum value among all filters) of a 3×3 filter. In the top two panels of Figure 5a, the vertical and horizon axis represent the height and width of the weight matrix, respectively. For the third panel, we arrange the weight matrix of Lds-DenseNet-BC-100 in this way for alignment. Figure 5b shows the curve between 48 3 × 3 convolutional layers in dense blocks and the average norm of weights for three models. The results suggest that the picking stage indeed reduces the redundancy in the weight matrix and picks up more valuable filters. We additionally experiment with Dw-DenseNet-BC-100 and Lds-DenseNet-BC-100 ($k = 2$) without additional convolutions (AC) in the final part of Table 1. Without additional convolutions, the combing stage becomes the common optimization one. The accuracy dramatically drops down indicating that the combing stage is indispensable. Furthermore, additional convolutions arrange the sparsified convolutions into standard depthwise separable convolutions improving the computation cost at test time. Besides, Dw-DenseNet-BC-100 shows lower accuracy and non negligible gap in the convergence speed compared with the baseline in Figure 5c. On the contrary, Lds-DenseNet-BC-100 trained with the two-stage training framework owns a better curve of convergence speed which is near to the baseline.

4.1.8. Results on Other Models

To evaluate the effectiveness of the proposed LdsConv with the situation discussed in the above in different networks, we choose currently popular models as the baselines including ResNet [1], DenseNet [42], MobileNet [9], and SE-ResNet [43]. For all experiments, we set the pruning factor k to 2, the stage factor s to 4 and the balance loss function active. In DenseNet, we set its pruning cardinality N_O to the same value as the number of output channels. In other networks, we set the pruning cardinality N_O to 8. The experimental results are shown in Table 2. After using our modules to replace the convolutions in the original models, these networks generally achieve the effect of reducing the FLOPs and the number of parameters, meanwhile maintaining or even improving the accuracy. It shows that our method can effectively reduce the redundancy in convolutional filters. It also suggests that the LdsConv can perform well without too many adjustments on hyper-parameters.

Table 2. The table shows the results for different models on CIFAR-100. '*' refers to the LdsConv using the balance loss. With the setting obtained from the ablation study, we can simply improve the performance of the model by replacing the standard 3 × 3 convolution with our proposed LdsConv.

Model	Accuracy (%)	GFLOPs	Params (M)
MobileNet [9]	77.1	0.62	3.31
Lds-MobileNet *	78.0	0.51	2.74
ResNet50 [1]	80.2	4.46	23.71
Lds-ResNet50 *	80.9	2.86	14.97
ResNet152 [1]	81.7	14.20	58.34
Lds-ResNet152 *	82.0	8.66	35.53
SE-ResNet50 [43]	81.2	4.46	26.22
Lds-SE-ResNet50 *	81.5	2.87	16.54
DenseNet-BC-100 [42]	77.7	0.30	0.79
Lds-DenseNet-BC-100 *	77.7	0.23	0.64

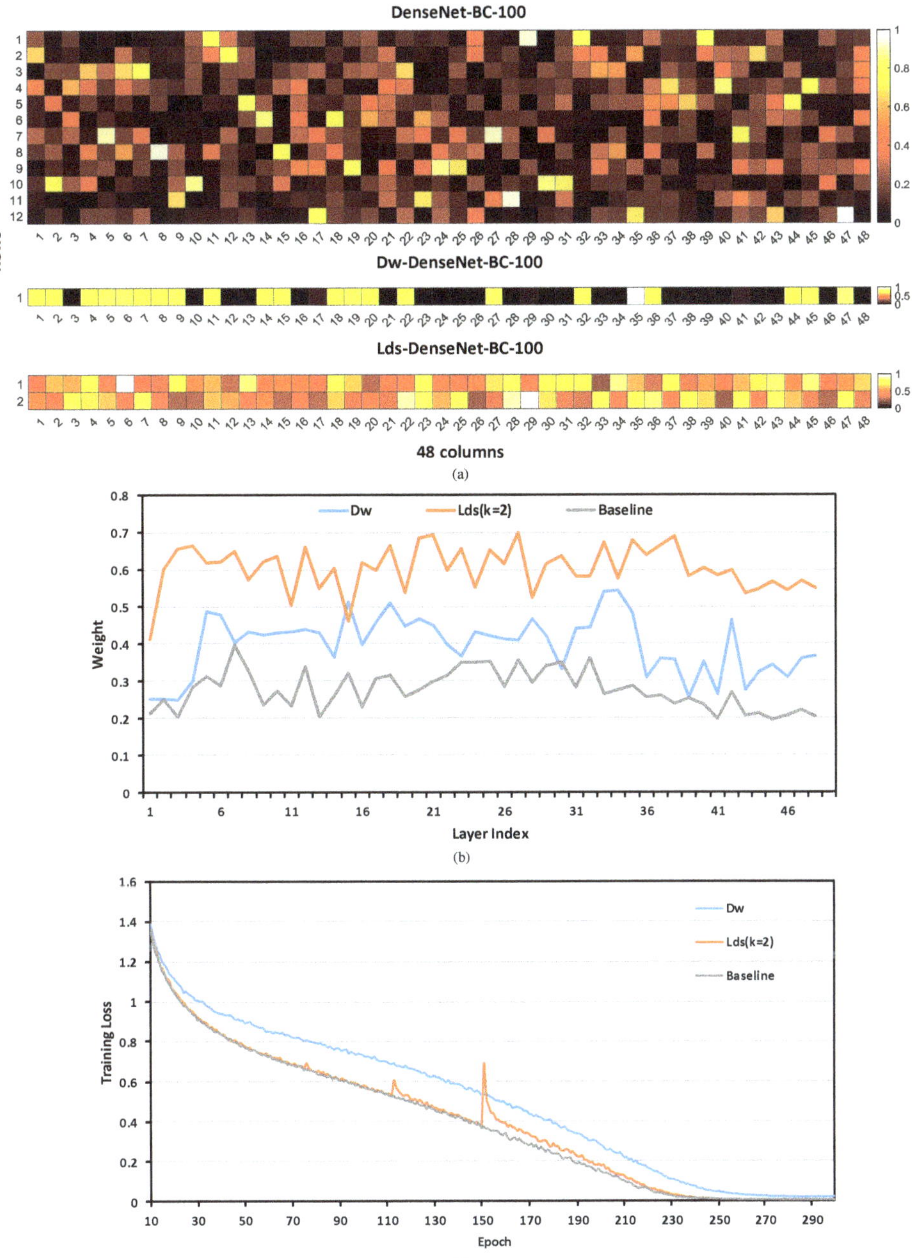

Figure 5. (**a**): Norm of weights of three models in Cifar 100. The block with darker color has less value of norm. The vertical and horizon axis represent the height and width of the weight matrix expect the Lds-DenseNet-BC-100 in which we arrange it in this way for alignment. (**b**) The curve between 48 3×3 convolutional layers in dense blocks and the average norm of weights for three models. (**c**) The curve of convergence speed for three models.

4.2. Results on ImageNet

In a set of experiments, we test LdsConv filters on the ImageNet dataset.

4.2.1. Training Details

We use the SGD method to train all the models and adopt Nesterov momentum with a momentum weight of 0.9 without dampening using a weight decay of $1e^{-4}$. We use 135 as the total training epochs, in which the picking stage takes 45 epochs, the combining stage involves 90 epochs. The learning rate change strategy is shown in Figure 3. For MobileNet, we choose to simply increase the training epochs rather than adjusting hyper-parameters to the best. Thus, we use 300 as the total training epochs, in which the epoch size of the picking stage and combining stage is set as 100 and 200, respectively. The initial learning rate is 0.045, and its weight decay is $4e^{-5}$.

4.2.2. Model Configurations

In the experiments on ImageNet, we set the balance loss function active, the pruning factor k to 2 and the stage factor s to 4. Except for DenseNet, we set the group cardinality N_O to 8. In DenseNet, we still set its group cardinality to the same value as the number of output channels ensuring the group to be 1.

4.2.3. Comparison on ImageNet

We continue to use ResNet [1], DenseNet [42], MobileNet [9], and SE-ResNet [43] as the baseline for comparison, and the results are shown in Table 3. All results of baselines come from their original papers. In MobileNet, we can slightly reduce parameters and FLOPs, and highly increase the accuracy by 2.3%. For other networks using standard convolution originally, we not only improve the accuracy but also obviously reduce the number of parameters and FLOPs. What's more, our modules can coexist with SE-modules to further improve the efficiency of the model.

Table 3. The table shows the results for different models on ImageNet. '*' refers to the LdsConv using the balance loss. By simply replacing the standard 3×3 convolutional filters with our proposed LdsConv filters, we can not only improve the accuracy but also reduce the FLOPs and the number of parameters a lot. For the case of MobileNet, we highly increase the accuracy by 2.3% which is a pretty considerable improvement.

Model	Error% (Top-1)	GFLOPs	Params (M)
MobileNet [9]	29.0	0.57	4.2
Lds-MobileNet *	26.7	0.49	3.7
ResNet50 [1]	24.7	3.86	25.6
Lds-ResNet50 *	22.9	2.71	16.8
ResNet152 [1]	23.0	11.30	60.2
Lds-ResNet152 *	21.2	7.14	37.4
SE-ResNet50 [43]	23.3	3.87	28.1
Lds-SE-ResNet50 *	22.0	2.71	17.1
SE-ResNet152 [43]	21.6	11.32	66.8
Lds-SE-ResNet152 *	20.7	7.15	38.2
DenseNet121 [42]	25.0	2.88	8.0
Lds-DenseNet121 *	24.2	1.99	6.5
DenseNet264 [42]	22.2	5.86	33.3
Lds-DenseNet264 *	21.7	4.72	29.9

4.2.4. Comparison with Model Compression Methods

To investigate the compressing ability of our proposed LdsConv, we adjust the bottleneck block with LdsConv in the ResNet to a extreme state as shown in Figure 6. To this end, we remove the Bn and Relu layers after the 3×3 group convolutional layer before training. When the combination stage begins, we integrate the additional pointwise convolution (AC) with the sequent 1×1 convolution by the matrix multiply operation since no non-linear operation exists between them. When the model formally enters the combing stage, we only train one 1×1 convolution after every LdsConv. In Table 4, we compare the LdsConv with the existing compression methods including ThiNet [30], NISP [56] and FPGM [57]. We use ResNet50 as the baseline, replace the standard convolution with the LdsConv, and reduce the number of parameters further by setting the pruning factor to 1 and combing the additional pointwise convolution with the sequent 1×1 convolution. We also set $s = 6$ and $E_G = E_P/4$, which lengthens the training epochs, in order to relieve the negative effect of extremely compressing. Compared with these pruning methods, our method, denoted as Lds-ResNet50-extreme, not only improves the accuracy outperforming all other compared methods but also reduces the FLOPs by 40.9%. Furthermore, the real inference speed of Lds-ResNet50-extreme is 42 batches (16 images per batch) per second with the practical evaluation on GPU Nvidia RTX 2080 compared with the 28.9 batches per second on the baseline of ResNet50. We can obtain nearly $1.5\times$ speed up without special hardware support.

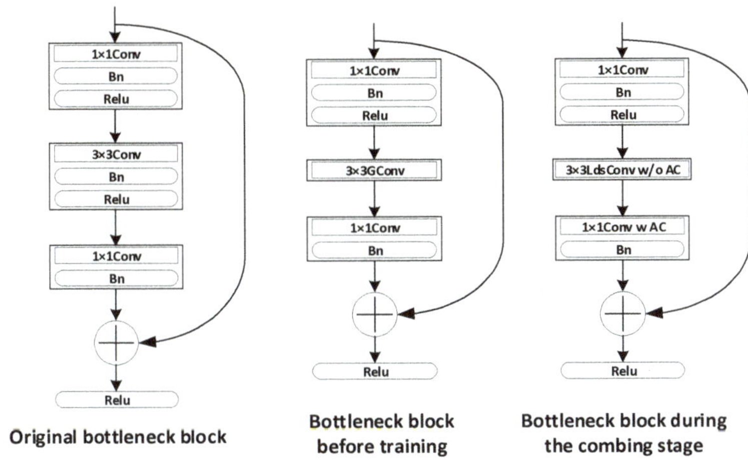

Figure 6. The extreme state of LdsConv in the ResNet. We remove the Bn and Relu layer after the 3×3 convolution and combine the additional convolution with the sequent 1×1 convolution by the matrix multiply operation. Finally the standard convolution is replaced with only depthwise convolution. The 3×3 LdsConv w/o AC means the depthwise part in LdsConv. The sequent 1×1 Conv w AC means the combing result of the additional convolution and original sequent 1×1 convolution.

Table 4. The table shows the comparison with existing compression methods for ResNet50 on ImageNet. Our Lds-ResNet50-extreme outperforms all other methods in terms of accuracy and still has a comparable reduction on FLOPs.

Model	Error% (Top-1)	GFLOPs	FLOPs↓ (%)
ThiNet-70 [30]	27.9	-	36.8
NISP [56]	27.3	-	27.3
FPGM-only 30% [57]	24.4	-	42.2
Lds-ResNet50-extreme	23.4	2.28	40.9

4.3. Comparison with Similar Works

To further verify the effectiveness of our approach, we do several experiments using three different networks, namely, ND [50], FLGC [48] and GDConv [49] as well as the proposed model. A comparison of the four models is shown in Table 5. These methods perform similarly when they transform a regular convolution into a depthwise/groupwise convolution. To fairly evaluate the performance of each method, we reimplement these methods in ResNet50 since they have different baselines in their original papers. FLGC mainly transforms the 1×1 convolution into groupwise one and thus can reduce the FLOPs a great deal. However, FLGC also sacrifices the accuracy a lot in order to reach such a reduction on computational cost. On the contrary, our proposed LdsConv mainly transforms the 3×3 convolution into the depthwise separable one and make a sweet balance between the FLOPs and the accuracy. ND decomposes the regular convolution into the accumulation of several depthwise separable convolutions. While our approach aims to replace the standard convolution with a single depthwise separable convolution. Further more, our Lds-ResNet50-extreme replaces with only one depthwise convolution (w/o separable one) resulting a extreme reduction on computation cost which can be never transcended by ND. The goal of DGConv is to construct a groupwise convolution with dynamic groups. While our approach is to construct a depthwise (Lds-ResNet50-extreme) or depthwise separable convolution with most valuable filters. Our Lds-ResNet50-extreme plays a role as the upper bound of reduction on FLOPs for DGConv-ResNet50 and our Lds-ResNet50* simply surpasses the accuracy with fewer extra FLOPs. As shown in Table 5, our Lds-ResNet50* outperforms other methods in terms of accuracy and still has a considerable reduction on FLOPs and number of parameters. Our Lds-ResNet50-extreme remains a comparable accuracy with strong compression on the model.

Table 5. The table shows the comparison with similar methods for ResNet50 on ImageNet. '*' refers to the LdsConv using the balance loss.

Model	Error% (Top-1)	GFLOPs	Params (M)
FLGC-ResNet50 [48]	34.2	1.0	7.6
ND-ResNet50 [50]	26.7	3.12	20.6
DGConv-ResNet50 [49]	23.3	2.46	14.6
Lds-ResNet50 *	22.9	2.71	16.8
Lds-ResNet50-extreme	23.4	2.28	14.3

4.4. Network Visualization with Grad-CAM

We further apply the Grad-CAM [58] to models using images from the ImageNet validation set. Grad-CAM uses gradients to calculate the importance of the spatial locations in convolutional layers. As the gradients are calculate with respect to a specific class, Grad-CAM results show attended regions clearly. By visualizing the importance map for the network, we are able to understand which part the network is interested in and how the network is making use of the features for predicting a class. We compare the visualization results between our proposed Lds-ResNet50 and baseline (ResNet50) in Figure 7.

From Figure 7 it can be clearly seen that the Grad-CAM results of Lds-ResNet50 cover the target regions better than those of the original ResNet50. It suggests that LdsConv-integrated network learns well to exploit information in target regions and aggregate features from them.

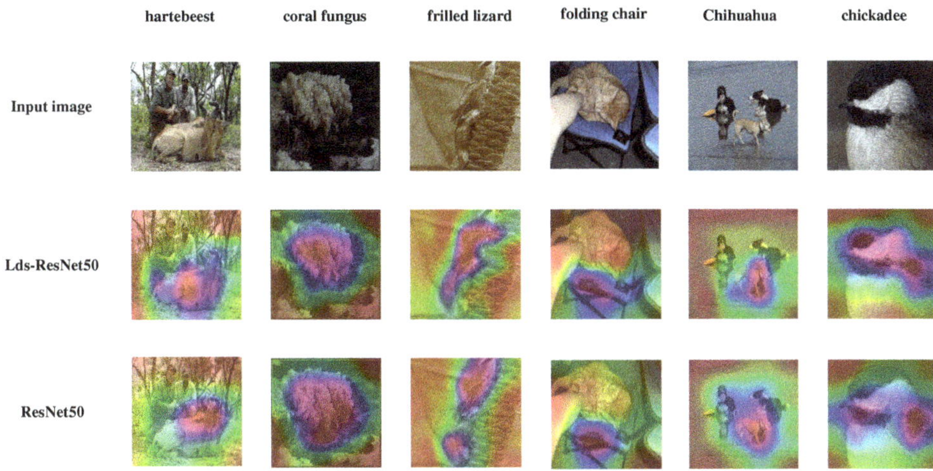

Figure 7. Grad-CAM [58] visualization results. We compare the visualization results between our Lds-ResNet50 and ResNet50. The Grad-CAM visualization is calculated for the last convolutional outputs. The ground-truth label is shown on the top of each input image.

5. Conclusions

In this work, we propose a new type of convolution called LdsConv. We have compared our proposed convolutional filters with the original convolutional filters on various existing architectures. Experimental results show that our LdsConv is more efficient than existing convolutions in these models. We also have compared the LdsConv method with the FLOPs compression methods and similar motivated works. Results from our experiments show that the proposed method produces the overall best accuracy while still having competitive FLOPs.

Author Contributions: Conceptualization, W.L. and Y.D.; investigation, Y.D., H.-L.W. and X.P.; methodology, W.L., Y.D. and X.P.; project administration, Y.D.; software, W.L.; supervision, Y.D. and H.-L.W.; validation, Y.Z.; visualization, W.L.; writing original draft, W.L.; writing review and editing, Y.D., H.-L.W., X.P. and Y.Z. All authors have read and agreed to the published version of the manuscript.

Funding: This research received no external funding.

Conflicts of Interest: The authors declare no conflict of interest.

References

1. He, K.; Zhang, X.; Ren, S.; Sun, J. Deep residual learning for image recognition. In Proceedings of the IEEE Conference on Computer Vision and Pattern Recognition, Las Vegas, NV, USA, 27–30 June 2016; pp. 770–778.
2. Krizhevsky, A.; Sutskever, I.; Hinton, G.E. Imagenet classification with deep Convolutional Neural Networks. In Proceedings of the Advances in Neural Information Processing Systems, Lake Tahoe, NV, USA, 3–6 December 2012; pp. 1097–1105.
3. Simonyan, K.; Zisserman, A. Very deep Convolutional Networks for large-scale image recognition. *arXiv* **2014**, arXiv:1409.1556.
4. Du, M.; Ding, Y.; Meng, X.; Wei, H.L.; Zhao, Y. Distractor-aware deep regression for visual tracking. *Sensors* **2019**, *19*, 387. [CrossRef] [PubMed]
5. Lyu, J.; Bi, X.; Ling, S.H. Multi-level cross residual network for lung nodule classification. *Sensors* **2020**, *20*, 2837. [CrossRef] [PubMed]
6. Xia, H.; Zhang, Y.; Yang, M.; Zhao, Y. Visual tracking via deep feature fusion and correlation filters. *Sensors* **2020**, *20*, 3370. [CrossRef] [PubMed]
7. Hwang, Y.J.; Lee, J.G.; Moon, U.C.; Park, H.H. SSD-TSEFFM: New SSD using trident feature and squeeze and extraction feature fusion. *Sensors* **2020**, *20*, 3630. [CrossRef]

8. Liang, S.; Gu, Y. Towards robust and accurate detection of abnormalities in musculoskeletal radiographs with a multi-network model. *Sensors* **2020**, *20*, 3153. [CrossRef]

9. Howard, A.G.; Zhu, M.; Chen, B.; Kalenichenko, D.; Wang, W.; Weyand, T.; Andreetto, M.; Adam, H. Mobilenets: Efficient Convolutional Neural Networks for mobile vision applications. *arXiv* **2017**, arXiv:1704.04861.

10. Chollet, F. Xception: Deep learning with depthwise separable convolutions. In Proceedings of the IEEE Conference on Computer Vision and Pattern Recognition, Honolulu, HI, USA, 21–26 July 2017; pp. 1251–1258.

11. Iandola, F.N.; Han, S.; Moskewicz, M.W.; Ashraf, K.; Dally, W.J.; Keutzer, K. SqueezeNet: AlexNet-level accuracy with 50× fewer parameters and <0.5 MB model size. *arXiv* **2016**, arXiv:1602.07360.

12. Zhang, X.; Zhou, X.; Lin, M.; Sun, J. Shufflenet: An extremely efficient Convolutional Neural Network for mobile devices. In Proceedings of the IEEE Conference on Computer Vision and Pattern Recognition, Salt Lake City, UT, USA, 18–23 June 2018; pp. 6848–6856.

13. Szegedy, C.; Vanhoucke, V.; Ioffe, S.; Shlens, J.; Wojna, Z. Rethinking the inception architecture for computer vision. In Proceedings of the IEEE Conference on Computer Vision and Pattern Recognition, Las Vegas, NV, USA, 27–30 June 2016; pp. 2818–2826.

14. Szegedy, C.; Ioffe, S.; Vanhoucke, V.; Alemi, A.A. Inception-v4, Inception-ResNet and the impact of residual connections on learning. In Proceedings of the Thirty-First AAAI Conference on Artificial Intelligence, San Francisco, CA, USA, 4–9 February 2017.

15. Wei, B.; Hamad, R.A.; Yang, L.; He, X.; Wang, H.; Gao, B.; Woo, W.L. A deep-learning-driven light-weight phishing detection sensor. *Sensors* **2019**, *19*, 4258. [CrossRef]

16. Ying, C.; Klein, A.; Christiansen, E.; Real, E.; Murphy, K.; Hutter, F. Nas-Bench-101: Towards reproducible neural architecture search. In Proceedings of the International Conference on Machine Learning, Long Beach, CA, USA, 10–15 June 2019; pp. 7105–7114.

17. Real, E.; Aggarwal, A.; Huang, Y.; Le, Q.V. Regularized evolution for image classifier architecture search. In Proceedings of the AAAI Conference on Artificial Intelligence, Honolulu, HI, USA, 27 January–1 February 2019; Volume 33, pp. 4780–4789.

18. Zoph, B.; Vasudevan, V.; Shlens, J.; Le, Q.V. Learning transferable architectures for scalable image recognition. In Proceedings of the IEEE Conference on Computer Vision and Pattern Recognition, Salt Lake City, UT, USA, 18–23 June 2018; pp. 8697–8710.

19. Liu, C.; Zoph, B.; Neumann, M.; Shlens, J.; Hua, W.; Li, L.J.; Li, F.-F.; Yuille, A.; Huang, J.; Murphy, K. Progressive neural architecture search. In Proceedings of the European Conference on Computer Vision (ECCV), Munich, Germany, 8–14 September 2018; pp. 19–34.

20. Zoph, B.; Le, Q.V. Neural architecture search with reinforcement learning. *arXiv* **2016**, arXiv:1611.01578.

21. Pham, H.; Guan, M.Y.; Zoph, B.; Le, Q.V.; Dean, J. Efficient neural architecture search via parameter sharing. *arXiv* **2018**, arXiv:1802.03268.

22. Hutter, F.; Kotthoff, L.; Vanschoren, J. *Automated Machine Learning: Methods, Systems, Challenges*; Springer Nature: Berlin, Germany, 2019.

23. Singh, P.; Verma, V.K.; Rai, P.; Namboodiri, V.P. Hetconv: Heterogeneous kernel-based convolutions for deep CNNs. In Proceedings of the IEEE Conference on Computer Vision and Pattern Recognition, Long Beach, CA, USA, 16–20 June 2019; pp. 4835–4844.

24. Chen, Y.; Fang, H.; Xu, B.; Yan, Z.; Kalantidis, Y.; Rohrbach, M.; Yan, S.; Feng, J. Drop an Octave: Reducing spatial redundancy in Convolutional Neural Networks with Octave Convolution. *arXiv* **2019**, arXiv:1904.05049.

25. Liao, S.; Yuan, B. CircConv: A structured Convolution with low complexity. In Proceedings of the AAAI Conference on Artificial Intelligence, Honolulu, HI, USA, 27 January–1 February 2019; Volume 33, pp. 4287–4294.

26. Vanhoucke, V. Learning visual representations at scale. *ICLR Invit. Talk* **2014**, *1*, 2.

27. Szegedy, C.; Liu, W.; Jia, Y.; Sermanet, P.; Reed, S.; Anguelov, D.; Erhan, D.; Vanhoucke, V.; Rabinovich, A. Going deeper with Convolutions. In Proceedings of the IEEE Conference on Computer Vision and Pattern Recognition, Boston, MA, USA, 7–12 June 2015; pp. 1–9.

28. Han, S.; Mao, H.; Dally, W.J. Deep compression: Compressing deep Neural Networks with pruning, trained quantization and Huffman coding. *arXiv* **2015**, arXiv:1510.00149.

29. Zhu, L.; Deng, R.; Maire, M.; Deng, Z.; Mori, G.; Tan, P. Sparsely aggregated Convolutional Networks. In Proceedings of the European Conference on Computer Vision (ECCV), Munich, Germany, 8–14 September 2018; pp. 186–201.
30. Luo, J.H.; Wu, J.; Lin, W. Thinet: A filter level pruning method for deep Neural Network compression. In Proceedings of the IEEE International Conference on Computer Vision, Venice, Italy, 22–29 October 2017; pp. 5058–5066.
31. Singh, P.; Kadi, V.S.R.; Verma, N.; Namboodiri, V.P. Stability based filter pruning for accelerating deep CNNs. In Proceedings of the 2019 IEEE Winter Conference on Applications of Computer Vision (WACV), Waikoloa Village, HI, USA, 9–11 January 2019; pp. 1166–1174.
32. He, Y.; Zhang, X.; Sun, J. Channel pruning for accelerating very deep Neural Networks. In Proceedings of the IEEE International Conference on Computer Vision, Venice, Italy, 22–29 October 2017; pp. 1389–1397.
33. Li, H.; Kadav, A.; Durdanovic, I.; Samet, H.; Graf, H.P. Pruning filters for efficient ConvNets. *arXiv* **2016**, arXiv:1608.08710.
34. He, Y.; Kang, G.; Dong, X.; Fu, Y.; Yang, Y. Soft filter pruning for accelerating deep Convolutional Neural Networks. *arXiv* **2018**, arXiv:1808.06866.
35. Singh, P.; Manikandan, R.; Matiyali, N.; Namboodiri, V. Multi-layer pruning framework for compressing single shot multibox detector. In Proceedings of the 2019 IEEE Winter Conference on Applications of Computer Vision (WACV), Waikoloa Village, HI, USA, 7–11 January 2019; pp. 1318–1327.
36. Singh, P.; Verma, V.K.; Rai, P.; Namboodiri, V.P. Leveraging filter correlations for deep model compression. *arXiv* **2018**, arXiv:1811.10559.
37. Rastegari, M.; Ordonez, V.; Redmon, J.; Farhadi, A. Xnor-Net: ImageNet classification using binary Convolutional Neural Networks. In Proceedings of the European Conference on Computer Vision (ECCV), Amsterdam, The Netherlands, 8–16 October 2016; Springer: Cham, Switzerland, 2016; pp. 525–542.
38. Park, E.; Yoo, S.; Vajda, P. Value-aware quantization for training and inference of Neural Networks. In Proceedings of the European Conference on Computer Vision (ECCV), Munich, Germany, 8–14 September 2018; pp. 580–595.
39. Zhang, D.; Yang, J.; Ye, D.; Hua, G. LQ-Nets: Learned quantization for highly accurate and compact deep Neural Networks. In Proceedings of the European Conference on Computer Vision (ECCV), Munich, Germany, 8–14 September 2018; pp. 365–382.
40. Ren, S.; He, K.; Girshick, R.; Sun, J. Faster R-CNN: Towards real-time object detection with region proposal Networks. In Proceedings of the Advances in Neural Information Processing Systems, Montreal, QC, Canada, 7–12 December 2015; pp. 91–99.
41. Cai, Z.; Vasconcelos, N. Cascade R-CNN: delving into high quality object detection. In Proceedings of the IEEE Conference on Computer Vision and Pattern Recognition, Salt Lake City, UT, USA, 18–23 June 2018; pp. 6154–6162.
42. Huang, G.; Liu, Z.; Van Der Maaten, L.; Weinberger, K.Q. Densely connected Convolutional Networks. In Proceedings of the IEEE Conference on Computer Vision and Pattern Recognition, Honolulu, HI, USA, 21–26 July 2017; pp. 4700–4708.
43. Hu, J.; Shen, L.; Sun, G. Squeeze-and-excitation networks. In Proceedings of the IEEE Conference on Computer Vision and Pattern Recognition, Salt Lake City, UT, USA, 18–23 June 2018; pp. 7132–7141.
44. He, K.; Zhang, X.; Ren, S.; Sun, J. Identity mappings in deep residual networks. In Proceedings of the European Conference on Computer Vision (ECCV), Amsterdam, The Netherlands, 8–16 October 2016; Springer: Cham, Switzerland, 2016; pp. 630–645.
45. Xie, S.; Girshick, R.; Dollár, P.; Tu, Z.; He, K. Aggregated residual transformations for deep Neural Networks. In Proceedings of the IEEE Conference on Computer Vision and Pattern Recognition, Honolulu, HI, USA, 21–26 July 2017; pp. 1492–1500.
46. Woo, S.; Park, J.; Lee, J.Y.; So Kweon, I. CBAM: Convolutional block attention module. In Proceedings of the European Conference on Computer Vision (ECCV), Munich, Germany, 8–14 September 2018; pp. 3–19.
47. Hu, M.; Lin, H.; Fan, Z.; Gao, W.; Yang, L.; Liu, C.; Song, Q. Learning to recognize chest-Xray images faster and more efficiently based on multi-kernel depthwise convolution. *IEEE Access* **2020**, *8*, 37265–37274. [CrossRef]

48. Wang, X.; Kan, M.; Shan, S.; Chen, X. Fully learnable group convolution for acceleration of deep Neural Networks. In Proceedings of the IEEE Conference on Computer Vision and Pattern Recognition, Long Beach, CA, USA, 16–20 June 2019; pp. 9049–9058.

49. Zhang, Z.; Li, J.; Shao, W.; Peng, Z.; Zhang, R.; Wang, X.; Luo, P. Differentiable learning-to-group channels via groupable Convolutional Neural Networks. In Proceedings of the IEEE International Conference on Computer Vision, Seoul, Korea, 27 October–2 November 2019; pp. 3542–3551.

50. Guo, J.; Li, Y.; Lin, W.; Chen, Y.; Li, J. Network decoupling: From regular to depthwise separable convolutions. *arXiv* **2018**, arXiv:1808.05517.

51. Huang, G.; Liu, S.; Van der Maaten, L.; Weinberger, K.Q. Condensenet: An efficient densenet using learned group convolutions. In Proceedings of the IEEE Conference on Computer Vision and Pattern Recognition, Salt Lake City, UT, USA, 18–23 June 2018; pp. 2752–2761.

52. Huang, G.; Li, Y.; Pleiss, G.; Liu, Z.; Hopcroft, J.E.; Weinberger, K.Q. Snapshot ensembles: Train 1, get m for free. *arXiv* **2017**, arXiv:1704.00109.

53. Loshchilov, I.; Hutter, F. SGDR: Stochastic gradient descent with warm restarts. *arXiv* **2016**, arXiv:1608.03983.

54. Deng, J.; Dong, W.; Socher, R.; Li, L.J.; Li, K.; Li, F.-F. Imagenet: A large-scale hierarchical image database. In Proceedings of the 2009 IEEE conference on computer vision and pattern recognition, Miami, FL, USA, 20–25 June 2009; pp. 248–255.

55. Krizhevsky, A.; Hinton, G. *Learning Multiple Layers of Features from Tiny Images*; Technical Report; Citeseer: Princeton, NJ, USA, 2009.

56. Yu, R.; Li, A.; Chen, C.F.; Lai, J.H.; Morariu, V.I.; Han, X.; Gao, M.; Lin, C.Y.; Davis, L.S. NISP: Pruning networks using neuron importance score propagation. In Proceedings of the IEEE Conference on Computer Vision and Pattern Recognition, Salt Lake City, UT, USA, 18–23 June 2018; pp. 9194–9203.

57. He, Y.; Liu, P.; Wang, Z.; Hu, Z.; Yang, Y. Filter pruning via geometric median for deep Convolutional Neural Networks acceleration. In Proceedings of the IEEE Conference on Computer Vision and Pattern Recognition, Long Beach, CA, USA, 16–20 June 2019; pp. 4340–4349.

58. Selvaraju, R.R.; Cogswell, M.; Das, A.; Vedantam, R.; Parikh, D.; Batra, D. Grad-CAM: Visual explanations from deep networks via gradient-based localization. In Proceedings of the IEEE international conference on computer vision, Venice, Italy, 22–29 October 2017; pp. 618–626.

Article

SlimDeblurGAN-Based Motion Deblurring and Marker Detection for Autonomous Drone Landing

Noi Quang Truong, Young Won Lee, Muhammad Owais , Dat Tien Nguyen, Ganbayar Batchuluun, Tuyen Danh Pham * and Kang Ryoung Park

Division of Electronics and Electrical Engineering, Dongguk University, 30 Pildong-ro 1-gil, Jung-gu, Seoul 04620, Korea; noitq.hust@gmail.com (N.Q.T.); lyw941021@dongguk.edu (Y.W.L.); malikowais266@gmail.com (M.O.); nguyentiendat@dongguk.edu (D.T.N.); ganabata87@gmail.com (G.B.); parkgr@dongguk.edu (K.R.P.)
* Correspondence: phamdanhtuyen@gmail.com; Tel.: +82-10-9264-4449; Fax: +82-2-2277-8735

Received: 11 June 2020; Accepted: 12 July 2020; Published: 14 July 2020

Abstract: Deep learning-based marker detection for autonomous drone landing is widely studied, due to its superior detection performance. However, no study was reported to address non-uniform motion-blurred input images, and most of the previous handcrafted and deep learning-based methods failed to operate with these challenging inputs. To solve this problem, we propose a deep learning-based marker detection method for autonomous drone landing, by (1) introducing a two-phase framework of deblurring and object detection, by adopting a slimmed version of deblur generative adversarial network (DeblurGAN) model and a You only look once version 2 (YOLOv2) detector, respectively, and (2) considering the balance between the processing time and accuracy of the system. To this end, we propose a channel-pruning framework for slimming the DeblurGAN model called SlimDeblurGAN, without significant accuracy degradation. The experimental results on the two datasets showed that our proposed method exhibited higher performance and greater robustness than the previous methods, in both deburring and marker detection.

Keywords: unmanned aerial vehicle; autonomous landing; deep-learning-based motion deblurring and marker detection; network slimming; pruning model

---

1. Introduction

Unmanned aerial vehicles (UAVs) or drones are successfully used in several industries. They have a wide range of applications such as surveillance, aerial photography, infrastructural inspection, and rescue operations. These applications require that the onboard system can sense the environment, parse, and react according to the parsing results. Scene parsing is a function that enables the system to understand the visual environment, such as recognizing the type of objects, place of objects, and regions of object instances in a scene. These problems are the main topics in computer vision—classification, object detection, and object segmentation. Object detection is a common topic and has attracted the most interest in recent studies. In object detection, traditional handcrafted feature-based methods showed limited performance [1–12]. A competitive approach is to apply deep-learning-based methods, which have gained popularity in recent years [13–16]. However, deploying deep learning models to a UAV onboard system raises new challenges—(1) difficulties of scene parsing in cases of low-resolution or motion-blurred input, (2) difficulties of deploying the model to an embedded system with limited memory and computation power, and (3) balancing between model accuracy and execution time. Autonomous landing is a core function of an autonomous drone, and it has become an urgent problem to be solved in autonomous drone applications. Recently, deploying deep learning models to UAV systems has become more feasible, because of both the growth in computing power and the extensive studies of deep neural networks, which have achieved significant results in scene parsing tasks, such as object detection tasks (e.g., faster region-based convolutional neural network (R-CNN) [17] and single-shot multibox detector (SSD) [18]. Therefore, the topic of autonomous drone landing has attracted much research interest, and the trend is toward autonomous landing by using deep-learning-based methods for tracking a guiding marker. Several state-of-the-art (SOTA) object detectors, based on convolutional neural networks (CNNs) have been proposed and deployed successfully for marker detection in marker tracking tasks. You only look once (YOLO) models might be the most popular deep object detectors in practical applications, because the detection accuracy and execution time are well balanced. Nevertheless, those systems have low robustness and are prone to failure when dealing with low-resolution [16] or motion-blurred images [19]. Such inputs need to be preprocessed before being fed to the detector. Thus, using a combination of a few networks as a pipeline is a promising approach to achieve this goal. In addition, drone landing causes motion of the attached camera. Even if a drone has an antivibration damper gimbal, the recorded frames are affected by motion blurring, especially in the case of high-speed landing [20]. For this reason, marker detection with motion-blurred input is a critical problem that necessarily needs to be addressed.

Therefore, we propose to learn efficient motion deblurring marker detection for autonomous drone landing, through a combination of motion deblurring and object detection, and apply a slimming deblurring model to balance the system speed with accuracy, on embedded edge devices. To this end, we trained the DeblurGAN network on our synthesized dataset and then pruned the model to obtain the slimmed version, SlimDeblurGAN. Moreover, we trained a variant of the YOLO detector on our synthesized dataset. Finally, we stacked SlimDeblurGAN and the detector, and then evaluated the system on a Desktop PC and a Jetson board TX2 environment. This research was novel compared to previous studies, in the following four ways:

- This is one of the first studies on simultaneous deep-learning-based motion deblurring and marker detection for autonomous drone landing.
- The balance of accuracy and processing speed is critical when deploying a marker tracking algorithm on an embedded system with limited memory and computation power. By proposing a dedicated framework for pruning the motion deblurring model, our proposed SlimDeblurGAN acquires the real-time speed on embedded edge devices, with high detection accuracy.
- Through an iterative performance of channel pruning and fine-tuning, our proposed SlimDeblurGAN showed a lower computing complexity but a higher accuracy of marker detection, compared to the state-of-the-art methods, including original DeblurGAN. The SlimDeblurGAN generator uses batch normalization, instead of instance normalization and imposes sparsity regularization. By performing channel pruning on the convolutional layers of the generator, SlimDeblurGAN has a more compact and effective channel configuration of the convolutional layers. Furthermore, it has a smaller number of trainable parameters than DeblurGAN. Thus, its inference time is shorter than that of the original DeblurGAN, with a small degradation of accuracy.
- The codes of pruning framework for slimming DeblurGAN, SlimDeblurGAN, and YOLOv2, two synthesized motion-blurred datasets, and trained models are available to other researchers through our website (Dongguk drone motion blur datasets and the pretrained models. http://dm.dgu.edu/link.html), for fair comparisons.

2. Related Works

There are numerous studies on autonomous drone landing, which can be classified into two types—those not considering motion blurring and those considering motion blurring.

Not considering motion blurring: At the initial stages, researchers considered objects on the runway with a lamp to guide the UAV, to determine a proper landing area. Gui et al. [1] proposed a vision-based navigation method for UAV landing, by setting up a system in which a near-infrared (NIR) light camera was integrated with a digital signal processing processor and a 940-nm optical filter was used to detect NIR light-emitting diode (LED) lamps on the runway. Their method had a significant advantage, i.e., it could not only work well in the daytime but also at nighttime. However, this method required a complicated setup of four LEDs on the runway. In addition, this method could only be performed in a wide area. Therefore, it failed to operate in narrow urban landing areas. Forster et al. [2] proposed a landing method including generating a 3D terrain depth map from the images captured by a downward-facing camera, and determining a secure area for landing.

It was completely proven that this method could work well in both indoor and outdoor environments. Nevertheless, the depth estimation algorithm was only tested at a maximum range of 5 m, and this method exhibited a slow processing speed. Two limitations of markerless methods are the difficulty of spotting a proper area for landing and the requirement of complicated setups for the landing area.

To solve these problems, marker-based methods were proposed. According to the type of features used, marker-based methods could be categorized into two kinds—handcrafted feature-based and deep feature-based methods. One of the handcrafted feature-based approaches that was robust to low-light conditions adopted a thermal camera-based method. These methods have high performance, even in nighttime scenarios, by using the emission of infrared light from a target on the ground. However, such methods require the drone to carry an additional thermal camera, as thermal cameras are not available in conventional drone systems. Other handcrafted marker-based approaches are based on visible-light cameras. Lin et al. [4] proposed a method to track the relative position of the landing area, using a single visible-light camera-based method. They used an international H-pattern marker to guide a drone landing in a cluttered shipboard environment. The characteristic of this method was that it could restore the marker from partial occlusion, and correctly detect the marker from complicated backgrounds. Moreover, they adopted the Kalman filter to fuse the vision measurement with the inertial measurement unit (IMU) sensor outputs, to obtain a more accurate estimate. Following that approach, Lange et al. [5] introduced a method to control the landing position of autonomous multirotor UAVs. They also proposed a new hexagonal pattern of landing pads, including concentric white rings on a black background and an algorithm to detect the contour rings from the landing pads. In addition, they used auxiliary sensors such as the SRF10 sonar sensor (Robot Electronics, Norfolk, UK), which accurately measured the current altitude above the ground, and the Avago ADNS-3080 optical flow sensor (Broadcom Inc., San Jose, CA, USA), which output the UAV's current velocity. These methods have the same disadvantage as the previous one, mandatorily carrying additional hardware, such as the IMU sensor, sonar sensor, and optical flow sensor. Some previous studies investigated UAV landing on a moving platform [6,20]. These studies take account of the six-degrees of freedom (6-DOF) pose of the marker, by using special landing pads, like fiducial markers. They also investigated a landing scenario in which the markers were positioned on the deck of a ship or placed on a moving platform. Other than landing on a fixed area, this method not only solved the marker-tracking problem but also tackled the more challenging obstacle. However, it requires more calculations and the estimation of relative position between the UAV and the moving target. Hence, they used SOTA computer vision methods, including multisensor fusion, tracking, and motion prediction of landing target on the moving platform. Consequently, the limitation of such methods is the short working range, due to the limited working range of the hardware employed. In particular, a previous study adopted the fiducial AprilTag [21] marker as the landing pad, owing to its robustness in difficult situations, such as severe rotation, heavy occlusion, light variation, and low image resolution. Although this study successfully tracked the marker in daytime conditions, the maximum distance between the landing target and the UAV was only approximately 7 m.

Araar et al. [7] proposed a new solution for multirotor UAV landing, using a new landing pad and relative-pose-estimation algorithm. In addition, they adopted two filters (an extended Kalman filter and an extended H_∞) to fuse the estimated pose and the inertial measurement. Although their method was highly accurate, it required information on the inertial measurements. Additionally, only indoor environment experiments were conducted, and the maximum working range was limited, owing to the drawback of the employed AprilTag marker. A novel idea was adopted in another study, taking advantage of cloud computing to overcome the limitation of the onboard hardware [11]. Specifically, the heavy computation tasks of computer vision were transferred to a cloud-based system, and the onboard system of the UAV only handled the returned results. Barták et al. [8] introduced an adequate handcrafted marker-based method for drone landing. Handcrafted feature-based techniques, such as blob pattern recognition, were adopted to identify and recognize the landing target. Control algorithms were also employed to navigate the drone to the appropriate target area. In this way, this method worked well in real-world environments. Nevertheless, their experiments were conducted only during daytime, and the maximum detection range was limited to 2 m. In an attempt to address autonomous UAV landing on a marine vehicle, Venugopalan et al. [9] proposed a method that adopted handcrafted feature-based techniques, like color detection, shape detection, pattern recognition, and image recognition, to track the landing target. Additionally, a searching and landing algorithm and a state machine-based method, were proposed. Their method worked well, with a success rate of over 75%, even in some difficult environmental conditions like oscillatory motion associated with the landing target or wind disturbance. However, the testing distance between the landing target and the UAV in their experiments was close. Wubben et al. [10] proposed the method for accurate landing of unmanned aerial vehicles, based on ground pattern recognition. In their method, a UAV equipped with a low-cost camera could detect ArUco markers sized 56×56 cm, from an altitude of up to 30 m. When the marker was detected, the UAV changed its flight behavior in order to land on the accurate position where the marker was located. Through experiments, they confirmed an average offset of only 11 cm from the target position, which vastly enhanced the landing accuracy, compared to the conventional global positioning system (GPS)-based landing, which typically deviated from the intended target by 1 to 3 m. Some researchers studied the autonomous landing of micro aerial vehicles (MAVs), using two visible-light cameras [12]. They performed a contour-based ellipse detection algorithm to track a circular landing pad marker in the images obtained from the forward-facing camera. When the MAV was close to the target position, the downward-facing camera was used because the fixed forward-facing camera view was limited. By using two cameras to extend the field of view of the MAV, the system could search for the landing pad even when it was not directly below the MAV. However, this method was only tested in an indoor scenario, which limited the working range.

In order to overcome the performance limitations of the handcrafted feature-based methods, deep feature-based methods were introduced, which exhibited high accuracy and increased detection range. Nguyen et al. [13] proposed a marker tracking method for autonomous drone landing, based on a visible-light camera on a drone. They proposed a variant of YOLOv2 named lightDenseYOLO to predict the marker location, including its center and direction. In addition, they introduced Profile Checker V2 to improve accuracy. As a result, their method could operate with a maximum range of 50 m. Similarly, Yu et al. [14] introduced a deep-learning-based method for MAV autonomous landing systems, and they adopted a variant of the YOLO detector to detect landmarks. The system achieved high accuracy of marker detection and exhibited robustness to various conditions, such as variations in landmarks under different lighting conditions and backgrounds. Despite achieving high performance in terms of detection range and accuracy, these methods did not consider input images under conditions like low-resolution and motion-blurred images. In another study, Polvara et al. [15] proposed a method based on deep reinforcement learning to solve the autonomous landing problem. Specifically, they adopted a hierarchy of double-deep Q-networks that were used as high-level control policies to reach the landing target. Their experiments, however, were only conducted in indoor environments.

Recently, Truong et al. [16] proposed a super-resolution reconstruction (SR) marker detection method for autonomous drone landing, by using a combination of SR and marker-detection deep CNNs, to track the marker location. Their proposed method successfully handled the obstacle of low-resolution input. Moreover, they introduced a cost-effective solution for autonomous drone landing, as their system required only a low-cost, low-resolution camera sensor, instead of expensive, high-resolution cameras. Furthermore, their proposed system could operate on an embedded system and acquired a real-time speed. However, they did not consider the case of motion blurring in the captured image. A low-resolution image was acquired by a low-resolution camera, including the small number of pixels from the camera sensor, but the motion blurring was caused by the f-number of the camera lens and the camera exposure time. A small f-number and a large exposure time caused a large amount of motion blurring in the captured image. It is often the case that motion blurring frequently occurred in the captured image by drone camera, because the image was captured while the drone was moving or landing. Therefore, we propose a new method of motion deblurring and marker detection for drone landing, which is completely different from the previous work [16], which considers only SR of the low-resolution image by drone camera, without motion deblurring. In addition, we propose a new network of SlimDeblurGAN for motion deblurring (that is different from previous work [16]), which used deep CNN with a residual net skip connection and network-in-network (DCSCN) for SR. Considering the motion blur method: All previous methods exhibited promising solutions for autonomous landing. They conducted experiments based on various scenarios like indoor, outdoor, daytime, and nighttime, as well as difficult conditions like low light and low resolution of the input. However, the input images under the motion blur effect, which frequently occur due to the movement of the drone, were not considered in their studies. Therefore, we propose a deep-learning-based motion deblurring and marker detection method for drone landing. These research studies [13,14] were about marker detection by a drone camera and did not consider the motion blurring in the captured image, which was different from our research considering motion deblurring. The research in [20] dealt with the motion blurring in the captured image by UAV, but they did not measure the accuracy of marker detection and the processing speed on the actual embedded system for the drone. Different from this research, we measured the accuracies of marker detection by our method and compared them with the state-of-the-art methods. In addition, we measured the processing speed of marker detection by our method on the actual embedded system for the processing on the drone and compared them with the state-of-the-art methods. The research [19] studied the detection of motion-blurred vehicle logo. However, its target was only for logo detection, which was different from our research of marker detection by a drone camera. Although the method in [13,14,21] achieved a 99% accuracy for landmark or marker, based on field experiments, they assumed only the slow movement or landing of drone, which did not generate the motion blurring. However, in the actual case of drone movement or landing at normal speed, motion blurring occurred frequently, as mentioned in [20]. Table 1 presents a comparison of the proposed and previous methods.

Table 1. Summary of theoretical comparisons between the proposed method and previous studies on vision-based drone landing.

Category		Type of Feature	Type of Camera	Description	Strength	Weakness
Not considering motion deblurring	Without marker	Handcrafted features	NIR light camera [1]	Detect the position of four infrared light-emitting diode (LED) lamps on the runway by using a camera attached below the head of the UAV	• The maximum detection range is 450 m • Works in both day and nighttime • Realtime processing	Setting up the NIR lamps on the ground is difficult in various places
			Single down-facing visible-light camera [2]	• Generate a local 3D elevation map of the ground environment by processing the input images from the camera sensor • Estimate a safe landing spot by a probabilistic method	Can find the landing spot in an emergency case without a marker	• The testing environment was ideal • The maximum height for testing was 4–5 m
	With marker	Handcrafted features	Thermal images [3]	Detect letter-based marker on thermal images	Can handle various illumination challenges in marker detection	Requires an expensive thermal camera
			Single visible-light camera [4–11]	Use line segments or contour detectors to detect the marker	Requires only a single visible-light camera.	• Marker detection only works in the daytime • Limited detection range
			Two visible-light cameras [12]	Vision-based method for cost-effective autonomous quadrotor landing by using a landing pad	Can detect the landing pad even in the case that only a part of landing pad is visible in the camera image	Indoor experiment and limited space

Table 1. *Font.*

Category	Type of Feature	Type of Camera	Description	Strength	Weakness
Considering motion deblurring	Deep features	Single visible-light camera	Use lightDenseYOLO CNN to detect the marker, and Profile Checker v2 to detect the center and direction of marker [13]	• Realtime processing on the embedded system • The maximum range is 50 m	• The system hardware requires deep CNN support • Requires a high-resolution camera
			Use a deep learning-based method for MAV autonomous landing system by detecting landmarks based on a variance of YOLO detector [14]	• High-accuracy marker detection • Robust to various conditions.	
			Marker detection by using double-deep Q-networks, and it navigates the drone to reach the target simultaneously [15]	Use deep reinforcement learning to solve the autonomous landing problem	Tested only in an indoor environment
		Low-resolution single visible-light camera [16]	Robust marker tracking by using a combination of SR and object detection	• Requires only a single cheap and low-resolution camera • Realtime processing	The system hardware requires deep CNN support
		Single visible-light camera (**proposed method**)	Robust marker tracking by using a two-phase framework of motion deblurring and marker detection.	• Can handle the non-uniform motion-blurred input images • Realtime processing	The system hardware requires deep CNN support

3. Proposed Method

3.1. A. Proposed Two-Phase Framework of Motion Deblurring and Marker Detection for Autonomous Drone Landing

Our goal was to propose a model M to accurately detect a marker object in a motion-blurred image x^{blur}. The factor blur indicated that the image x was affected by motion blur. For that, a framework was considered, which combined two models, including a motion deblurring model that acted as a preprocessing model (P) to predict the sharp image $\hat{y}^{sharp} = P(x^{blur}, \theta_P)$, and a follow-up marker detection model (S) that predicted the marker object based on the predicted sharp image $\hat{y} = S(\hat{y}^{sharp}, \theta_S)$. Here, θ_P is the set of trainable parameters of the preprocessing model (P), and θ_S is the set of trainable parameters of the marker detection model (S). This framework was promising because the motion deblurring model helped to recover the blurred input image to the sharp image, on which the detector could easily act. In addition, it had several advantages like separate independent training, guaranteed model convergence in the framework, and leveraging the SOTA models. Therefore, we proposed a two-phase framework for motion deblurring and marker detection, as shown in Figure 1. Phase I is a motion deblurring preprocessor P that uses our proposed SlimDeblurGAN model, and Phase II is the marker detector S that uses a YOLOv2 detector, which intakes the motion deblurred output from Phase I and outputs the predicted bounding box of the marker. The remainder of this section on the proposed method is organized according to the two phases.

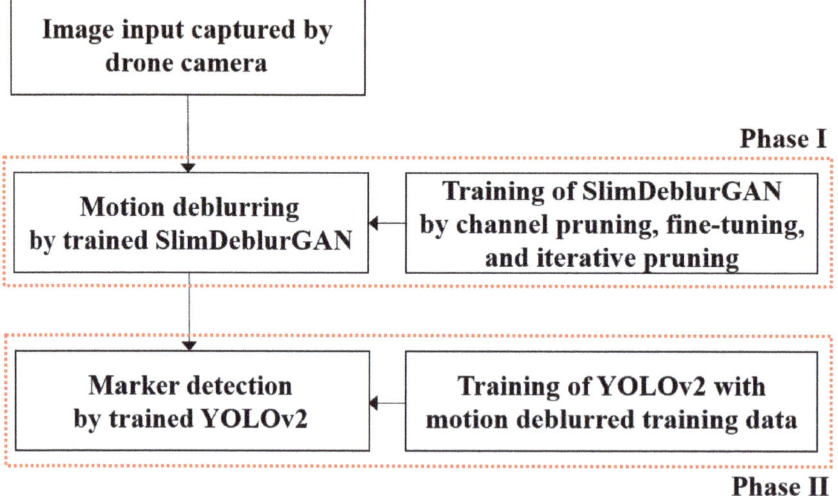

Figure 1. A block diagram of proposed motion deburring and marker detection system.

3.2. Phase I: Blind Motion Deblurring by SlimDeblurGAN

Motion deblurring is a method of sharpening the blurring of an image caused by the motion of object or camera during the exposure time. Such methods are categorized into two kinds—blind and nonblind deblurring. Nonblind deblurring methods assume that the blur source is known, whereas blind deblurring methods suppose that blur source is unknown, and they estimate both a latent sharp image and blur kernels. Kupyn et al. proposed DeblurGAN [22], which is a blind motion deblurring method that achieved the SOTA performance, while being faster than its closest competitor, DeepDeblur [23], by a factor of five. In this study, we did not directly use DeblurGAN in our framework; instead, we used a slim version that was pruned from the base model DeblurGAN. The pruning process is described in Section 3.2.2. Section 3.2.1 briefly explains the original DeblurGAN.

3.2.1. Blind Motion Deblurring by DeblurGAN

The family of conditional generative adversarial network (cGAN) [24] was successfully applied in some image translation applications such as super-resolution [25], style transfer [26], and motion deblurring. DeblurGAN was designed as a cGAN using the Wasserstein GAN gradient penalty (WGAN-GP) [27] as the critic function. Training GAN models required the procedure of finding a Nash equilibrium of a noncooperative two-player game [28]. Sometimes the gradient descent does this and at others, it does not, and no good equilibrium-finding algorithm was reported yet. These difficulties led to a novel idea, WGAN, which used an alternative objective function—using the Wasserstein distance instead of the traditional Jensen–Shannon distance, because it helped to increase the training stability [29]. Gulrajani et al. [27] then proposed WGAN-GP, which was an updated version, robust to the choice of generator architecture. For this crucial reason, DeblurGAN adopted WGAN-GP as a critic function, which allowed DeblurGAN to use a lightweight CNN architecture as a generator. The DeblurGAN architecture included a generator network and a critic network, as shown in Figure 2.

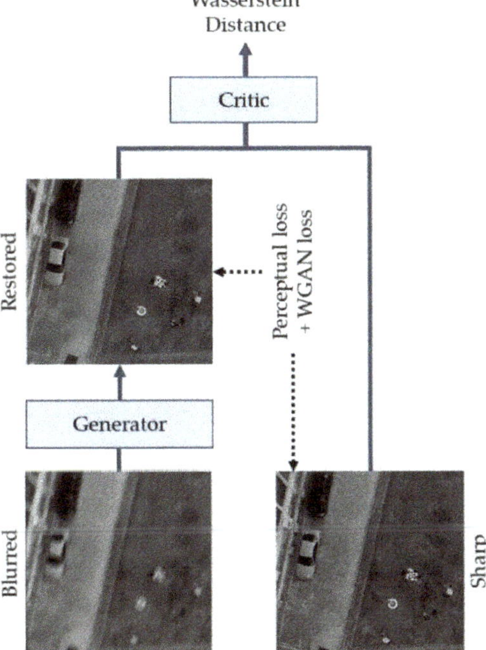

Figure 2. DeblurGAN architecture. The generator recovers the latent sharp image from the blurred image. The critic outputs the distance between the restored and sharp images. The total loss consists of perceptual loss and WGAN loss from the critic. After training and channel pruning, only the generator is used in Phase I.

The generator was the same as that proposed by Johnson et al. [30] for style transfer tasks. It contained two convolution blocks, nine residual blocks [31] (ResBlocks), and two transposed convolution blocks. Each ResBlock had a convolution layer, instance normalization layer [32], and rectified linear unit (ReLU) [33] activation. In contrast to the original one proposed by Johnson et al. [30], the DeblurGAN generator had an additional global skip connection, which was referred to as ResOut. The detailed information of the generator architecture is shown in Table 2. The critic network architecture was identical to that of PatchGAN [34].

Table 2. DeblurGAN generator architecture (CL means convolutional layer).

Layer Type	The Number of Filters	Size of Kernel (Height × Width)	The Number of Strides	Size of Output (Height × Width × Channel)
Input layer (Res)	-	-	-	$256 \times 256 \times 3$
1st CL	64	7×7	1×1	$256 \times 256 \times 64$
2nd CL	128	3×3	2×2	$256 \times 256 \times 128$
3rd CL	256	3×3	2×2	$256 \times 256 \times 256$
ResBlock $\begin{bmatrix} 3 \times 3\ conv,\ 1 \\ 3 \times 3\ conv,\ 1 \end{bmatrix} \times 9$	256	3×3	1×1	$256 \times 256 \times 256$
1st transpose CL	128	3×3	1×1	$256 \times 256 \times 128$
2nd transpose CL	64	3×3	1×1	$256 \times 256 \times 64$
Last CL	3	7×7	1×1	$256 \times 256 \times 3$
Output [Res + Last CL]	-	-	-	$256 \times 256 \times 3$

DeblurGAN loss included content loss and adversarial loss, as shown in Equation (1):

$$\mathcal{L} = \mathcal{L}_{GAN} + \lambda \cdot \mathcal{L}_X, \tag{1}$$

where the total loss $\mathcal{L}$ is the sum of the adversarial loss $\mathcal{L}_{GAN}$ and content loss $\mathcal{L}_X$; the coefficient λ denotes the balance between the two types of losses and it was set to 100 in all experiments.

Adversarial loss was described as:

$$\mathcal{L}_{GAN} = \sum_{n=1}^{N} -D_{\theta_D}\left(G_{\theta_G}\left(X^{blur}\right)\right). \tag{2}$$

D_{θ_D} and G_{θ_G} are the discriminator and generator, respectively. θ_D and θ_G. are the trainable parameters of the discriminator and generator, respectively.

Content loss was the perceptual loss [30], which was defined as:

$$\mathcal{L}_X = \frac{1}{W_{i,j}H_{i,j}} \sum_{x=1}^{W_{i,j}} \sum_{y=1}^{H_{i,j}} \left(\phi_{i,j}\left(X^{sharp}\right)_{x,y} - \phi_{i,j}\left(G_{\theta_G}\left(X^{blur}\right)\right)_{x,y}\right)^2, \tag{3}$$

where $\phi_{i,j}$ is the feature map obtained by the ith convolution within the VGG19 network, pretrained on ImageNet [35], and $W_{i,j}$ and $H_{i,j}$ are the width and height of the feature maps, respectively. θ_G is the set of trainable parameters of the generator (G_{θ_G}).

The authors proved experimentally that without this perceptual loss or without replacing the perceptual loss with a simple mean square error (MSE), the network did not converge to a meaningful state [22].

3.2.2. Proposed SlimDeblurGAN

As we adopted a two-phase process, the execution time of the proposed framework $t_{overall}$ was the sum of the time for each model element in two phases, including the processing time of motion deblurring t_P, and that of marker detection t_D, as illustrated in Equation (4).

$$t_{overall} = t_{Phase\ I} + t_{Phase\ II} = t_P + t_D. \tag{4}$$

In Equation (4), the processing time of detection was much shorter than that of the motion deblurring model. Informatively, the processing time of the YOLOv2 detector was shorter than that of

DeblurGAN, by almost 17 times. Therefore, a slimmed deblurring model P was crucial to reduce the execution time, and thus increased the execution speed of the overall system.

Considering the recently proposed methods for network lightening, such as using MobileNet [36] as a backbone, manually reducing the number of layers, network slimming [37], knowledge distillation [38], and dynamic computation [39], we settled on the network slimming proposed by Liu et al. [37]. This was a novel learning scheme for learning efficient convolutional networks, which reduced the model size, decreased the run-time memory footprint, and lowered the number of computing operations, without compromising accuracy. Essentially, the network slimming method is a technique to learn more compact CNNs. It directly imposed sparsity-induced regularization on the scaling factors in batch normalization layers, and the unimportant channels could thus be automatically identified during training, which could then be pruned. It is conceptually easy to understand; however, proposing a framework that can prune well for every network is challenging, as each network has its different components and irregular network architecture. Liu et al. applied a network slimming method to prune image classifier CNNs [37]. Zhang et al. [40] then extended the scheme to a coarse-grained method and successfully applied it to a slim YOLOv3 network. Inspired by the works of Liu and Zhang et al. [37,40], we proposed a model pruning procedure for pruning the DeblurGAN model to obtain SlimDeblurGAN, as shown in Figure 3 and Table 3.

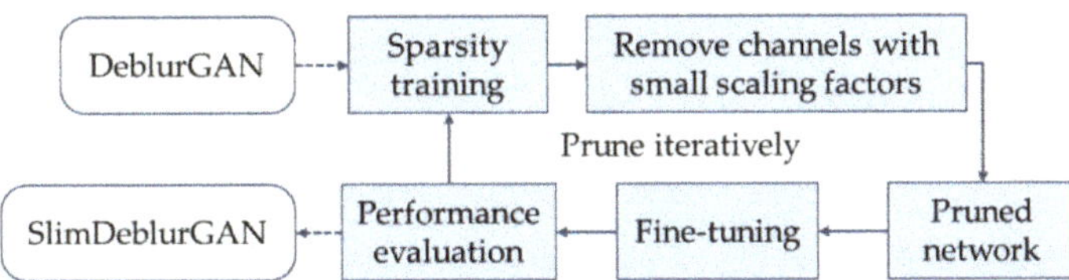

Figure 3. Iterative procedure of pruning of the DeblurGAN model through sparsity training and channel pruning for SlimDeblurGAN.

Adapting DeblurGAN for model pruning. Our goal was to reduce the processing time of the proposed system by reducing the execution time of Phase I. This phase was a motion deblurring task that could be performed by the generator of DeblurGAN. In addition, only the generator was kept and employed in the testing time. Therefore, we conducted the process of training and pruning DeblurGAN to obtain SlimDeblurGAN. To this end, we proposed a slimming framework for pruning only the generator, while keeping the remaining part of the DeblurGAN. We adapted DeblurGAN for the pruning process, by modifying the generator architecture. In more detail, the original DeblurGAN generator used the instance normalization layer; however, we replaced all instance normalization layers with batch normalization (BN) layers and imposed L1 regularization on the BN layers.

Channel-level sparsity training of DeblurGAN. Sparsity could be implemented at different levels, such as the weight level, kernel level, layer level, or channel level. Among these levels, the channel level provided the best tradeoff between flexibility and ease of implementation. The idea of channel-level sparsity training was to adopt a scaling factor γ for each channel, where $|\gamma|$ denoted the channel importance, and then to jointly train the network weights and the scaling factors. As there are some identical properties between desired architecture and the BN architecture, the implementation of channel-level sparsity could leverage the BN layer.

Specifically, the BN layer was formulated as shown in the following equations:

$$\hat{z} = \frac{z_{in} - \mu_{\mathrm{B}}}{\sqrt{\sigma_{\mathrm{B}}^2 + \epsilon}}, \tag{5}$$

$$z_{out} = \gamma\hat{z} + \beta, \tag{6}$$

where z_{in}, μ_B, and σ_B^2 are respectively the input features, mean, and variance of input features in a minibatch, and γ and β denote the trainable scale factor and bias (scale and shift), respectively. The trainable scale factor γ in the BN layer could be adopted as an indicator of channel importance. To impose sparsity regularization, a sparsity-induced penalty was added to the training objective ($loss_{network}$), which was given as

$$L = loss_{network} + \lambda \sum_{\gamma} f(\gamma), \tag{7}$$

where $f(\gamma) = |\gamma|$ and λ denotes the penalty factor. γ is the trainable scale factor.

Table 3. SlimDeblurGAN: DeblurGAN generator architecture after the model pruning process (CL—convolutional layer).

Layer Type		The Number of Filters	Size of Kernel (Height × Width)	The Number of Strides	Size of Output (Height × Width × Channel)
Input layer (Res)		-	-	-	256 × 256 × 3
1st CL		10	7 × 7	1 × 1	256 × 256 × 10
2nd CL		69	3 × 3	2 × 2	256 × 256 × 69
3rd CL		251	3 × 3	2 × 2	256 × 256 × 251
ResBlock 1	4th CL	107	3 × 3	1 × 1	256 × 256 × 107
	5th CL	251	3 × 3	1 × 1	256 × 256 × 251
ResBlock 2	6th CL	59	3 × 3	1 × 1	256 × 256 × 59
	7th CL	251	3 × 3	1 × 1	256 × 256 × 251
ResBlock 3	8th CL	24	3 × 3	1 × 1	256 × 256 × 24
	9th CL	251	3 × 3	1 × 1	256 × 256 × 251
ResBlock 4	10th CL	9	3 × 3	1 × 1	256 × 256 × 9
	11th CL	251	3 × 3	1 × 1	256 × 256 × 251
ResBlock 5	12th CL	3	3 × 3	1 × 1	256 × 256 × 3
	13th CL	251	3 × 3	1 × 1	256 × 256 × 251
ResBlock 6	14th CL	3	3 × 3	1 × 1	256 × 256 × 3
	15th CL	251	3 × 3	1 × 1	256 × 256 × 251
ResBlock 7	16th CL	3	3 × 3	1 × 1	256 × 256 × 3
	17th CL	251	3 × 3	1 × 1	256 × 256 × 251
ResBlock 8	18th CL	13	3 × 3	1 × 1	256 × 256 × 13
	19th CL	251	3 × 3	1 × 1	256 × 256 × 251
ResBlock 9	20th CL	20	3 × 3	1 × 1	256 × 256 × 20
	21st CL	251	3 × 3	1 × 1	256 × 256 × 251
1st transpose CL		128	3 × 3	1 × 1	256 × 256 × 128
2nd transpose CL		64	3 × 3	1 × 1	256 × 256 × 64
Last CL		3	7 × 7	1 × 1	256 × 256 × 3
Output [Res + Last CL]		-	-	-	256 × 256 × 3

Channel pruning. To achieve this goal, we adopted an expected pruning ratio r, which was an expected ratio of the number of expected pruned channels to the overall feature channels. Based on r and the sorted list of all $|\gamma|$, a global threshold $\hat{\gamma}$ was experimentally obtained, which determined whether a channel of the feature map was to be pruned. Feature channels, whose scaling factors were smaller than the threshold $\hat{\gamma}$ were pruned.

Fine-tuning SlimDeblurGAN. After channel pruning, model fine-tuning was necessary to compensate for temporary accuracy loss, which showed an even higher accuracy than the model without fine-tuning.

Iterative pruning. The repetition of the pruning procedure (as shown in Figure 3) helped to avoid over-pruning, which caused the pruned model degradation and could not be recovered by fine-tuning or performing more pruning steps.

3.2.3. Summarized Differences between the Original DeblurGAN and Proposed SlimDeblurGAN

We summarized the differences between the original DeblurGAN and the proposed SlimDeblurGAN, as follows.

- SlimDeblurGAN generator uses batch normalization instead of instance normalization and imposes sparsity regularization.
- By performing channel pruning on convolutional layers of the generator, SlimDeblurGAN has a more compact and effective channel configuration of the convolutional layers and it has a smaller number of trainable parameters than DeblurGAN. Thus, its inference time is shorter than that of the original DeblurGAN, with a small degradation of accuracy, as shown in the experimental section.

3.3. Phase II: Marker Detection by YOLOv2 Detector

Deep object detectors have attracted much interest in recent years. Several SOTA deep object detectors were proposed, including Fast R-CNN [41], Faster R-CNN [17], R-FCN [42], RetinaNet [43], SSD [18], and the YOLO series (YOLO [44], YOLOv2 [45], and YOLOv3 [46]). According to the adoption of extra region proposal modules, these could be categorized into two classes—two-stage and one-stage detectors. In particular, the YOLO series, which were one-stage detectors, were widely adopted in practical applications [44–46], because the accuracy and speed were well-balanced. Therefore, we adopted YOLOv2 using Darknet19 of Figure 4 as the main part of the feature extraction, as a marker detector in Phase II.

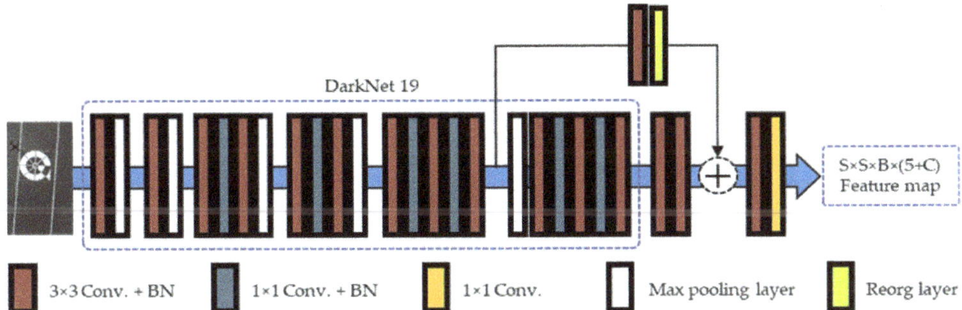

Figure 4. YOLOv2 backbone convolutional neural networks (CNN) architecture. The backbone network is a combination of the first 18 layers of Darknet19 and the YOLOv2 header. The header includes four convolutional layers and a Reorg layer. The Reorg layer refers to reorganization, which manipulates the output feature map of the 13th convolution of Darknet19 to obtain a reorganized feature map. The Concat layer concatenates the reorganized feature map and the output feature map of the 20th convolution. The final output is the feature map of shape $S \times S \times B \times (5 + C)$.

In YOLOv2 [45], the Reorg layer was used to reshape the feature map, so that the width and height of the input feature map matched the other output feature map, and these two feature maps could be concatenated together. However, our marker datasets were quite different, compared to other object detection datasets like COCO [47], because the number of classes to be detected was only one (the marker), and its size varied from small to large, according to the height of the drone above the landing area. Therefore, it was necessary to adapt YOLOv2 to remove the redundant computations. First, we considered the anchor boxes. As the marker was a circle-based shape, the ground truth bounding boxes were theoretically square. However, the image size of our dataset was 1280 × 720 pixels, and in training and testing, the input was resized to 320 × 320 pixels by bilinear interpolation [48]. Hence, the height-to-width ratio of the marker was changed to a certain ratio. Therefore, instead of choosing the anchor boxes by hand or using anchor boxes obtained from other datasets, we normalized all ground truth bounding boxes of the training dataset and clustered them through K-means clustering with a distance metric, to obtain the proper anchor boxes for our dataset. The number of anchor boxes and their sizes could be determined by the elbow curve method on the graph of the number clustering and the IoU threshold. Second, we set the input size for the backbone network of 320 × 320 pixels as close to the output of Phase I, 256 × 256 pixels.

According to the YOLOv2 network [45], the output image was represented as S × S grids, and S was defined as 10. Therefore, the output feature map of the S × S grids was 10 × 10. B was the number of anchor boxes (in our experiment, B was defined as 3). In detail, each grid could have three anchor boxes for representing the detected object [45]. The output shape of the feature map of the YOLOv2 network was S × S × B × (5 + C) [45]. Here, "5" meant center x, center y, width, height, and confidence of one anchor box. In addition, C was the number of class probability (in our case, C was one because there was only one marker class) of one anchor box. Consequently, the output shape of the feature map of the YOLOv2 network (S × S × B × (5 + C)) became 10 × 10 × 3 × 6 in our case.

4. Experimental Results

4.1. Experimental Environment and Datasets

As there was no open dataset of blurred images acquired from landing drones, we synthesized images from the Dongguk drone camera database version 2 (DDroneC-DB2) [13] as the synthesized motion blur drone database 1 (SMBD-DB1), and we also obtained the real motion blur drone database 1 (RMBD-DB1), which contained real motion blur in the wild.

Training and testing were performed based on a two-fold cross-validation method on these two databases. All subsets were equally distributed to training and testing per fold. In the 1st fold validation, half of the images in the SMBD-DB1 were used for training, and the other half was for testing. This procedure was repeated by exchanging the training and testing data with each other in the 2nd fold validation. The average accuracy from the two-fold validations was determined as the final accuracy. The same rule was also applied to RMBD-DB1. The details of these two datasets are presented in the following section.

SMBD-DB1: This dataset was generated by following the idea proposed by Kupyn et al. [22], which was based on random trajectory generation. More specifically, motion-blurred images were generated by applying the motion-blurring kernels to the original images. The motion-blurring kernels were created by applying subpixel interpolation to the trajectory vector. Each trajectory vector, which was a complex-valued vector, corresponded to the discrete positions of an object undergoing 2D random motion in a continuous domain. Figure 5 illustrates four samples of SMBD-DB1, including the motion-blurring kernels, original images, and obtained blurred images. SMBD-DB1 contains 10,642 images generated from three sub-datasets (acquired in the morning, afternoon, and evening) of DDroneC-DB2, as shown in Table 4.

RMBD-DB1: For validating our method in a real-world environment, we used an additional dataset that contained real motion blur in the wild. We obtained six drone-landing videos and obtained 2991 images. The details of RMBD-DB1 are shown in Table 5. Therefore, the RMBD-DB1 included real motion of captured images by a real drone camera, and our method was expected to be performed on a real drone. The third case of Figure 6 shows the real motion blurring when the drone was fast landing.

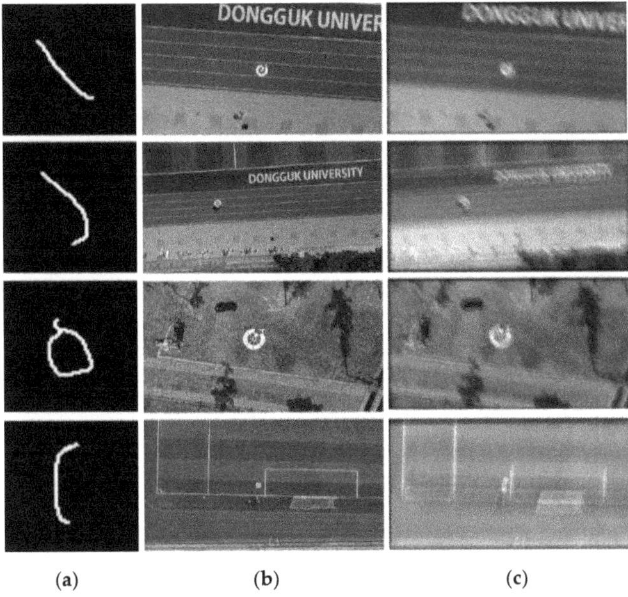

(a)   (b)   (c)

Figure 5. Examples of SMBD-DB1. (**a**) Synthetically generated kernels, (**b**) ground truth (original images), and (**c**) motion-blurred images.

(a)   (b)

Figure 6. Examples of RMBD-DB1. (**a**) Ground truth (original images), and (**b**) motion-blurred images.

Table 4. Descriptions of SMBD-DB1.

Sub-Dataset	The Number of Images
Morning	4154
Afternoon	2640
Evening	3848
Total	10,642

Table 5. Descriptions of RMBD-DB1.

Sub-Dataset	The Number of Images
Morning	502
Afternoon	1557
Evening	932
Total	2991

4.2. Training the Proposed Method

The training of our method included two parts to train SlimDeblurGAN and YOLOv2, as explained in the following sections. All experiments, including both training and testing, were performed on a desktop computer with an Intel®Core™ i7-3770K CPU 3.5 GHz (4 cores) (Intel Corp., Santa Clara, CA, USA), 8 GB of main memory, and an NVIDIA GeForce 1070 graphics processing unit (GPU) card (1920 compute unified device architecture (CUDA) cores, and 8 GB of graphics memory) (NVIDIA Corp., Santa Clara, CA, USA).

4.2.1. Training SlimDeblurGAN

We conducted a model pruning process, as mentioned in the previous section, to obtain SlimDeblurGAN. First, we created the base model and performed sparse training. Second, we repeated the iterative process of pruning and fine-tuning, until the resulting model showed a critical degradation of accuracy. We chose a model from among all pruned models throughout the pruning process, which had the best balance between accuracy and model size. Aiming to generate the base model, we replaced all instance normalizations by batch normalizations in the generator network for model pruning adaptation. As a result, we could increase the batch size to accelerate the training process. The training batch size of the DeblurGAN base model was chosen as 8, as it was the maximum batch size for which the model could be loaded in our training environment. We retained this batch size in the fine-tuning of the iterative pruning process. The number of parameters and accuracies of the resulting models obtained from the iterative channel pruning process are presented in Table 6. In this table, the peak signal-to-noise ratio (PSNR) [48] was widely used for mathematical measurements of image quality, based on the mean square error (MSE) between the pixels of ground truth image (Im(i,j)) and motion-deblurred image (Res(i,j)), as shown in Equations (8) and (9). The structural similarity measure (SSIM) index [49] could also predict the perceived quality of images. In detail, SSIM was the index showing the similarity between two images based on means, standard deviations, and the covariance of the two images.

$$\mathrm{MSE} \;=\; \frac{\left(\sqrt{\sum_{j=1}^{M}\sum_{i=1}^{N}\left(\mathrm{Im}(i,j)-\mathrm{Res}(i,j)\right)^{2}}\right)^{2}}{MN},\tag{8}$$

$$\mathrm{PSNR} \;=\; 10log_{10}\!\left(\frac{255^{2}}{MSE}\right),\tag{9}$$

where M and N represent the width and height of image, respectively.

Table 6. DeblurGAN channel pruning iterations.

Iteration	Model	The Number of Parameters of Generator	PSNR/SSIM
0	Base model (DeblurGAN)	11.39×10^6	20.59/0.49
1	1st pruned model	2.47×10^6	21.46/0.39
2	2nd pruned model (SlimDeblurGAN)	1.64×10^6	20.92/0.34
3	3rd pruned model	1.14×10^6	18.16/0.26

The base model had 11.39 million parameters in the generator, and we performed sparse training from scratch with this model, which showed successful convergence with a PSNR of 20.59 and an SSIM of 0.49 on SMBD-DB1. We further performed channel pruning and fine-tuning with this trained base model. As a result, we obtained the first pruned model with the number of parameters in the generator reduced to 2.47 million. The fine-tuning yielded a PSNR of 21.46 and an SSIM of 0.39. Likewise, we performed the next channel pruning iterations, in which the base model was replaced with the resulting model from the previous iteration. The number of parameters in the generator after the second and third iterations decreased to 1.64 million and 1.14 million, respectively. Meanwhile, the accuracies of the resulting models degraded to 20.92 (PSNR) and 0.34 (SSIM) after the second iteration, followed by 18.16 (PSNR) and 0.26 (SSIM) after the third iteration. As shown in Table 6, the number of remaining parameters of the generator was dramatically reduced by 4.6 times after the first iteration, 6.9 times after the second iteration, and almost 10 times after the third iteration, compared to that of the base model. Although the accuracy increased after the first iteration, it decreased slightly after the second iteration, and critically degraded after the third iteration. We stopped the pruning process after the third iteration, as the observed degradation indicated over-pruning. We considered that the second pruned model had a good balance between the number of parameters and accuracy. Hence, we applied this slimmed model to the motion-deblurring phase of our system and referred to it as SlimDeblurGAN. The successful PSNR training with fine-tuning of the SlimDeblurGAN for 70 epochs appears in Figure 7.

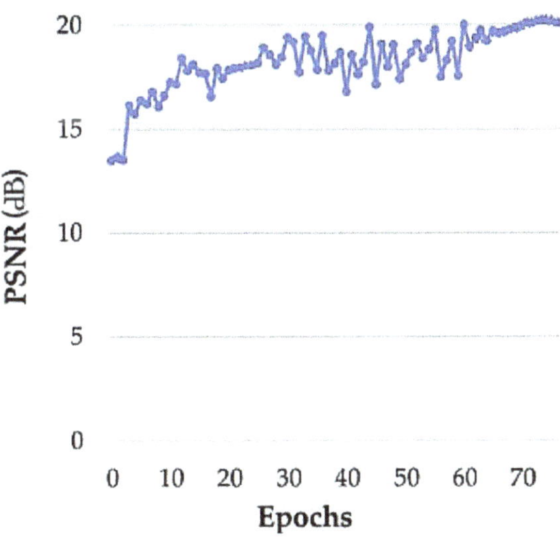

Figure 7. Training PSNR of SlimDeblurGAN.

4.2.2. Training Marker Detection CNN

In an attempt to facilitate the training process of the marker detection CNN, we considered the distribution of object bounding boxes in the training dataset, in order to generate a set of anchor boxes used by YOLOv2.

Generating anchor boxes: We performed K-means clustering on the bounding boxes of SMBD-DB1, based on the mean IoU distance with K from 2 to 9. As shown in Figure 8, we could determine the number of clusters by the elbow curve method. As a result, 3 was the best candidate for the number of clusters. In detail, the case of K from 2 to 3 showed the largest increment in the mean IoU and we chose 3 for the number of clusters in YOLOv2.

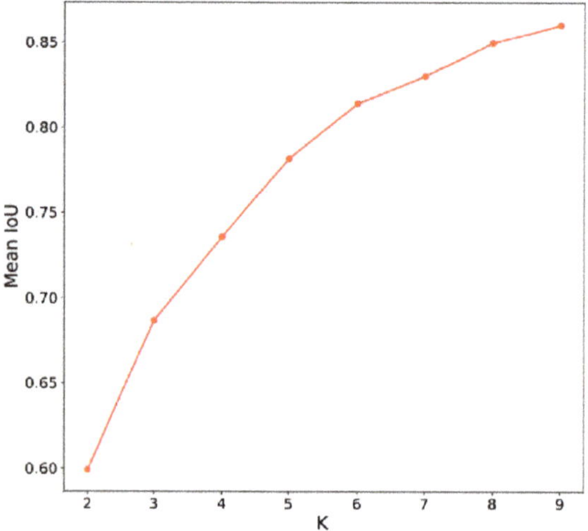

Figure 8. Mean IoU with respect to K for the optimal K determination.

The details of the three selected anchor boxes are shown in Table 7, and these anchor boxes are visualized in Figure 9. Notably, the actual size of the anchor boxes used in YOLOv2 depended on the grid size of the output feature map of the YOLOv2 backbone. In our experiments, we designed the backbone network to generate an output feature map grid of size 10×10. Therefore, the size of the anchor boxes was 10 times larger than the normalized size. The normalized bounding boxes are detailed in Table 7.

Table 7. Size of the three selected anchor boxes (* The ranges of normalized height and width are from 0 to 1, respectively).

Anchor Box	Normalized *		Size
	Height	Width	
Box 1	0.072	0.040	Small
Box 2	0.152	0.086	Medium
Box 3	0.371	0.212	Large

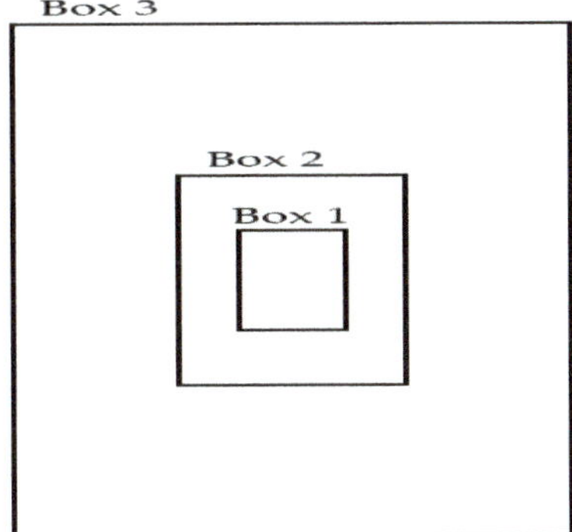

Figure 9. Anchor boxes.

4.3. Testing the Proposed Method

4.3.1. Accuracy of Motion Deblurring of the Proposed SlimDeblurGAN

We conducted training and channel pruning processes to obtain SlimDeblurGAN on one fold and tested on the other fold of SMBD-DB1. In addition to measuring the testing accuracy, we measured the number of floating-point operations (FLOPs) to show the effectiveness of our proposed SlimDeblurGAN in terms of computation, compared to DeepDeblur [23], DeblurGAN [22], and DeblurGAN, using MobileNet as the backbone nets [50]. All parameters for others [22,23,50] were optimally selected by us with training data. The average results of the measurements from the two folds are presented in Table 8. The comparison of SlimDeblurGAN and the SOTA methods is also illustrated in Figure 10. As shown in this figure and table, DeblurGAN [22] showed the highest PSNR of 21.6; however, it also had a very high number of operations, at 99.3 Giga FLOPs. DeepDeblur [23] showed the lowest PSNR and highest FLOPs. Both the SlimDeblurGAN and the DeblurGAN model using MobileNet as a backbone had a small number of FLOPS, nearly one-sixth as much as that of DeblurGAN. However, DeblurGAN using MobileNet failed to maintain accuracy with a PSNR of 19.5, whereas SlimDeblurGAN had a slightly decreased accuracy, with a PSNR of 20.9. Therefore, we confirmed that our channel pruning process successfully generated a compact version of DeblurGAN with fewer FLOPs, yet high accuracy.

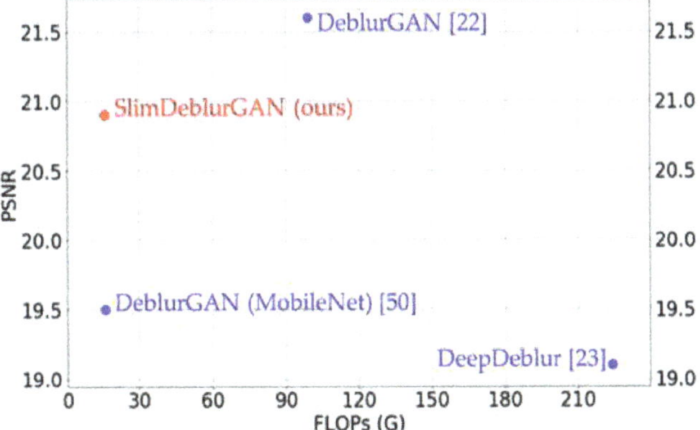

Figure 10. Comparison of DeepDeblur, DeblurGAN, DeblurGAN (MobileNet), and our proposed model of SlimDeblurGAN, in terms of motion deblurring accuracy and the number of FLOPs.

Table 8. Comparison of the number of FLOPs and accuracies of Deep Deblur [23], DeblurGAN [22], DeblurGAN using MobileNet [50], and our proposed model, SlimDeblurGAN.

Model	FLOPs (Giga)	SSIM/PSNR
DeepDeblur [23]	224.1	0.41/19.1
DeblurGAN [22]	99.3	0.40/21.6
DeblurGAN (MobileNet) [50]	16.1	0.32/19.5
SlimDeblurGAN (ours)	16	0.34/20.9

Figure 11 presents examples of motion deblurring by four methods, i.e., DeepDeblur [23], DeblurGAN [22], DeblurGAN (MobileNet) [50], and SlimDeblurGAN. As shown in the figure, the results by DeblurGAN (MobileNet) and by DeepDeblur [23] were worse than those of the other methods, because the marker was still blurred and had the ghost effect as in the motion-deblurred image. However, the results obtained by DeblurGAN and SlimDeblurGAN showed sharp, non-ghost effects and recognizable markers, even by the human eye. Although the accuracy of DeblurGAN was slightly higher than that of SlimDeblurGAN, as presented in Table 8, the results obtained from both methods were almost the same in terms of perceptual comparisons.

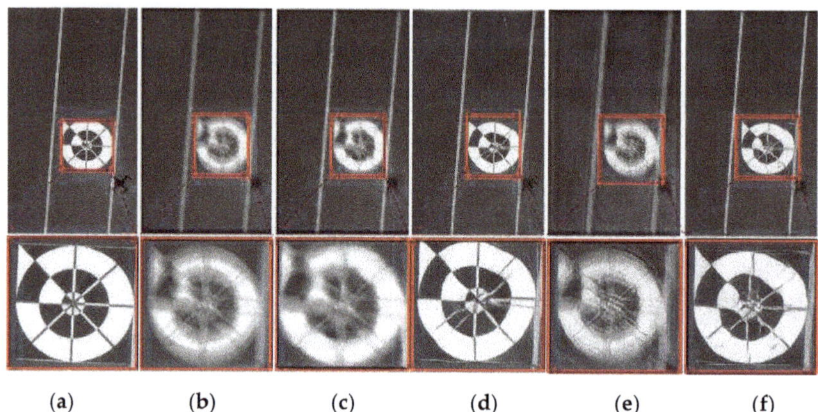

(a)   (b)   (c)   (d)   (e)   (f)

Figure 11. Examples of motion deblurring on SMBD-DB1. (**a**) Ground truth original image, (**b**) motion-blurred image, motion deblurring by (**c**) DeepDeblur, (**d**) DeblurGAN, (**e**) DeblurGAN (MobileNet), and (**f**) SlimDeblurGAN (ours).

4.3.2. Accuracy of Marker Detection

As the most common method of evaluating the performance of object detection system is to analyze the precision, recall, and F1 score at different IoU thresholds, we also evaluated our system in this way. These metrics were based on true positive (TP), false positive (FP), true negative (TN), and false negative (FN). In our study, TP, FP, TN, and FN could be determined by the following case studies of detection results:

Case 1: The system could not detect the marker on the image. We considered this case to be FN, as presented in Figure 12a.

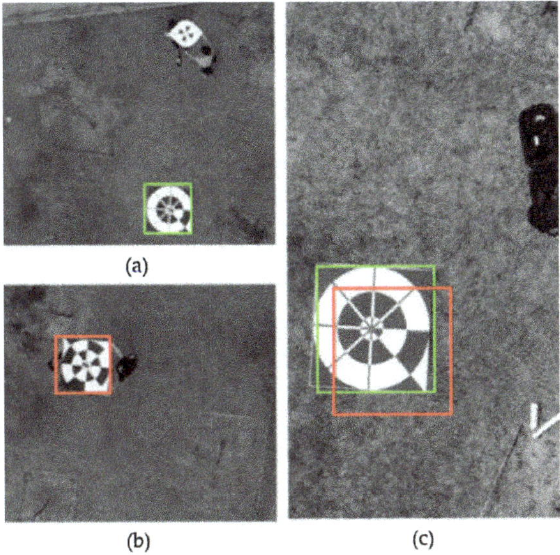

Figure 12. Examples of the detected results. The green boxes are ground truth labels; the red boxes are detected results. (**a**) The model cannot detect the maker on the image, and thus this case is considered as FN. (**b**) The detected result is not the marker but rather a marker-like object, and therefore this case is considered as FP. (**c**) The model can correctly detect the marker on the image; this case is considered as TP if the IoU is greater than or equal to the predefined threshold; otherwise, it is considered as FN and FP.

Case 2: The detected object was not the marker but rather a marker-like object (i.e., wrong detection). We considered this case to be FP, as shown in Figure 12b.

Case 3: The marker was detected by the system, as illustrated in Figure 12c; we considered the IoU between the detected bounding box and the ground truth bounding box. We compared the IoU score with the predefined threshold. If the IoU was greater than or equal to the predefined threshold, this case could be considered as TP; otherwise, it could be considered to be FP and FN.

Following these definitions, we counted the number of true positives (Num. of TP), the number of false positives (Num. of FP), and the number of false negatives (Num. of FN) on the testing dataset. As a result, the accuracies could be calculated based on Equations (10)–(12).

$$\text{Precision} = \frac{\text{Num. of TP}}{\text{Num. of TP} + \text{Num. of FP}}, \tag{10}$$

$$\text{Recall} = \frac{\text{Num. of TP}}{\text{Num. of TP} + \text{Num. of FN}}, \tag{11}$$

$$\text{F1 score} = 2 \cdot \frac{\text{Precision} \cdot \text{Recall}}{\text{Precision} + \text{Recall}}. \tag{12}$$

Precision reflected the proportion of correct detections out of the total number of detections, whereas recall indicated the proportion of correct detections to the total number of ground truth marker boxes in the testing dataset. There was a trade-off between precision and recall. If the model learned to predict with high precision, it tended to overfit, which caused a reduction in the recall. In contrast, if the model learned to be able to predict all markers in the dataset for a high recall, it could be more general in marker detection, which caused underfitting and degraded precision. The higher the precision, the lower recall, and vice versa. The F1 score turned out to be a better metric, which was based on precision and recall, as shown in Equation (12). In our experiments, we evaluated the detection performance on five methods—(1) YOLOv2, (2) a combination of DeepDeblur and YOLOv2, (3) a combination of DeblurGAN and YOLOv2, (4) a combination of a modified version of DeblurGAN that used MobileNet as the backbone, and YOLOv2, and (5) our proposed method—a combination of SlimDeblurGAN and YOLOv2. The measurements were conducted at different IoU thresholds on both synthesized motion-blur datasets. All parameters for other methods [21–23,45,50] were optimally selected by us with training data.

Testing accuracies on SMBD-DB1: The marker detection results on SMBD-DB1 are shown in Tables 9–11 and Figure 13. The precision, recall, and F1 scores of the five methods were very high at the low IoU thresholds and decreased with increasing IoU. At first glance, we could see that without any motion deblurring preprocessing, YOLOv2 exhibited low accuracies of marker detection on the motion-blurred input, as shown by the blue curves in Figure 13. The reason for this was that the motion blur strongly distorted the feature, pattern, and shape of markers in the images, making them difficult to detect. Obviously, the ghost effects in motion-blurred images were obstacles to accurately detecting the markers. Therefore, YOLOv2 could only detect the marker in images that were less affected by motion blur; otherwise, it failed. In an attempt to overcome this problem, YOLOv2 was trained directly on motion-blurred images, which could help YOLOv2 to increase its recall by learning to generalize its detection. This approach, however, decreased the precision, owing to the aforementioned tradeoff between them. Hence, the system was less accurate in distinguishing markers between marker-like objects, which caused a reduction in precision. However, with the help of highly accurate motion deblurring models such as DeblurGAN or SlimDeblurGAN, YOLOv2 obtained high precision, recall, and F1 score. Apart from DeblurGAN and SlimDeblurGAN, using DeblurGAN (MobileNet) made YOLOv2 yield lower detection accuracy than the other methods, including YOLOv2, without any deblurring preprocessing method. The failure of the DeblurGAN (MobileNet) indicated that adopting the MobileNet architecture in the backbone of the SOTA model to reduce the computation cost was not always effective. Our proposed SlimDeblurGAN showed a higher detection accuracy than the other methods, confirming that we successfully generated a compact and highly accurate motion-deblurring model of SlimDeblurGAN. In detail, the detection result by SlimDeblurGAN + YOLOv2 was slightly higher than that by the DeblurGAN + YOLOv2, as shown in Tables 9–11 and Figure 13. However, the number of parameters and FLOPs of SlimDeblurGAN were less by factors of 10 and 6, respectively, than those of DeblurGAN, as shown in Tables 6 and 8.

Table 9. Comparisons of precision on SMBD-DB1.

Methods	IoU Threshold														Average
	0.3	0.35	0.4	0.45	0.5	0.55	0.6	0.65	0.7	0.75	0.8	0.85	0.9	0.95	
YOLOv2 [45]	0.980	0.969	0.953	0.927	0.890	0.840	0.776	0.681	0.563	0.407	0.247	0.117	0.029	0.002	0.599
DeepDeblur [23] + YOLOv2	0.971	0.954	0.927	0.889	0.840	0.772	0.682	0.581	0.453	0.331	0.211	0.103	0.032	0.002	0.553
DeblurGAN [22] + YOLOv2	0.961	0.951	0.937	0.921	0.895	0.857	0.810	0.747	0.653	0.540	0.385	0.213	0.070	0.008	0.639
DeblurGAN (MobileNet) [50] + YOLOv2	0.958	0.939	0.911	0.868	0.801	0.701	0.573	0.416	0.280	0.164	0.087	0.039	0.010	0.000	0.482
DCSCN + lightDenseYOLO [21]	0.971	0.969	0.951	0.919	0.905	0.861	0.805	0.733	0.654	0.533	0.379	0.207	0.070	0.007	0.640
SlimDeblurGAN + YOLOv2 (ours)	0.972	0.964	0.952	0.933	0.908	0.870	0.821	0.748	0.658	0.542	0.382	0.208	0.068	0.009	0.645

Table 10. Comparisons of recall on SMBD-DB1.

Methods	IoU Threshold														Average
	0.3	0.35	0.4	0.45	0.5	0.55	0.6	0.65	0.7	0.75	0.8	0.85	0.9	0.95	
YOLOv2 [45]	0.820	0.811	0.798	0.776	0.745	0.703	0.649	0.570	0.471	0.341	0.207	0.098	0.024	0.001	0.501
DeepDeblur [23] + YOLOv2	0.827	0.812	0.789	0.757	0.715	0.657	0.580	0.495	0.386	0.282	0.180	0.088	0.027	0.002	0.471
DeblurGAN [22] + YOLOv2	0.942	0.933	0.919	0.903	0.877	0.840	0.794	0.732	0.640	0.529	0.377	0.209	0.069	0.008	0.627
DeblurGAN (MobileNet) [50] + YOLOv2	0.902	0.884	0.857	0.817	0.753	0.660	0.539	0.392	0.264	0.155	0.082	0.037	0.009	0.000	0.454
DCSCN + lightDenseYOLO [21]	0.942	0.921	0.908	0.901	0.882	0.852	0.783	0.728	0.638	0.519	0.362	0.198	0.054	0.005	0.621
SlimDeblurGAN + YOLOv2 (ours)	0.966	0.958	0.946	0.927	0.902	0.865	0.816	0.744	0.654	0.538	0.379	0.207	0.068	0.009	0.641

Table 11. Comparisons of the F1 Score on SMBD-DB1.

Methods	IoU Threshold														Average
	0.3	0.35	0.4	0.45	0.5	0.55	0.6	0.65	0.7	0.75	0.8	0.85	0.9	0.95	
YOLOv2 [45]	0.893	0.883	0.868	0.844	0.811	0.766	0.707	0.621	0.513	0.371	0.225	0.107	0.026	0.001	0.545
DeepDeblur [23] + YOLOv2	0.893	0.878	0.853	0.817	0.772	0.710	0.627	0.535	0.417	0.305	0.194	0.095	0.029	0.002	0.509
DeblurGAN [22] + YOLOv2	0.951	0.942	0.928	0.912	0.886	0.849	0.802	0.739	0.646	0.535	0.381	0.211	0.069	0.008	0.633
DeblurGAN (MobileNet) [50] + YOLOv2	0.929	0.911	0.883	0.841	0.776	0.680	0.556	0.404	0.272	0.159	0.084	0.038	0.009	0.000	0.467
DCSCN + lightDenseYOLO [21]	0.956	0.944	0.929	0.910	0.893	0.856	0.794	0.730	0.645	0.526	0.370	0.202	0.061	0.006	0.630
SlimDeblurGAN + YOLOv2 (ours)	0.969	0.961	0.949	0.930	0.905	0.867	0.818	0.746	0.656	0.540	0.381	0.208	0.068	0.009	0.643

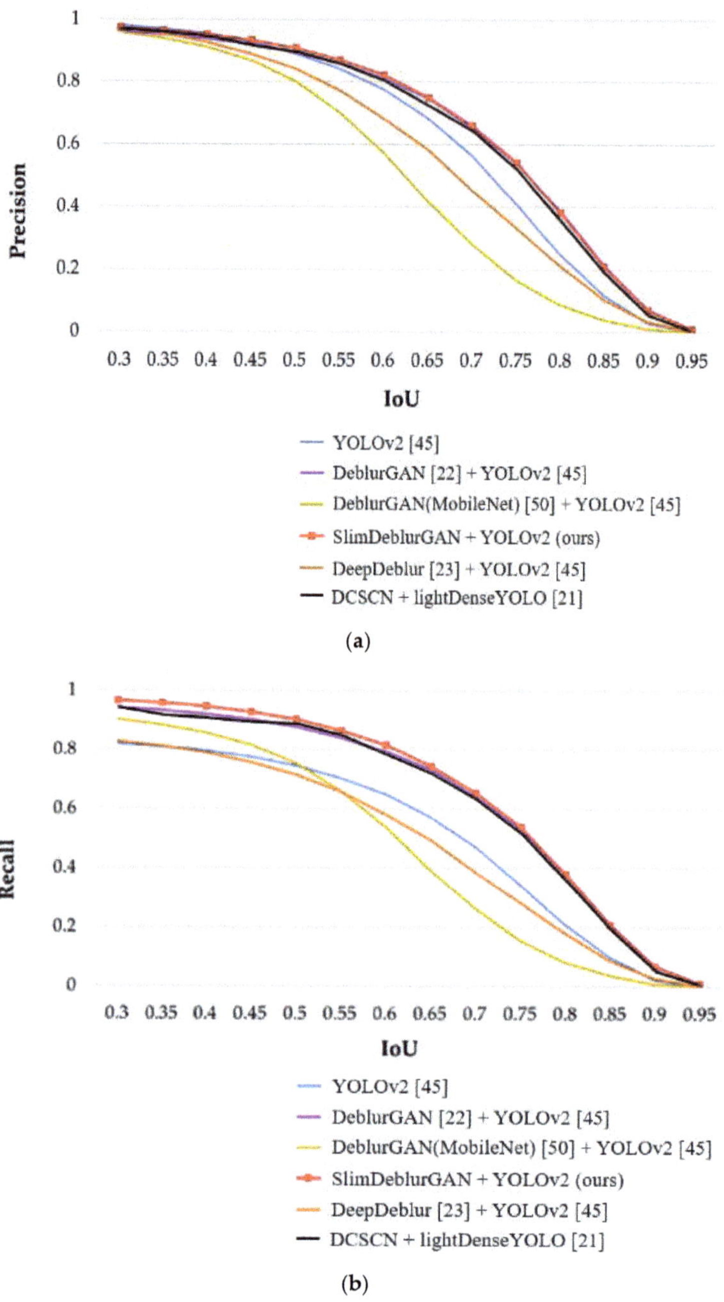

(**a**)

(**b**)

Figure 13. *Cont.*

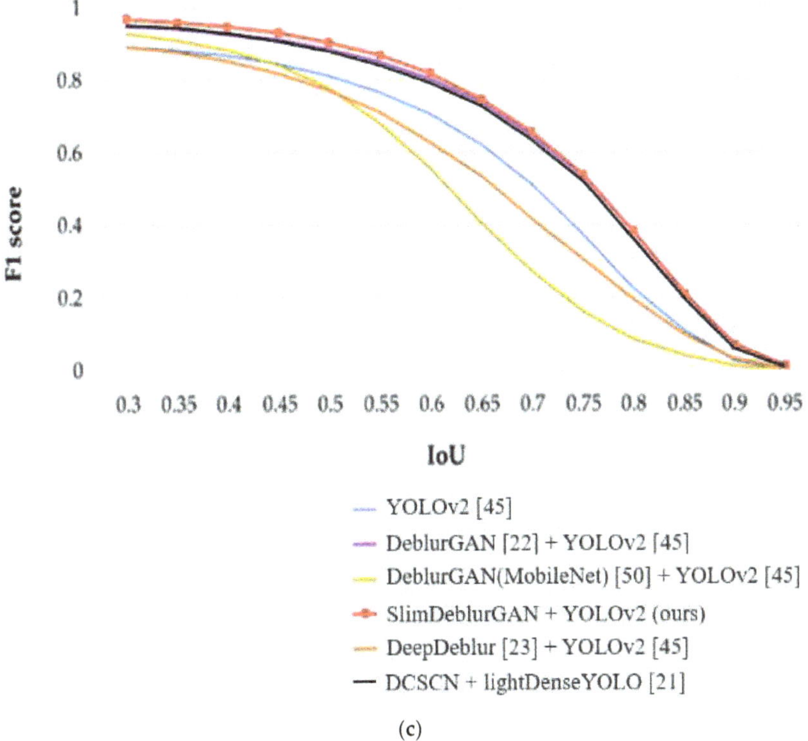

Figure 13. Marker detection results on SMBD-DB1 for the five methods at different IoU thresholds: (**a**) precision, (**b**) recall, and (**c**) F1 score.

Figure 14 presents examples of the detection result on SMBD-DB1; the green boxes show the ground truths, and the red boxes show the detected boxes. As shown in this figure, the marker in the motion-blurred image could not be detected by YOLOv2 (without importing the additional motion deblurring stage) and could not be recognized even by the human eye. The images in Figure 14b–e were detection results on the resulting images of different deblurring methods—DeepDeblur, DeblurGAN, DeblurGAN (MobileNet), and SlimDeblurGAN, respectively.

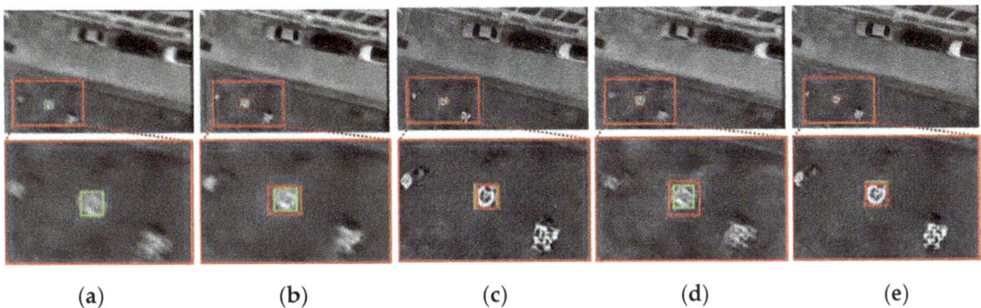

Figure 14. Examples of detection results on SMBD-DB1. The green boxes represent the ground truth bounding boxes and the red boxes represent the boxes detected by (**a**) YOLOv2, (**b**) DeepDeblur and YOLOv2, (**c**) a combination of DeblurGAN and YOLOv2, (**d**) a combination of DeblurGAN (MobileNet) and YOLOv2, and (**e**) a combination of SlimDeblurGAN and YOLOv2 (ours).

Although the detection results of Figure 14b–e showed that the marker in the image could be successfully detected, the detected bounding boxes by different methods were different. Specifically, the detected boxes by DeblurGAN (c) and by SlimDeblurGAN (e) were closer to the ground truth than those by DeepDeblur (b) and by DeblurGAN (MobileNet) (d). From these results, we could confirm that the motion deblurring method can overcome the challenge of motion-blurred input to an object-detection system. The more accurate the motion-deblurring method, the more accurate the detection result becomes.

Testing accuracies on RMBD-DB1: Table 12, Table 13, Table 14 present the detection accuracies of precision, recall, and F1 score, respectively, on the RMBD-DB1 dataset. In addition, the comparative graphs of the experimental results are presented in Figure 15. As shown in these tables and figure, the methods combining motion deblurring and marker detection showed better detection results than the method without motion deblurring, and our proposed method yielded a better result than the SOTA methods. By testing on the RMBD-DB1 realistic motion blur dataset, we could confirm that our proposed system can work well in the real-world environment.

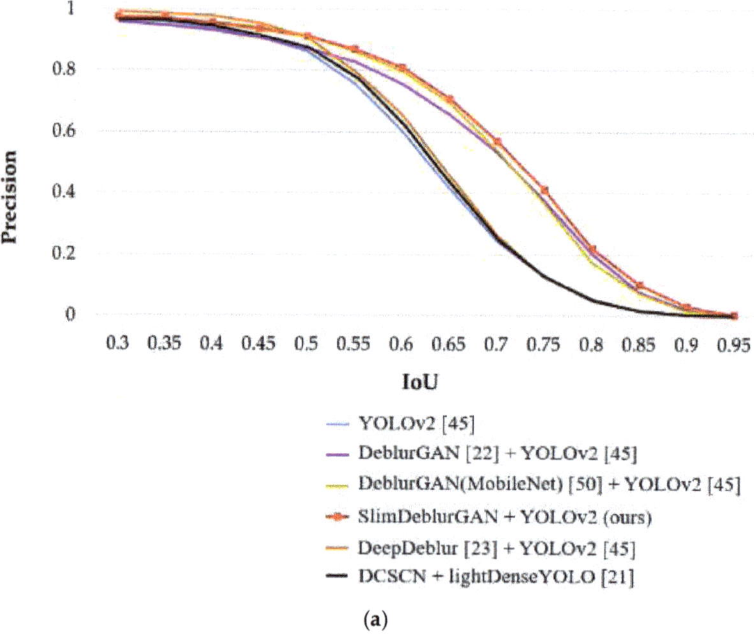

(a)

Figure 15. *Cont.*

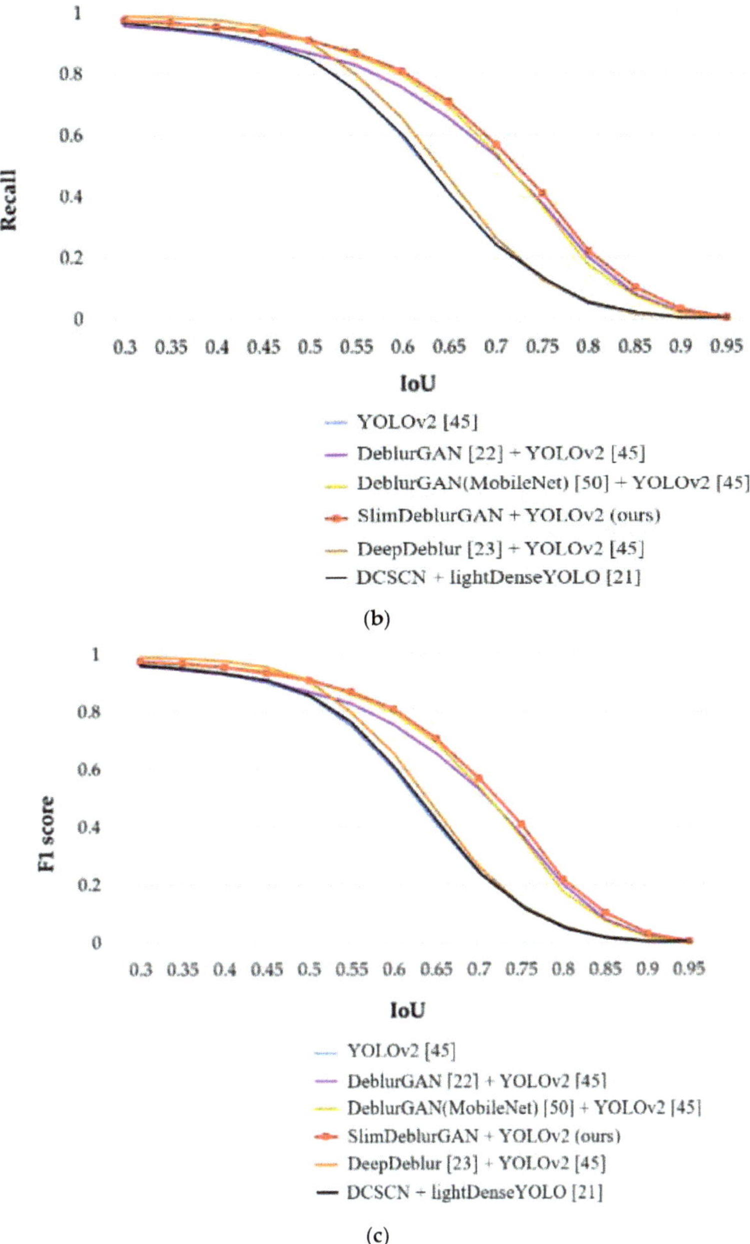

Figure 15. Marker detection results on RMBD-DB1 for five methods at different IoU thresholds: (**a**) precision, (**b**) recall, and (**c**) F1 score.

Table 12. Comparisons of precision on RMBD-DB1.

Methods	IoU Threshold														Average
	0.3	0.35	0.4	0.45	0.5	0.55	0.6	0.65	0.7	0.75	0.8	0.85	0.9	0.95	
YOLOv2 [45]	0.972	0.959	0.940	0.910	0.862	0.758	0.604	0.416	0.245	0.131	0.050	0.019	0.003	0.000	0.491
DeepDeblur [23] + YOLOv2	0.992	0.987	0.978	0.956	0.908	0.799	0.657	0.459	0.264	0.127	0.057	0.016	0.005	0.001	0.515
DeblurGAN [22] + YOLOv2	0.961	0.949	0.933	0.908	0.871	0.830	0.757	0.657	0.534	0.376	0.203	0.080	0.021	0.002	0.577
DeblurGAN (MobileNet) [50] + YOLOv2	0.974	0.966	0.957	0.943	0.916	0.862	0.799	0.695	0.541	0.367	0.176	0.075	0.018	0.003	0.592
DCSCN + lightDenseYOLO [21]	0.971	0.956	0.939	0.909	0.870	0.771	0.625	0.431	0.262	0.125	0.055	0.018	0.004	0.001	0.496
SlimDeblurGAN + YOLOv2 (ours)	0.976	0.970	0.956	0.936	0.912	0.870	0.811	0.710	0.571	0.411	0.221	0.103	0.032	0.005	0.606

Table 13. Comparisons of recall on RMBD-DB1.

Methods	IoU Threshold														Average
	0.3	0.35	0.4	0.45	0.5	0.55	0.6	0.65	0.7	0.75	0.8	0.85	0.9	0.95	
YOLOv2 [45]	0.959	0.947	0.928	0.898	0.851	0.749	0.596	0.410	0.241	0.129	0.050	0.019	0.003	0.000	0.484
DeepDeblur [23] + YOLOv2	0.991	0.986	0.978	0.956	0.907	0.799	0.657	0.459	0.264	0.127	0.057	0.016	0.005	0.001	0.514
DeblurGAN [22] + YOLOv2	0.960	0.949	0.933	0.907	0.871	0.829	0.756	0.657	0.534	0.376	0.203	0.080	0.021	0.002	0.577
DeblurGAN (MobileNet) [50] + YOLOv2	0.973	0.966	0.957	0.942	0.915	0.861	0.798	0.695	0.540	0.367	0.176	0.075	0.018	0.003	0.592
DCSCN + lightDenseYOLO [21]	0.960	0.947	0.929	0.899	0.852	0.749	0.597	0.409	0.242	0.128	0.051	0.019	0.004	0.000	0.485
SlimDeblurGAN + YOLOv2 (ours)	0.975	0.969	0.955	0.935	0.911	0.870	0.810	0.709	0.570	0.411	0.220	0.103	0.032	0.005	0.605

Table 14. Comparisons of F1 Score on RMBD-DB1.

Methods	IoU Threshold														Average
	0.3	0.35	0.4	0.45	0.5	0.55	0.6	0.65	0.7	0.75	0.8	0.85	0.9	0.95	
YOLOv2 [45]	0.965	0.953	0.934	0.904	0.856	0.753	0.600	0.413	0.243	0.130	0.050	0.019	0.003	0.000	0.487
DeepDeblur [23] + YOLOv2	0.991	0.986	0.978	0.956	0.908	0.799	0.657	0.459	0.264	0.127	0.057	0.016	0.005	0.001	0.515
DeblurGAN [22] + YOLOv2	0.960	0.949	0.933	0.908	0.871	0.829	0.757	0.657	0.534	0.376	0.203	0.080	0.021	0.002	0.577
DeblurGAN (MobileNet) [50] + YOLOv2	0.973	0.966	0.957	0.942	0.915	0.862	0.798	0.695	0.540	0.367	0.176	0.075	0.018	0.003	0.592
DCSCN + lightDenseYOLO [21]	0.965	0.951	0.934	0.904	0.861	0.760	0.611	0.419	0.252	0.126	0.053	0.018	0.004	0.000	0.489
SlimDeblurGAN + YOLOv2 (ours)	0.975	0.970	0.956	0.936	0.911	0.870	0.810	0.709	0.570	0.411	0.220	0.103	0.032	0.005	0.606

Figure 16 presents examples of detection results on RMBD-DB1 performed by five methods, i.e., YOLOv2, a combination of DeblurGAN and YOLOv2, a combination of DeblurGAN (MobileNet) and YOLOv2, a combination of DeepDeblur and YOLOv2, and our proposed framework combining SlimDeblurGAN and YOLOv2. The motion blur was due to the downward movement of the drone, which caused the failure of object detection by YOLOv2 (without importing the additional motion deblurring stage), as shown in Figure 16a. However, the combinations of motion deblurring models and the YOLOv2 detector successfully detected the marker, as shown in Figure 16b–d, and our method showed more accurate results of marker detection than the other methods, as shown in Figure 16e.

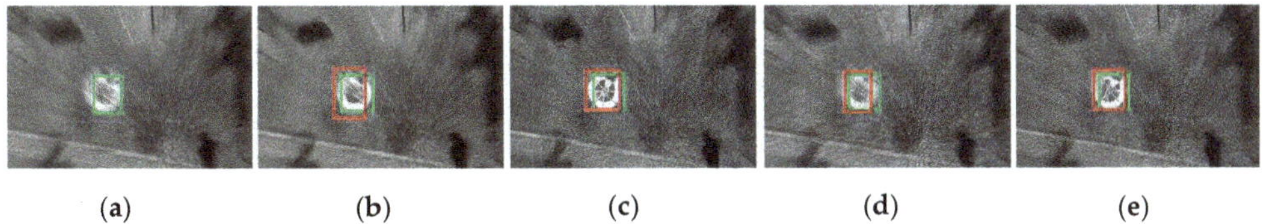

(a)   (b)   (c)   (d)   (e)

Figure 16. Examples of detection results on RMBD-DB1. The green boxes represent the ground truth bounding boxes and the red boxes represent the boxes detected by (**a**) YOLOv2, (**b**) DeepDeblur and YOLOv2, (**c**) a combination of DeblurGAN and YOLOv2, (**d**) a combination of DeblurGAN (MobileNet) and YOLOv2, and (**e**) a combination of SlimDeblurGAN and YOLOv2 (ours).

4.3.3. Comparisons on Processing Speed and Discussion

We measured the processing speed of our method on both an embedded system and a desktop computer. The specifications of the desktop computer are explained in Section 4.2, and a Jetson TX2 system was used as the embedded system, as shown in Figure 17. A Jetson TX2 embedded system is a fast, power-efficient device optimized for artificial intelligence (AI). It includes an NVIDIA Pascal™-family GPU (256 CUDA cores) with 8 GB of memory and features various standard hardware interfaces that facilitate integration into a wide range of products like UAVs and autonomous vehicle [51]. As the board was pre-flashed with a Linux development environment, we installed the Ubuntu 16.04 operating system, which was a convenient environment for training and testing deep learning models, as recommended by NVIDIA. Our proposed SlimDeblurGAN and YOLOv2, including the comparative algorithms were implemented in desktop computer by TensorFlow 1.14 [52], CUDA® toolkit (ver. 10.0) [53], and NVIDIA CUDA® deep neural network library (CUDNN) (ver. 7.6.2) [54]. These were also implemented in the Jetson TX2 system by TensorFlow 1.12 [52], CUDA® toolkit (ver. 9.0) [53], and NVIDIA CUDNN (ver. 7.3) [54]. The full specifications of the Jetson TX2 embedded system are presented in Table 15.

Figure 17. Jetson TX2 embedded system.

Table 15. Specifications of the Jetson TX2 embedded system.

Jetson TX2 Embedded System	
CPU	HMP Dual Denver (2 MB L2) + Quad ARM® A57 (2 MB L2)
GPU	NVIDIA Pascal™, 256 CUDA cores
Memory	8 GB
Data Storage	32 GB
Operating System	Linux for Tegra R28.1 (L4T 28.1)
Dimensions	50 mm × 87 mm

We measured the processing time of each phase, separately. Table 16 presents the processing time per image and the FPS of four motion deblurring methods—DeepDeblur, DeblurGAN, DeblurGAN (MobileNet), and SlimDeblurGAN. The processing speed of SlimDeblurGAN on the desktop computer was extremely fast, at about 98 FPS, and it was approximately 54.6 FPS on the Jetson TX2 system. SlimDeblurGAN had the highest processing speed in both the desktop and Jetson TX2 environments.

Table 16. Processing time (ms) per image with FPS of motion deblurring networks on two platforms.

Model	Execution Time/FPS	
	Desktop Computer	**Jetson TX2 Board**
DeepDeblur [23]	52/19.2	398/2.5
DeblurGAN [22]	37/27	349/2.9
DeblurGAN (MobileNet) [50]	26/38.5	215/4.7
DCSCN [21]	101/10	188/5.3
SlimDeblurGAN (ours)	10.2/98	18.3/54.6

In addition, we measured the total processing time per image of our method, including the YOLOv2 detector, as shown in Table 17. In the desktop environment, YOLOv2 archived the fast speed at 50 FPS, and it was still fast on the Jetson TX2 board with approximately 32.3 FPS. As a result, the total processing speed of our proposed method could be 33.1 FPS on the desktop computer and 20.3 FPS on the Jetson TX2 embedded system. In addition, our method was faster than the previous method, as shown in Tables 17 and 18. YOLOv2 [45] and our previous method [21] were already applied to autonomous drone landing, and our method outperformed these algorithms, as shown in Tables 9–14, Tables 16 and 18 and Figures 13 and 15, which confirmed the necessity of motion blur restoration by the proposed method, for accurate marker detection. In [21], DSCN + YOLOv2 was compared with DSCN + lightDenseYOLO, which confirmed that DSCN + lightDenseYOLO proposed in [21] outperformed DSCN + YOLOv2. Therefore, we compared our method (SlimDeblurGAN + YOLOv2) with DSCN + lightDenseYOLO proposed in [21]. In Figure 14, the green boxes represent the ground truth bounding boxes, and the red boxes represent the boxes detected. There was no red box in the lower image of Figure 14a, which meant that YOLOv2 could not detect the marker in the motion blurred image. The same result could also be observed in Figure 16a. As shown in Tables 9–14, and Figures 13 and 15, YOLOv2 without the motion deblurring method showed lower accuracies of marker detection than our proposed method.

Table 17. Total processing time (ms) per image by our method on two platforms.

Platform	SlimDeblurGAN	YOLOv2	Total/FPS
Desktop computer	10.2	20	30.2/33.1
Jetson	18.3	31	49.3/20.3

Table 18. Total processing time (ms) per image by the previous method [21] on two platforms.

Platform	DCSCN	lightDenseYOLO	Total/FPS
Desktop computer	101	20	121/8.3
Jetson	190	35	225/4.4

Although the improvement in precision by the proposed method with SMBD-DB1 was 0.5%, compared to previous work [21] (shown in Table 9), those of recall and the F1 Score by the proposed method were, respectively, 2% and 1.3% compared to previous work [21] as shown in Tables 10 and 11. In addition, with RMBD-DB1, the improvement of precision, recall, and F1 Score of the proposed method were respectively, 11%, 12%, and 11.7%, compared to previous work [21], as shown in Tables 12–14.

5. Conclusions

We introduced a deep-learning-based marker detection method for autonomous drone landing, which considered motion deblurring, by proposing a two-phase framework system. To the best of our knowledge, this study was the first to consider the performance of a combination of motion deblurring and marker detection for autonomous drone landing. In addition, we considered the balance between accuracy and execution time by adopting our proposed motion-deblurring network and the real-time object detector of YOLOv2. To this end, we proposed a SlimDeblurGAN by channel pruning, to lighten the pretrained DeblurGAN model without significant degradation of accuracy, which was significantly faster than the original version. We adopted such models to our system by training from scratch and testing on our two synthesized motion-blurred datasets acquired from landing drones. We confirmed experimentally that our system could be operated well on non-uniform motion-blurred input, and it could be applied to an embedded system with low processing power. For our future work, we plan to combine the two networks of motion deblurring and marker detection into one model, including shallower layers and fewer parameters, which could reduce the processing time. In addition, we would apply our network to other applications of pedestrian detection at a distance, for intelligent surveillance camera environments, object detection in satellite images, small object detection, and moving object detection, etc.

Author Contributions: N.Q.T. and T.D.P. implemented the overall system of motion deblurring and marker detection, and wrote this paper. Y.W.L., M.O., D.T.N., G.B., and K.R.P. helped with the experiments and analyzed the results. All authors have read and agreed to the published version of the manuscript.

Acknowledgments: This work was supported in part by the National Research Foundation of Korea (NRF) funded by the Ministry of Education, through the Basic Science Research Program under Grant NRF-2018R1D1A1B07041921;

in part by the NRF funded by the Ministry of Science and ICT (MSIT), through the Basic Science Research Program under Grant NRF-2019R1A2C1083813; in part by the NRF funded by the MSIT, through the Bio and Medical Technology Development Program under Grant NRF-2016M3A9E1915855.

Conflicts of Interest: The authors declare no conflict of interest.

References

1. Gui, Y.; Guo, P.; Zhang, H.; Lei, Z.; Zhou, X.; Du, J.; Yu, Q. Airborne vision-based navigation method for UAV accuracy landing using infrared lamps. *J. Intell. Robot. Syst.* **2013**, *72*, 197–218. [CrossRef]
2. Forster, C.; Faessler, M.; Fontana, F.; Werlberger, M.; Scaramuzza, D. Continuous on-board monocular-vision-based elevation mapping applied to autonomous landing of micro aerial vehicles. In Proceedings of the IEEE International Conference on Robotics and Automation, Seattle, WA, USA, 26–30 May 2015; IEEE: New York City, NY, USA, 2015; pp. 111–118.
3. Xu, G.; Qi, X.; Zeng, Q.; Tian, Y.; Guo, R.; Wang, B. Use of land's cooperative object to estimate UAV's pose for autonomous landing. *Chin. J. Aeronaut.* **2013**, *26*, 1498–1505. [CrossRef]
4. Lin, S.; Garratt, M.A.; Lambert, A.J. Monocular vision-based real-time target recognition and tracking for autonomously landing an UAV in a cluttered shipboard environment. *Auton. Robots* **2017**, *41*, 881–901. [CrossRef]
5. Lange, S.; Sunderhauf, N.; Protzel, P. A vision based onboard approach for landing and position control of an autonomous multirotor UAV in GPS-denied environments. In Proceedings of the IEEE International Conference on Advanced Robotics, Munich, Germany, 22–26 June 2009; IEEE: New York City, NY, USA, 2009; pp. 1–6.
6. Polvara, R.; Sharma, S.; Wan, J.; Manning, A.; Sutton, R. Towards autonomous landing on a moving vessel through fiducial markers. In Proceedings of the European Conference on Mobile Robots, Paris, France, 6–8 September 2017; pp. 1–6.
7. Araar, O.; Aouf, N.; Vitanov, I. Vision based autonomous landing of multirotor UAV on moving platform. *J. Intell. Robot. Syst.* **2017**, *85*, 369–384. [CrossRef]
8. Barták, R.; Hraško, A.; Obdržálek, D. On autonomous landing of AR. Drone: Hands-on experience. In Proceedings of the 27th International Florida Artificial Intelligence Research Society Conference, Pensacola Beach, FL, USA, 21–23 May 2014; pp. 400–405.
9. Venugopalan, T.K.; Taher, T.; Barbastathis, G. Autonomous landing of an unmanned aerial vehicle on an autonomous marine vehicle. In Proceedings of the Oceans Conference, Hampton Roads, VA, USA, 14–19 October 2012; pp. 1–9.
10. Wubben, J.; Fabra, F.; Calafate, C.T.; Krzeszowski, T.; Marquez-Barja, J.M.; Cano, J.-C.; Manzoni, P. Accurate landing of unmanned aerial vehicles using ground pattern recognition. *Electronics* **2019**, *8*, 1532. [CrossRef]
11. Skoczylas, M. Vision analysis system for autonomous landing of micro drone. *Acta Mech. Autom.* **2014**, *8*, 199–203. [CrossRef]
12. Dotenco, S.; Gallwitz, F.; Angelopoulou, E. Autonomous approach and landing for a low-cost quadrotor using monocular cameras. In Proceedings of the European Conference on Computer Vision Workshops, Zurich, Switzerland, 6–12 September 2014; pp. 209–222.
13. Nguyen, P.H.; Arsalan, M.; Koo, J.H.; Naqvi, R.A.; Truong, N.Q.; Park, K.R. LightDenseYOLO: A fast and accurate marker tracker for autonomous UAV landing by visible light camera sensor on drone. *Sensors* **2018**, *18*, 1703. [CrossRef] [PubMed]
14. Yu, L.; Luo, C.; Yu, X.; Jiang, X.; Yang, E.; Luo, C.; Ren, P. Deep learning for vision-based micro aerial vehicle autonomous landing. *Int. J. Micro Air Veh.* **2018**, *10*, 171–185. [CrossRef]
15. Autonomous Quadrotor Landing Using Deep Reinforcement Learning. Available online: https://arxiv.org/abs/1709.03339 (accessed on 15 January 2020).
16. Truong, N.Q.; Nguyen, P.H.; Nam, S.H.; Park, K.R. Deep learning-based super-resolution reconstruction and marker detection for drone landing. *IEEE Access* **2019**, *7*, 61639–61655. [CrossRef]
17. Ren, S.; He, K.; Girshick, R.; Sun, J. Faster R-CNN: Towards real-time object detection with region proposal networks. IEEE Trans. *Pattern Anal. Mach. Intell.* **2017**, *39*, 1137–1149. [CrossRef] [PubMed]

18. Liu, W.; Anguelov, D.; Erhan, D.; Szegedy, C.; Reed, S.; Fu, C.-Y.; Berg, A.C. SSD: Single shot multibox detector. In Proceedings of the European Conference on Computer Vision, Amsterdam, The Netherlands, 11–14 October 2016; pp. 21–37.

19. Zhou, L.; Min, W.; Lin, D.; Han, Q.; Liu, R. Detecting motion blurred vehicle logo in IoV using filter-DeblurGAN and VL-YOLO. *IEEE Trans. Veh. Technol.* **2020**, *69*, 3604–3614. [CrossRef]

20. Wang, R.; Ma, G.; Qin, Q.; Shi, Q.; Huang, J. Blind UAV images deblurring based on discriminative networks. *Sensors* **2018**, *18*, 2874. [CrossRef] [PubMed]

21. Wang, J.; Olson, E. AprilTag 2: Efficient and robust fiducial detection. In Proceedings of the IEEE/RSJ International Conference on Intelligent Robots and Systems, Daejeon, Korea, 9–14 October 2016; pp. 4193–4198.

22. Kupyn, O.; Budzan, V.; Mykhailych, M.; Mishkin, D.; Matas, J. DeblurGAN: Blind motion deblurring using conditional adversarial networks. In Proceedings of the IEEE/CVF Conference on Computer Vision and Pattern Recognition, Salt Lake City, UT, USA, 18–23 June 2018; pp. 8183–8192.

23. Nah, S.; Kim, T.H.; Lee, K.M. Deep multi-scale convolutional neural network for dynamic scene deblurring. In Proceedings of the IEEE Conference on Computer Vision and Pattern Recognition, Honolulu, HI, USA, 21–26 July 2017; pp. 257–265.

24. Conditional Generative Adversarial Nets. Available online: https://arxiv.org/abs/1411.1784 (accessed on 15 January 2020).

25. Ledig, C.; Theis, L.; Huszár, F.; Caballero, J.; Cunningham, A.; Acosta, A.; Aitken, A.; Tejani, A.; Totz, J.; Wang, Z.; et al. Photo-realistic single image super-resolution using a generative adversarial network. In Proceedings of the IEEE Conference on Computer Vision and Pattern Recognition, Honolulu, HI, USA, 21–26 July 2017; pp. 105–114.

26. Li, C.; Wand, M. Precomputed real-time texture synthesis with markovian generative adversarial networks. In Proceedings of the European Conference on Computer Vision, Amsterdam, The Netherlands, 11–14 October 2016; pp. 702–716.

27. Improved Training of Wasserstein GANs. Available online: https://arxiv.org/abs/1704.00028 (accessed on 4 January 2020).

28. Salimans, T.; Goodfellow, I.; Zaremba, W.; Cheung, V.; Radford, A.; Chen, X.; Chen, X. Improved techniques for training GANs. In Proceedings of the 30th Conference on Neural Information Processing Systems, Barcelona, Spain, 5–10 December 2016; pp. 1–9.

29. Wasserstein GAN. Available online: https://arxiv.org/abs/1701.07875 (accessed on 5 January 2020).

30. Johnson, J.; Alahi, A.; Fei-Fei, L. Perceptual losses for real-time style transfer and super-resolution. In Proceedings of the European Conference on Computer Vision, Amsterdam, The Netherlands, 11–14 October 2016; pp. 694–711.

31. He, K.; Zhang, X.; Ren, S.; Sun, J. Deep residual learning for image recognition. In Proceedings of the IEEE Conference on Computer Vision and Pattern Recognition, Las Vegas, NV, USA, 27–30 June 2016; pp. 770–778.

32. Instance Normalization: The Missing Ingredient for Fast Stylization. Available online: https://arxiv.org/abs/1607.08022 (accessed on 5 January 2020).

33. Deep Learning Using Rectified Linear Units (ReLU). Available online: https://arxiv.org/abs/1803.08375 (accessed on 4 January 2020).

34. Isola, P.; Zhu, J.-Y.; Zhou, T.; Efros, A.A. Image-to-image translation with conditional adversarial networks. In Proceedings of the IEEE Conference on Computer Vision and Pattern Recognition, Honolulu, HI, USA, 21–26 July 2017; pp. 5967–5976.

35. Deng, J.; Dong, W.; Socher, R.; Li, L.-J.; Li, K.; Fei-Fei, L. ImageNet: A large-scale hierarchical image database. In Proceedings of the IEEE Conference on Computer Vision and Pattern Recognition, Miami, FL, USA, 20–25 June 2009; pp. 248–255.

36. MobileNets: Efficient Convolutional Neural Networks for Mobile Vision Applications. Available online: https://arxiv.org/abs/1704.04861 (accessed on 5 January 2020).

37. Liu, Z.; Li, J.; Shen, Z.; Huang, G.; Yan, S.; Zhang, C. Learning efficient convolutional networks through network slimming. In Proceedings of the IEEE International Conference on Computer Vision, Venice, Italy, 22–29 October 2017; pp. 2755–2763.

38. Distilling the Knowledge in a Neural Network. Available online: https://arxiv.org/abs/1503.02531 (accessed on 7 January 2020).

39. Deep Learning with Dynamic Computation Graphs. Available online: https://arxiv.org/abs/1702.02181 (accessed on 7 January 2020).

40. Zhang, P.; Zhong, Y.; Li, X. SlimYOLOv3: Narrower, faster and better for real-time UAV applications. In Proceedings of the International Conference on Computer Vision, Seoul, Korea, 27 October–2 November 2019; pp. 1–9.

41. Girshick, R. Fast R-CNN. In Proceedings of the IEEE International Conference on Computer Vision, Santiago, Chile, 7–13 December 2015; pp. 1440–1448.

42. Dai, J.; Li, Y.; He, K.; Sun, J. R-FCN: Object detection via region-based fully convolutional networks. In Proceedings of the 30th Conference on Neural Information Processing Systems, Barcelona, Spain, 5–10 December 2016; pp. 1–9.

43. Lin, T.-Y.; Goyal, P.; Girshick, R.; He, K.; Dollár, P. Focal loss for dense object detection. In Proceedings of the IEEE International Conference on Computer Vision, Venice, Italy, 22–29 October 2017; pp. 2999–3007.

44. Redmon, J.; Divvala, S.; Girshick, R.; Farhadi, A. You only look once: Unified, real-time object detection. In Proceedings of the IEEE Conference on Computer Vision and Pattern Recognition, Las Vegas, NV, USA, 27–30 June 2016; pp. 779–788.

45. Redmon, J.; Farhadi, A. YOLO9000: Better, faster, stronger. In Proceedings of the IEEE Conference on Computer Vision and Pattern Recognition, Honolulu, HI, USA, 21–26 July 2017; pp. 6517–6525.

46. YOLOv3: An Incremental Improvement. Available online: https://arxiv.org/abs/1804.02767 (accessed on 10 January 2020).

47. Lin, T.-Y.; Maire, M.; Belongie, S.; Hays, J.; Perona, P.; Ramanan, D.; Dollár, P.; Zitnick, C.L. Microsoft COCO: Common objects in context. In Proceedings of the European Conference on Computer Vision, Zurich, Switzerland, 6–12 September 2014; pp. 740–755.

48. Gonzalez, R.C.; Woods, R.E. *Digital Image Processing*, 3rd ed.; Prentice-Hall: Upper Saddle River, NJ, USA, 2010.

49. Wang, Z.; Bovik, A.C.; Sheikh, H.R.; Simoncelli, E.P. Image quality assessment: From error visibility to structural similarity. *IEEE Trans. Image Process.* **2004**, *13*, 600–612. [CrossRef] [PubMed]

50. Kupyn, O.; Martyniuk, T.; Wu, J.; Wang, Z. DeblurGAN-v2: Deblurring (orders-of-magnitude) faster and better. In Proceedings of the International Conference on Computer Vision, Seoul, Korea, 27 October–2 November 2019; IEEE: New York City, NY, USA, 2019; pp. 8878–8887.

51. Jetson TX2 Module. Available online: https://www.nvidia.com/en-us/autonomous-machines/embedded-systems-dev-kits-modules/ (accessed on 19 December 2019).

52. TensorFlow: The Python Deep Learning library. Available online: https://www.tensorflow.org/ (accessed on 23 January 2020).

53. CUDA. Available online: https://developer.nvidia.com/cuda-toolkit-archive (accessed on 23 January 2020).
54. CUDNN. Available online: https://developer.nvidia.com/cudnn (accessed on 23 January 2020).

Article

A Hough-Space-Based Automatic Online Calibration Method for a Side-Rear-View Monitoring System

Jung Hyun Lee and **Dong-Wook Lee** *

Department of Electronics and Electrical Engineering, Dongguk University, Seoul 04620, Korea;
jhlee36@dongguk.edu
* Correspondence: dlee@dongguk.edu; Tel.: +82-2-2260-3350

Received: 18 May 2020; Accepted: 11 June 2020; Published: 16 June 2020

Abstract: We propose an automatic camera calibration method for a side-rear-view monitoring system in natural driving environments. The proposed method assumes that the camera is always located near the surface of the vehicle so that it always shoots a part of the vehicle. This method utilizes photographed vehicle information because the captured vehicle always appears stationary in the image, regardless of the surrounding environment. The proposed algorithm detects the vehicle from the image and computes the similarity score between the detected vehicle and the previously stored vehicle model. Conventional online calibration methods use additional equipment or operate only in specific driving environments. On the contrary, the proposed method is advantageous because it can automatically calibrate camera-based monitoring systems in any driving environment without using additional equipment. The calibration range of the automatic calibration method was verified through simulations and evaluated both quantitatively and qualitatively through actual driving experiments.

Keywords: side-rear-view monitoring system; automatic online calibration; Hough-space

1. Introduction

In recent years, vision-based Advanced Driver Assistance Systems (ADAS) based on cameras have been developed continuously to provide safety and convenience to motorists. Vision-based ADAS employ the intrinsic and extrinsic parameters of cameras to provide a specific Field Of View (FOV). A Surround View Monitoring System, one of the vision-based ADAS, uses camera parameters to generate a bird's eye view image with a FOV that contains all of the information around the vehicle [1,2]. A panoramic rear-view system also uses camera parameters to stitch side-view image and rear-view image to generate a panoramic image [3,4]. These vision-based ADAS transforms the captured image using camera parameters to generate the desired FOV image.

A side-rear-view monitoring system should especially provide a reliable FOV so that the driver can glean adequate information, and most countries legally specify the reliability of FOV. However, even though the same model vehicles are equipped with the same side-rear-view monitoring system devices, each monitoring system provides a different FOV due to manufacturing tolerances. FOV also changes when the same monitoring system device is mounted on different vehicles. Therefore, a side-rear-view monitoring system has to calibrate camera to provide uniform FOV even when various factors change. To provide consistent FOV according to the laws and circumstances of each country, control over the intrinsic and extrinsic parameters of cameras is required.

Camera calibration is one of the most useful methods for estimating the intrinsic and extrinsic parameters of a camera [5–11]. Camera calibration can improve camera performance by overcoming manufacturing tolerance limitations. Looser manufacturing tolerances can allow for lower cost and higher yield. Additionally, estimated parameters by calibration can be used to transform pixel-based metrics to physically based ones. This geometrical transformation enables FOV control.

Camera calibration for ADAS can be categorized into offline, self, and online calibration. Offline calibration methods use photographed targets, such as checker-patterns on a floor or wall [12–18]. This method is inconvenient because the size and position of the targets should be regulated depending on the location, orientation, and FOV of the camera. To this end, automobile manufacturers need to secure specialized facilities. Online calibration does not use specific targets and requires a moving camera. Traditional online calibration methods employ additional devices, such as encoders, Light Detection and Ranging (LiDAR) systems, odometry devices, and Inertial Measurement Units (IMU) [19–22] to overcome absence of specific targets. Other online calibration methods called self calibration [23–25] do not use additional devices but requires specific information about the road surface, such as lane markers [26–29]. However, it is not always possible to obtain specific information in the natural driving environment. In addition, self calibration methods have a constraint that some camera parameters must be known. Therefore, offline calibration must precede because the vehicle will not be operating on roads with lanes before it is sold. Unquestionably, the primary purpose of traditional online calibration is recalibration.

We propose a side-rear-view camera calibration method that is possible even if we do not know any camera parameters. Therefore, it does not require an offline calibration preprocessing step. In this method, a vision-based ADAS camera mounted near the side surface of a vehicle constantly photographs the vehicle. We call the part of the captured vehicle "Reflected-Vehicle Area (RVA)", and we can extract the RVA not influenced by the driving environment. Therefore, the RVA is an essential prerequisite for our method.

A segmentation method with artificial intelligence such as deep learning is one of good solution to detect the RVA [30]. However, deep learning requires a huge amount of data based on the type of vehicle, camera parameters, and various driving environments. Collecting these data is very inconvenient and difficult. In order to overcome this inconvenience, we utilize widely known and uncomplicated image processing techniques to detect the RVA.

The proposed method detects the boundary of a reflected vehicle and computes the interior of the boundary as the RVA. The boundary of the reflected vehicle can be represented by any curve in the captured image. Random Sample Consensus (RANSAC) is a useful curve-fitting method. However, RANSAC is not always able to identify the optimal curve from moderately contaminated data [31]. Therefore, we eliminate contaminated data to the extent possible before utilizing RANSAC.

After the RVA is detected, the proposed method computes a similarity score between the detected RVA and a stored vehicle model. The similarity score can be calculated by the reprojection error minimization. To minimize reprojection error, we have to extract and match the features from the captured image and the stored image using Scale-Invariant Feature Transform (SIFT) or Speeded Up Robust Features (SURF) [32]. However, it is difficult to extract adequate numbers of feature points from two-dimensional vehicle images, which are required to apply image matching. The challenge is mainly due to the fact that the feature point is a corner where two straight lines with different slopes meet whereas RVA consists mostly of smooth curves. Image-template-matching methods can compute the similarity score without requiring feature point extraction. The template size should be small to facilitate the utilization of ring projection in conventional template-matching methods [33,34], but the RVA is too large from the viewpoint of applying ring projection. To solve this problem, Yang et al. studied large-scale rotation-invariant template matching [35]. This method uses color information, but the color of the RVA changes continuously because it reflects the surrounding environment. We propose a large-scale rotation-invariant template-matching method that computes the similarity score by using edge information instead of color information. The proposed algorithm utilizes the normalized 2D cross-correlation and the Hough space expressed in the Hesse normal form.

The rest of this paper is organized as follows: Section 2 reviews related works, and Section 3 describes the essential procedures of the proposed method. Section 4 presents simulation and experimental results. Finally, we conclude with a summary of the work in Section 5.

2. Related Works

This literature review focuses only on how to calibrate the parameters of a vehicle camera since camera calibration has been extensively researched for a long time in a wide range of fields. The previous methods can be classified as offline and online calibration according to which features are used. Additionally, online calibration can be categorized according to whether additional devices are used.

2.1. Offline Calibration

Offline calibration methods estimate camera parameters using special patterns consisting of edges, circles, or lines. Since these methods use precisely drawn patterns aligned with a camera-mounted vehicle, it is possible to accurately estimate the camera parameters. However, accurate calibration cannot be performed without special facilities aligning a vehicle and patterns in the precise location.

The A&G company [13] provides calibration facilities that can align the vehicle and the calibration patterns for highly accurate camera calibration. Xia et al. [14] used multiple patterns and cameras by minimizing the reprojection error. Mazzei et al. [15] also minimized the reprojection error of checkboard patterns' corner locations to calibrate extrinsic parameters of the front view camera. Hold et al. [16] used a similar method using circle patterns on the ground. This method minimized the reprojection error of the centers of the circles. Tan et al. [17] drew an H-shaped pattern that consisted of two parallel lines and one perpendicular line to the vehicle. Li et al. [18] also used an H-shaped pattern to calibrate a rear-view camera.

2.2. Online Calibration with Additional Devices

Online calibration can estimate camera parameters using various sensors to utilize natural features instead of artificial patterns while driving. However, in terms of side-rear-view monitoring system calibration, this method has several drawbacks. Since the side-rear-view camera is looking at the horizon behind a vehicle rather than the road surface and part of the captured image is obscured by the driver's vehicle, it is difficult to detect enough natural features. Therefore, feature-based algorithms are inappropriate for calibration of a side-rear-view monitoring system.

Wang et al. [19] proposed a camera-encoder system to estimate extrinsic parameters. They obtained the distance that camera traveled through the encoder and calculated the Euclidean distance between matched image feature points using feature extracting and matching algorithms. This method can estimate extrinsic parameters by comparing the Euclidean distance of matched image feature points with the camera movement distance. Schneider et al. [20] also utilized odometry, camera, and matched feature points for estimating intrinsic parameters. Chien et al. [21] used visual odometry and LiDAR for online calibration. Visual odometry determines equivalent odometry information using feature extracting and matching algorithms. Li et al. [22] utilized IMU to calibrate the camera. The data measured by the IMU is fed into a processor, which calculates the position.

2.3. Online Calibration without Additional Devices

Online calibration without additional devices extracts and matches natural feature points in image sequences. Since there is no other assistant equipment, these methods highly depend on feature extracting and matching algorithms. All of the introduced papers in this section utilized the road lanes as the feature points.

Xu et al. [26] and de Paula et al. [27] utilized the detection of two symmetrical lanes to calibrate cameras of the lane departure warning systems and augmented reality systems, respectively. Wang et al. [28] detected two symmetrical dotted lanes for online calibration. However, a side-rear-view camera captures few or no symmetrical lanes. Choi et al. [29] proposed the recalibration method for around view monitoring systems. This method can calibrate only when the road lanes around the vehicle are detected. However, road lanes near the vehicle are not captured by the side-rear-view camera.

3. Automatic Online Calibration

Automatic online calibration is a method that automatically calibrates the camera's orientation and location in natural driving environments. However, the calibration method cannot change the orientation of a fixed camera in monitoring systems, which leads to deformed images. Therefore, we have to convert camera parameters into image deforming parameters.

The camera parameters can be classified into intrinsic and extrinsic parameters. Intrinsic parameters describe the optical properties of the camera, and extrinsic parameters describe the orientation and location of the camera. Since the optical properties of the manufactured camera such as the image sensor size, image sensor resolution, and distance between image sensor and lens hardly change, we assume that intrinsic parameters are constant. Therefore, we assume that intrinsic parameters are constant. Online calibration focuses only on the camera orientation because the orientation has considerably more influence on the image than the camera position [29,36]. Therefore, we also exclude camera location parameters, which is one of extrinsic parameters, from variables as well.

The camera orientation can be expressed in terms of its roll, pitch, and yaw angles, as shown in Figure 1. When the camera rotates in the roll direction, the subject rotates in the image. When the camera rotates in the yaw and pitch directions, the subject moves in the horizontal and vertical directions, respectively, in the image. Therefore, the roll direction corresponds to image rotation, while the pitch and yaw directions correspond to image translation. By using this relationship, we can express camera orientation as image deforming parameters: image rotation and image translation parameters.

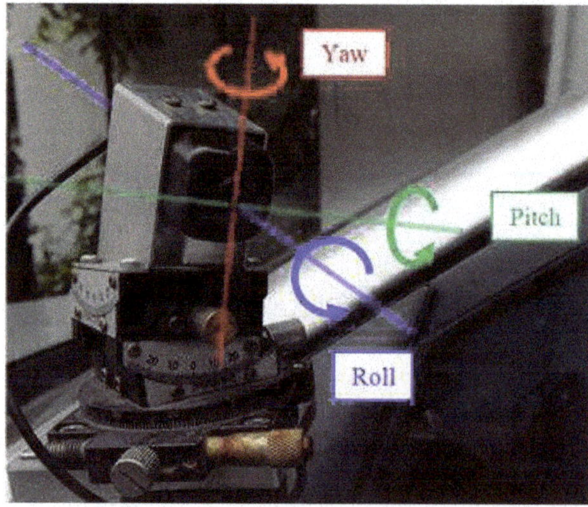

Figure 1. Parameterization of camera orientation.

We compare and analyze RVA of a pre-uploaded 3D vehicle model and RVA of a captured image in order to estimate the parameters. An RVA detection step has to be preceded for comparative analysis. RVA data in the image space is converted into the Hough space to estimate the image rotation parameters. We can estimate the image translation parameters using two RVA data with no image rotation difference. Figure 2 shows the procedure of the proposed automatic online calibration.

3.1. Reflected-Vehicle Area Detection

RVA is a part of the driver's vehicle photographed in the image. The algorithm for detecting RVA consists of two steps. The first step involves preprocessing to improve the accuracy of the second step and to eliminate, to the extent possible, the data that are not related to the reflected vehicle. In the second step, we utilized RANSAC to find the reflected-vehicle boundary and determine the inside of this boundary as the RVA. RANSAC is an iterative curve-fitting method for estimating the parameters of a mathematical model and classifying data into inliers and outliers. Inliers are the data whose distribution can be explained by some set of model parameters, and outliers are the data that do not fit the model. Therefore, outliers do not influence the estimated parameters. For this reason, RANSAC is used for outlier detection as well [37].

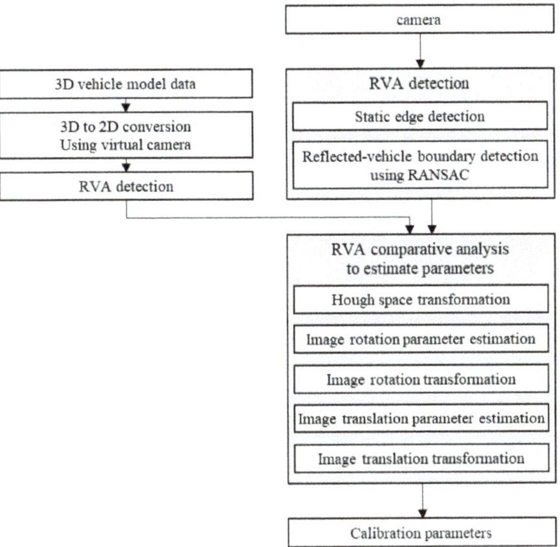

Figure 2. Block diagram of automatic online calibration.

In the first step, we eliminate outliers to improve the accuracy of RANSAC, which is inversely proportional to the number of outliers. We assume that the edge points of the reflected vehicle are inliers and all other points are outliers. The edge points of RVA always appear stationary because the moving speed and direction of the camera installed on the vehicle are the same as those of the vehicle. Therefore, we detected edges that do not change over time, and we call this process "static edge detection". To detect static edge points, we capture multiple images over a certain time period and detect the edges of each captured image using the Sobel filter. Thereafter, we could detect static edges by applying the logical and operation to each pixel coordinate. Figure 3 shows an example of the detection of static edges.

Edge detection using the Sobel filter does not guarantee robust results because it uses static parameters of filter to detect edges of dynamic images. However, the proposed static edge detection can overcome this problem by collecting lots of edge information from multiple images. Therefore, we must capture an adequate number of images to eliminate static edge points outside RVA to the extent possible. The static edge image in Figure 4a confirms that the static edge points inside RVA also form a curve as distinct as the reflected-vehicle boundary. Therefore, if there are several static edge points in each row of the image, only the leftmost static edge point is set as the candidate of the reflected-vehicle boundary. This process not only eliminates the static edge points inside the RVA but also represents candidates as a bijection function. Figure 4b shows the candidate points of the reflected-vehicle boundary. In this figure, most of the static edge points inside RVA are not candidates. After determining the candidate group, we utilize RANSAC to detect RVA in the second step.

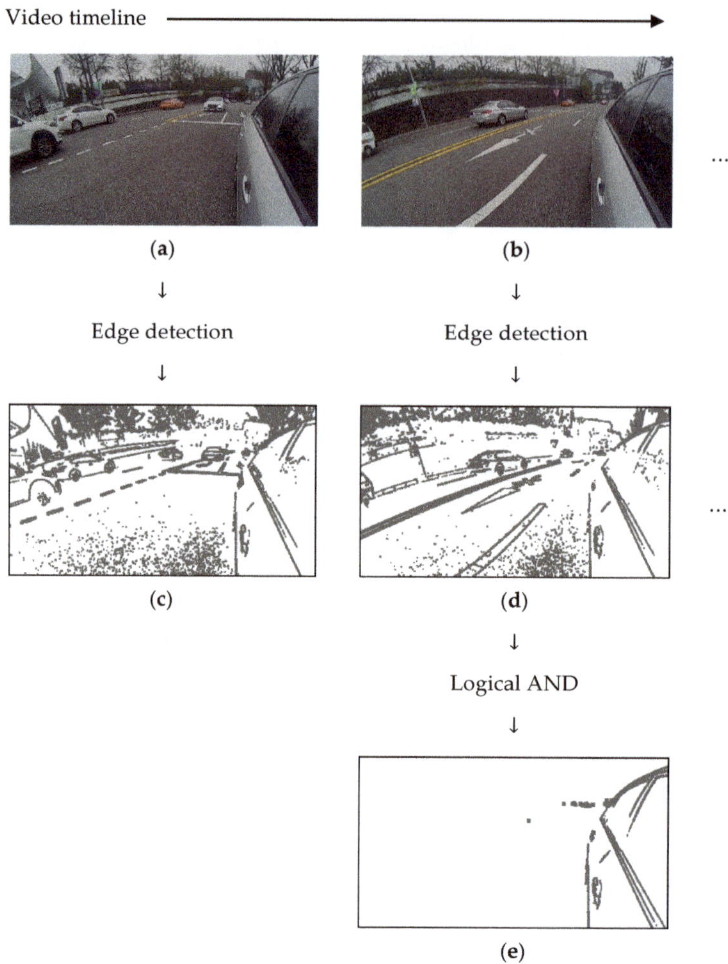

Figure 3. Concept of static edge detection. (**a**) First captured image; (**b**) second captured image; (**c**) edge image of (**a**); (**d**) edge image of (**b**); and (**e**) static edge image after logical and processing.

In the first step, we eliminated most of the static edge points, except for the points of the reflected-vehicle boundary. We utilized RANSAC to categorize the candidates into the reflected-vehicle boundary (inliers) and the others (outliers) in the second step. RANSAC is an iterative method involving two phases: hypothesis generation and hypothesis evaluation, as shown in Figure 5.

RANSAC generates the hypothesis for line fitting by randomly sampling data and estimating parameters using the sampled data. The highest score of hypothesis evaluation is given when all randomly picked-up data are inliers. Therefore, hypothesis generation and evaluation must be iterated so that all randomly picked-up data are inliers.

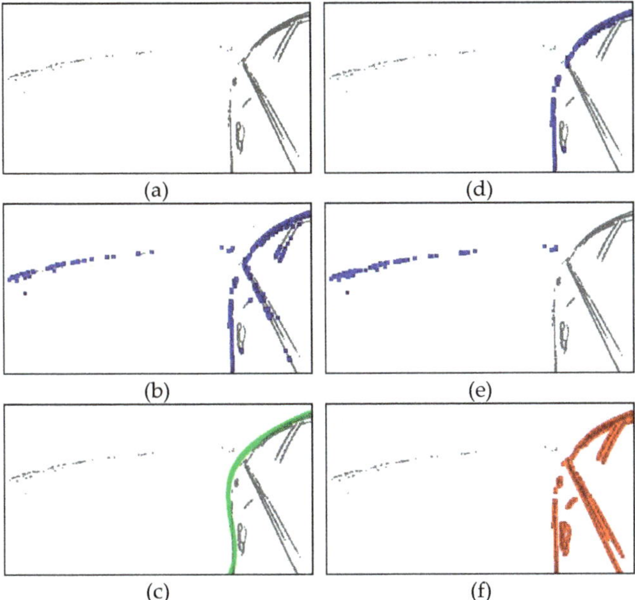

Figure 4. Example of reflected-vehicle area detection using Random Sample Consensus (RANSAC). (**a**) Static edge image; (**b**) determined candidates of the reflected-vehicle boundary; (**c**) estimated reflected-vehicle boundary using RANSAC; (**d**) identified reflected-vehicle boundary points from (**b**); (**e**) eliminated static edge points outside Reflected-Vehicle Area (RVA); and (**f**) static edge points inside RVA.

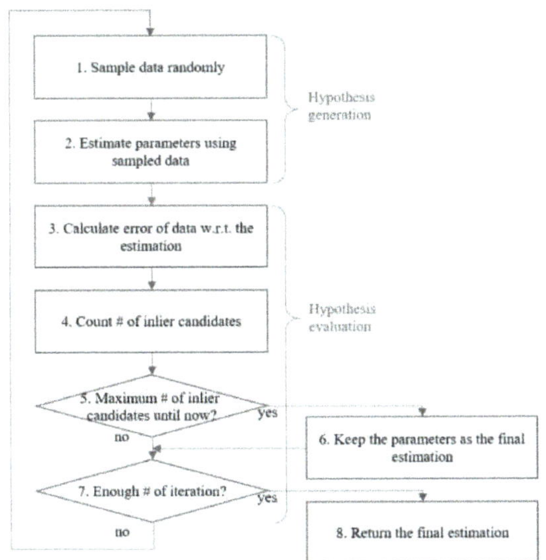

Figure 5. Flowchart of RANSAC.

The probability p that RANSAC will select all inlier samples at once is as follows.

$$p = 1 - (1 - \gamma^s)^N \tag{1}$$

where γ is the number of inliers divided by the number of points in the data, s is the number of samples selected each time, and N is the number of iterations. Equation (2) can be used to determine the number of iterations.

$$N = \frac{\log(1-p)}{\log(1-\gamma^s)} \tag{2}$$

We can determine the variables γ and s experimentally, but the probability p can only be determined empirically. After hypothesis generation, RANSAC calculates the error that the distance datum has to the estimated line and counts the number of inliers within a predefined threshold to evaluate the hypothesis. To predefine the threshold, we assumed that the error follows a normal distribution. In statistics, the empirical rule is expressed as follows: X is an observation from a normally distributed random variable, μ is the mean of the distribution, and σ is its standard deviation.

$$\begin{aligned}
\Pr(\mu - 1\sigma \leq X \leq \mu + 1\sigma) &\approx 0.6827 \\
\Pr(\mu - 2\sigma \leq X \leq \mu + 2\sigma) &\approx 0.9545 \\
\Pr(\mu - 3\sigma \leq X \leq \mu + 3\sigma) &\approx 0.9973
\end{aligned} \tag{3}$$

We obtained the standard deviation of the inliers σ and then predefine the threshold between 2σ and 3σ so that RANSAC can select inliers with a probability of 95% or higher. Finally, we could detect the reflected-vehicle boundary by using RANSAC.

Boundary of a reflected-vehicle can be represented by a smooth curve. However, the boundary changes depend on the vehicle type and camera parameters. Therefore, we used a third-order equation as a model of RANSAC as shown below.

$$f(v) = a_0 + a_1 v + a_2 v^2 + a_3 v^3 \tag{4}$$

where v is a horizontal direction coordinate of RVA point. Additionally, the order of the equation may increase as needed.

Figure 4c shows a reflected-vehicle boundary curve estimated using RANSAC. The blue points in Figure 4b are the candidates of a reflected-vehicle boundary, and these points are used as the input data for RANSAC. Figure 4c shows the curve with the most inliers, as estimated using RANSAC, and the blue points in Figure 4d are the candidates identified as inliers by RANSAC.

We assume that the interior of an estimated reflected-vehicle boundary is RVA. Figure 4e,f show the static edge points outside RVA and the static edge points inside RVA, respectively.

In the next section, we presented an automatic calibration method that employs these static edge points.

3.2. RVA Comparative Analysis to Estimate Parameters

We estimated the image rotation and translation parameters that represent the camera orientation by comparing RVA with the stored vehicle model. The stored vehicle model must be converted into an edge image to compare it with the RVA consisting of static edge points. Vehicle manufacturers may provide three-dimensional (3D) vehicle model data; if that is not the case, we can construct the data by using a 3D scanner. Then, we can regulate the camera position, orientation, and FOV and shoot a 3D vehicle in 3D virtual space. The Unity program is useful for regulating the virtual camera and for clicking pictures with it in 3D virtual space [38]. We applied edge detection to images captured using the virtual camera to obtain a reflected-vehicle edge image of the stored vehicle model. Figure 6 shows the process of converting a 3D vehicle model into an edge image by using the Unity program.

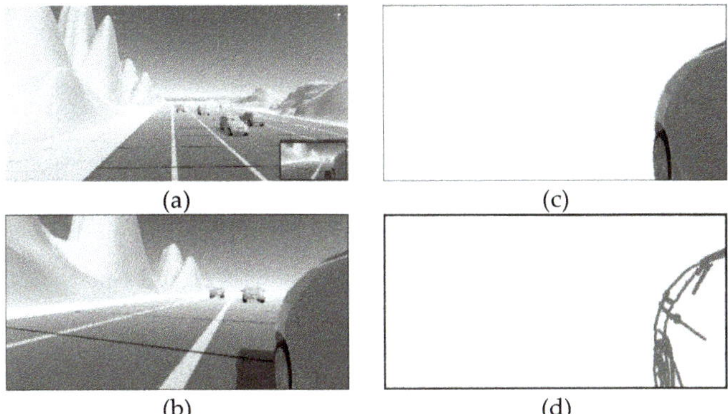

Figure 6. Process of converting a 3D vehicle model into an edge image by using the Unity program. (**a**) 3D virtual space of the Unity program; (**b**) photograph captured using a virtual camera; (**c**) reflected-vehicle area of image captured using a virtual camera; and (**d**) edge image of a 3D vehicle model in the reflected-vehicle area.

After converting the edge image from the 3D vehicle model, we utilized the Hough space to compare the converted edge image of the 3D vehicle model and the result of reflected-vehicle area detection in Section 2. The Hough space is a set of values that transform the edge points of the RVA into the Hesse normal form [39]. Equation (5) represents the Hesse normal form.

$$r = x \cos \theta + y \sin \theta \tag{5}$$

The coordinate (x, y) can be expressed as (r, θ) by using Equation (5), and we can visualize (r, θ) as a curve. Figure 7 shows a visualized curve corresponding to an image space point in the Hough space. We assumed that this curve can be expressed as $r = h(\theta)$.

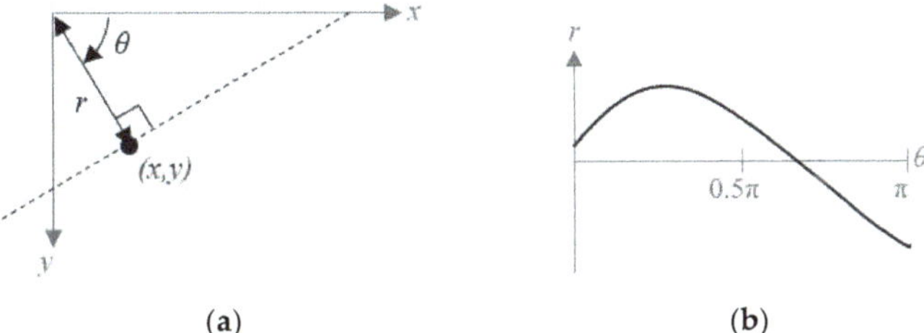

Figure 7. Parameterization of image space and Hough space. (**a**) A point in the image space, and (**b**) a curve corresponding to an image space point in the Hough space.

If the coordinate (x, y) is rotated by $\Delta\theta$ and moved to (x', y'), a degree-shift of $\Delta\theta$ occurs in the Hough space, and if $(\Delta x, \Delta y)$ image translation occurs, r-shifting occurs in the Hough space, as shown in Figure 8. This phenomenon indicates that the parameter θ and $(\Delta x, \Delta y)$ image translation are independent of each other. Therefore, the Hesse normal form can be re-expressed by considering that image rotation and translation occur simultaneously:

$$r + \Delta r = x'' \cos(\theta + \Delta\theta) + y'' \sin(\theta + \Delta\theta). \tag{6}$$

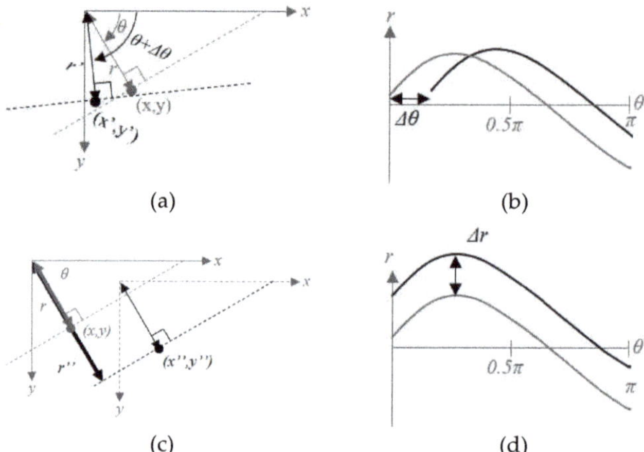

Figure 8. Parameterization of image space and Hough space by means of similarity transformation. (**a**) Rotation transformation in image space; (**b**) rotation transformation in Hough space; (**c**) translation transformation in image space; and (**d**) translation transformation in Hough space.

By using Equation (6), the Hough space curve $r = h(\theta)$ can be re-expressed as $r + \Delta r = h(\theta + \Delta\theta)$. We can estimate rotational similarity by comparing the difference between $h(\theta)$ and $h(\theta + \Delta\theta)$. $r = h(\theta)$ denotes a curve in the Hough space corresponding to one point in the image space. Many points exist in the image space, so we calculate the variance of $h(\theta)$ corresponding to each θ to solve this problem.

$$v(\theta) = \frac{1}{N-1} \sum_{i=1}^{N} \left| h_i(\theta) - \mu_h \right|^2 \tag{7}$$

where $v(\theta)$ is the variance of $h(\theta)$ corresponding to θ, N is the number of edge points, $h_i(\theta)$ is $h(\theta)$ corresponding to the ith edge point, and $\mu_h = \frac{1}{N} \sum_{i=1}^{N} h_i(\theta)$. Figure 9 shows an example of how (r, θ) of the Hough space and variance $v(\theta)$ are changed by image transformation. Figure 9g–i show that the variance $v(\theta)$ is shifted in the vertical direction owing to image rotation, and the amplitude of $v(\theta)$ is stretched in the horizontal direction owing to image scaling. Moreover, the effect of image translation is rarely seen in the Hough space. Therefore, we can estimate the rotational similarity between Figure 9a,b by computing the degree-shifting between variances $v(\theta)$, as shown in Figure 9g,h.

We utilized normalized cross-correlation to calculate the degree-shifting between $v_m(\theta)$ and $v_c(\theta)$, where $v_m(\theta)$ is a curve corresponding to the 3D vehicle model, and $v_c(\theta)$ is a curve corresponding to a static edge image of RVA. Normalization is applied to calculate the degree-shifting when the amplitude difference between two signals is large, as shown in Figure 9g,i. The normalized cross-correlation is one of the proper solutions for estimating the relationship between two signals, and it is expressed as follows:

$$R(\phi) = \frac{1}{K} \sum_k \frac{(v_m(\theta))^* v_c(\theta + \phi)}{\sigma_{v_m} \sigma_{v_c}}, \tag{8}$$

where σ_{v_m} is the variance of $v_m(\theta)$, σ_{v_c} is the variance of $v_c(\theta)$, $*$ is the complex conjugate, and K is the length of valid signals. Then, we can obtain the rotational similarity $\Delta\theta$ by using Equation (9).

$$\Delta\theta = \underset{\phi}{\mathrm{argmax}}(R(\phi)) \tag{9}$$

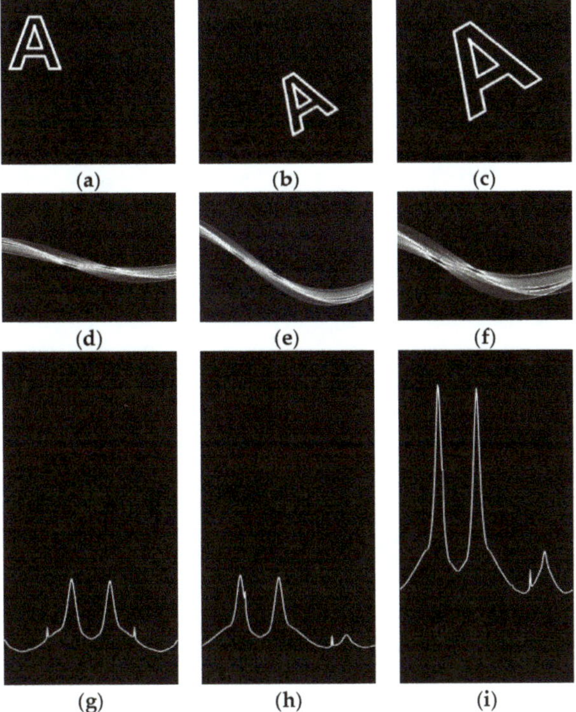

Figure 9. Example of Hough space data changed by image transformation. (**a**) A test image; (**b**) result of rotating and translating image (**a**); (**c**) result of scaling and translating image (**b**); (**d**) result of converting data (**a**) to the Hough space, where row is θ and column is r; (**e**) result of converting data (**b**); (**f**) result of converting data (**c**); (**g**) variance of (**d**) corresponding to θ, where row is θ, and column is variance $v(\theta)$; (**h**) variance of (**e**); and (**i**) variance of (**f**).

If the camera image is calibrated using the estimated rotational similarity score, only translational similarity remains to be determined. We can obtain translational similarity from the normalized 2D cross-correlation, which is widely used in computer vision [40].

$$\gamma(u,v) = \frac{\sum_{x,y}\left[I_m(x,y) - \mu_{I_m}\right]\left[\hat{I}_c(x-u,x-v) - \mu_{\hat{I}_c}\right]}{\left\{\sum_{x,y}\left[I_m(x,y) - \mu_{I_m}\right]^2 \sum_{x,y}\left[\hat{I}_c(x-u,x-v) - \mu_{\hat{I}_c}\right]^2\right\}^{0.5}} \tag{10}$$

where $\gamma(u,v)$ denotes the normalized 2D cross-correlation data at (u,v), I_m the edge image of the 3D vehicle model, $\hat{I}_c$ the rotation-corrected camera image, and μ_{I_m} and $\mu_{\hat{I}_c}$ the averages of I_m and $\hat{I}_c$, respectively. Then, we can compute the translational similarity by using Equation (11).

$$(\Delta x, \Delta y) = \underset{u,v}{\operatorname{argmax}}(\gamma(u,v)) \tag{11}$$

Finally, we can construct a similarity matrix from Δx, Δy, and $\Delta\theta$, and calibrate the image captured by the camera as follows:

$$H_S = \begin{bmatrix} \mathbf{R} & \mathbf{t} \\ 0 & 1 \end{bmatrix} = \begin{bmatrix} \cos\Delta\theta & -\sin\Delta\theta & \Delta x \\ \sin\Delta\theta & \cos\Delta\theta & \Delta y \\ 0 & 0 & 1 \end{bmatrix}, \tag{12}$$

where H_S is the similarity matrix, **R** is the image rotation matrix, and **t** is the image translation matrix instead of 3D camera orientation.

4. Simulation and Experimental Results

We performed several simulations and experiments to illustrate the performance of the proposed method. The purpose of the first experiment was to determine the number of captured images for static edge detection. The second experiment was performed and repeated to validate the effect of the driving environment on the automatic calibration. We compared our method with previous methods in the third experiment. The final experiment indicated the constraints of the proposed method. For these experiments, we installed High-Definition Low-Voltage Differential Signaling (HD LVDS) cameras with 60-degree FOV and a resolution of 1280 px × 720 px on the vehicle's left- and right-side mirror. These values were chosen due to their similarities to humans' angle of view. The camera was equipped with a three-axis goniometer to change its orientation, as shown in Figure 10a. We also produced and installed grabber equipment to acquire Controller Area Network (CAN) data and LVDS camera images, as shown in Figure 10b. The proposed algorithm was implemented in C++ on a portable PC. We analyzed the CAN data via the car' On-Board Diagnostic II (OBDII) port to detect vehicle speed and captured images only when the vehicle was in motion.

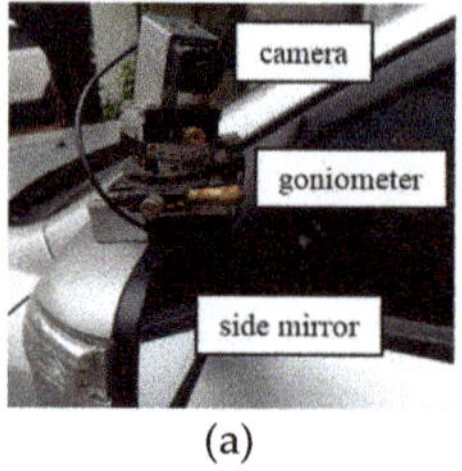

(a)

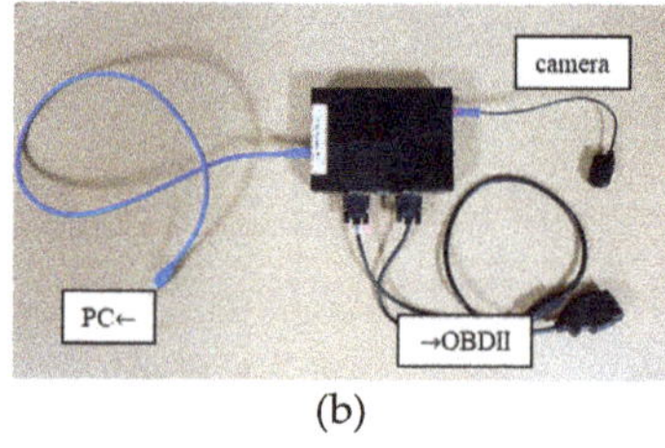

(b)

Figure 10. (**a**) Camera and goniometer used in the experiment and (**b**) grabber equipment for synchronizing and acquiring Controller Area Network (CAN) data and Low-Voltage Differential Signaling (LVDS) camera data.

4.1. Experiments for Determining An Appropriate Number of Captured Images

We compared the results of reflected-vehicle edge detection by changing the number of captured images to determine the appropriate number of captured images required for the purpose. Figure 11 shows the results of reflected-vehicle edge detection as a function of the number of captured images. The static edge points outside RVA were eliminated as the number of captured images increased. However, as the number of captured images increased, the time and memory costs increased as well. Due to this tradeoff relationship, we repeated this experiment in different driving environments and generalized the relationship between the number of captured images and the number of static edge points outside RVA.

Figure 12 shows the relationship between the number of static edge points outside RVA and the number of captured images. If more than 15 captured images were used, the ratio of the number of static edge points outside RVA to the number in the number of static edge points outside RVA converged to zero, as shown in Figure 12. Therefore, at least 15 captured images should be used so that the proportion of the number of static edge points outside RVA is less than 50%. Furthermore, we could eliminate a greater number of static edge points outside RVA by capturing more than 15 images, depending on the operating time and the computing power of the equipment.

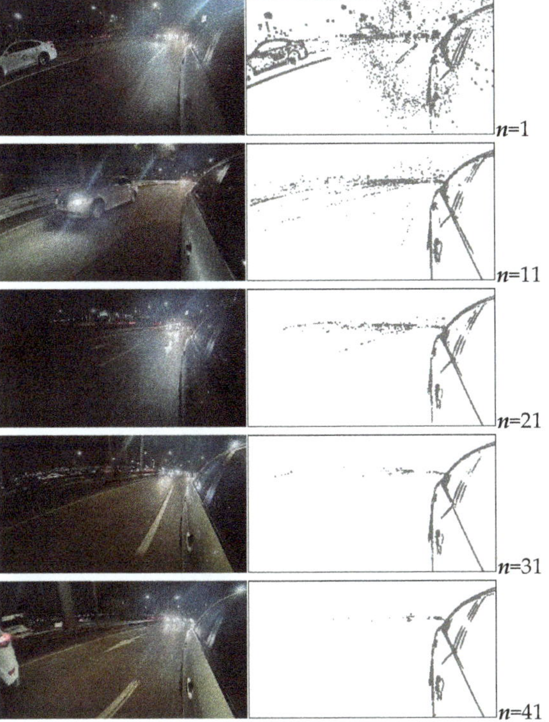

Figure 11. Left side: *n*th captured image, where *n* is the number of images captured; right side: static edge points of reflected-vehicle edge detection according to *n*.

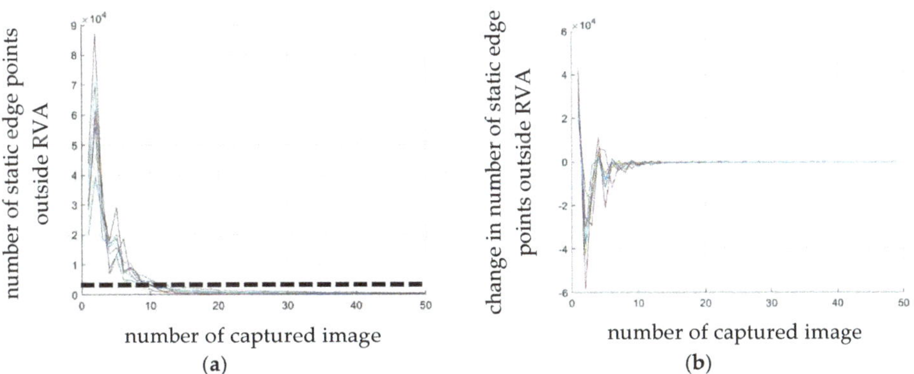

Figure 12. (a) Number of static edge points outside RVA according to the number of captured images, where the black dash line indicates that the ratio of the number of static edge points outside RVA to the number of static edge points is 0.5 and (b) change in the number of static edge points outside RVA according to the number of captured images.

4.2. Field Experiments for Quantitative and Qualitative Evaluation

We conducted experiments to verify each of the algorithms applied to the proposed method in natural driving environments. We used a goniometer and artificially regulated the camera orientation by 5° per axis. Figure 13 shows the process of the proposed method. Figure 13c shows the results of static edge detection and RVA detection. The static edge points outside RVA have been appropriately eliminated. Figure 13a shows one of the captured images, Figure 13b shows the result of automatic calibration of the captured images, and the green curves indicate the boundary of the 3D vehicle model. As shown in Figure 13b, the green curve almost matches the boundary of the reflected vehicle in the calibrated image. This result indicates that the proposed automatic calibration is apt for a side-rear-view monitoring system and that the proposed automatic calibration is accurate even when the camera orientation changes.

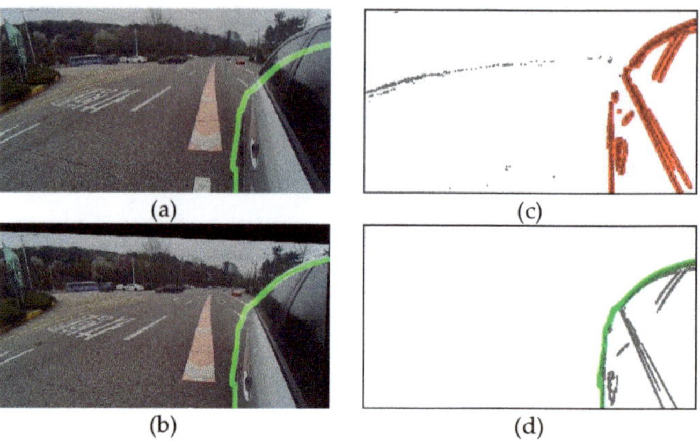

Figure 13. Field experiment result with speed limits of 80 km/h. (**a**) Captured image, where the green line denotes the boundary of the 3D vehicle model; (**b**) calibrated image, where the green line denotes the boundary of the 3D vehicle model; (**c**) RVA detection, where red points are static edge points inside RVA and gray points are static edge points outside RVA; and (**d**) edge image of 3D vehicle model, where the green line denotes the boundary of the 3D vehicle model.

We repeated the driving test in various environments without changing the camera orientation and the edge image of the 3D vehicle model to verify whether the proposed method can provide consistent results. We drove through a school campus with a speed limit of 20 km/h, city road with a speed limit of 50 km/h, speedway with a speed limit of 80 km/h, and an indoor parking lot during the day and night. Figure 14 shows the results of the experiment in each environment. The static edge points outside RVA appeared near the horizon when driving at the speed of 50 km/h or more. These static edges mostly indicate a horizontal vanishing line whose position hardly changed in the image. In the school campus or the underground parking lot, static edge points outside RVA were randomly scattered. Therefore, reflected-vehicle edge detection was essential when driving on a natural road with horizon views.

The static edge points inside RVA were more evident in the daytime than in the nighttime. Naturally, the edge of the car was more clearly visible in a bright environment than in a dark environment. Moreover, RVA detection results were evident in the dark when high-speed roads and parking lots were well lit. On the contrary, fewer static edge points inside RVA were detected in the campus during the night-time because of poorer lighting conditions compared to those in the other environments. Nevertheless, the proposed method took advantage of the Hough space (see Section 3) to ensure that the calibration can be performed even when there are only a few static edge points inside RVA.

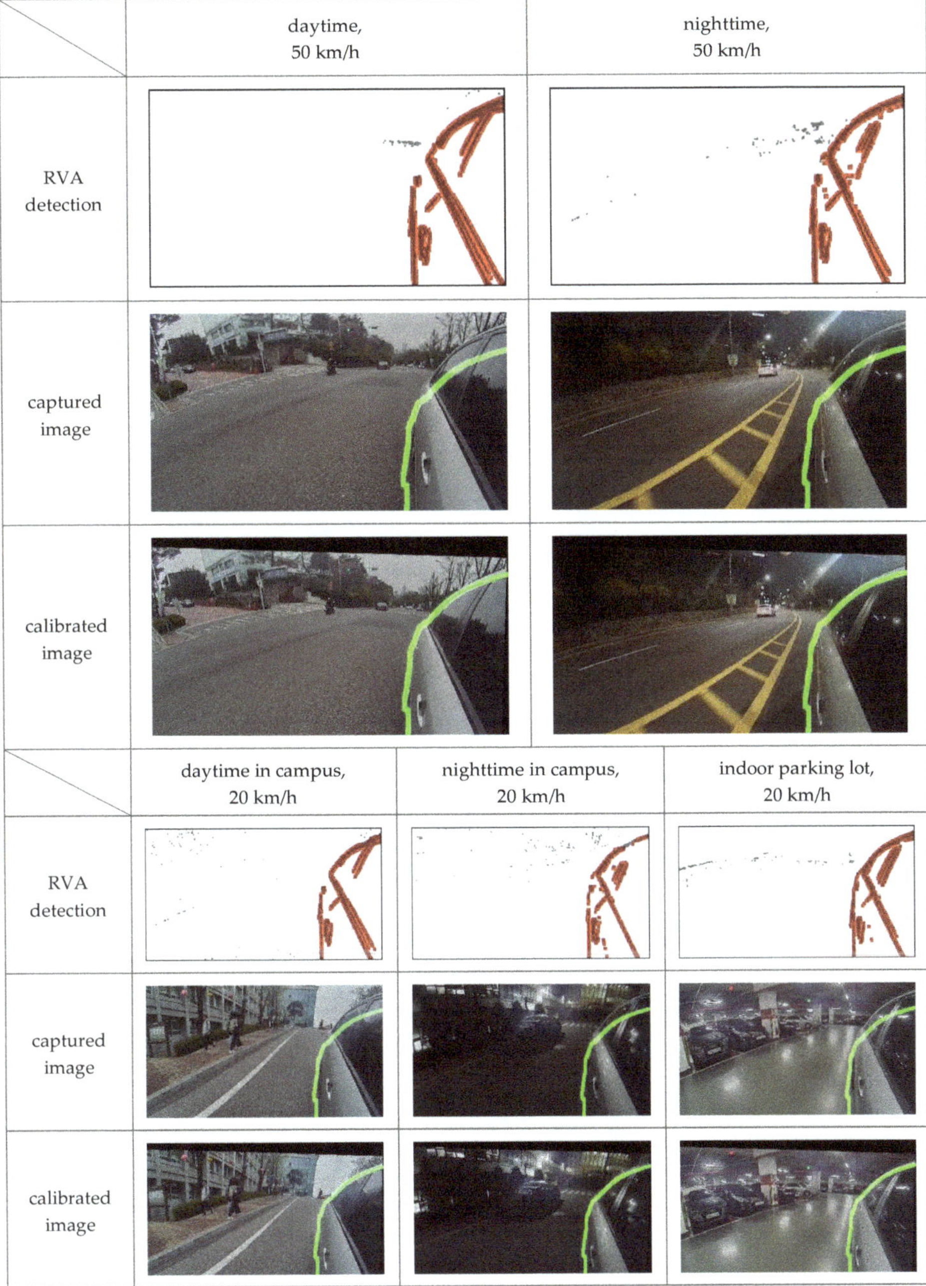

Figure 14. Results obtained using the proposed method in different experimental environments where the red points are the static edge points inside RVA, and the green line denotes the boundary of the 3D vehicle model.

We must know all camera parameters except rotation parameters to implement existing methods, whereas our method operates with unknown camera parameters. Moreover, the ground truths of the camera parameters installed in the vehicle were not available [19,41–43]. Therefore, we repeated this experiment 100 times and used precision, recall, and Root Mean Squared Error (RMSE) as the quantitative evaluation indexes. Furthermore, we experimented with a 150-degree FOV camera with lens distortion, a 115-degree FOV camera with lens distortion, and a 150-degree FOV camera without lens distortion to confirm the applicability of the algorithm to other ADAS cameras.

4.2.1. Precision, Recall, and RMSE

Precision and recall can be obtained by calculating true positive (TP), false positive (FP), false negative (FN), and true negative (TN) [44].

$$\text{precision} = \frac{TP}{TP+FP}$$
$$\text{recall} = \frac{TP}{TP+FN} \tag{13}$$

TP, FP, FN, and TN are defined in Table 1, where I_m denotes the edge image of the 3D vehicle model, $\hat{I}_c$ is a calibrated image, S_m is RVA of I_m, $(S_m)^c$ is area outside RVA of I_m, S_c is RVA of $\hat{I}_c$, and $(S_c)^c$ is area outside RVA of $\hat{I}_c$.

Table 1. Definitions of true positive (TP), false positive (FP), false negative (FN), and true negative (TN) for computing precision and recall.

		I_m	
		S_m	$(S_m)^c$
$\hat{I}_c$	S_c	TP	FP
	$(S_c)^c$	FN	TN

Figure 15 shows the TP, FP, FN, and TN areas visually. TP is the number of intersection points between S_m and S_c. The red region FN denotes the number of intersection points between S_m and $(S_c)^c$. We could calculate FP and TN as well in the same manner. The RMSE is defined as follows.

$$RMSE_{rot} = \sqrt{\frac{1}{N}\sum_n \left|\Delta\theta(n) - \mu_{\Delta\theta}\right|^2}, \text{ where } \left|\Delta\theta(n)\right| < 0.5\pi$$
$$RMSE_{tranx} = \sqrt{\frac{1}{N}\sum_n \left|\Delta x(n) - \mu_{\Delta x}\right|^2} \tag{14}$$
$$RMSE_{trany} = \sqrt{\frac{1}{N}\sum_n \left|\Delta y(n) - \mu_{\Delta y}\right|^2}$$

where N denotes the number of experiments; $\Delta\theta$, Δx, and Δy are parameters of a similarity matrix estimated using the proposed method, and $\mu_{\Delta\theta}$, $\mu_{\Delta x}$, and $\mu_{\Delta y}$ denote the average of $\Delta\theta$, Δx, and Δy, respectively. The range of $\Delta\theta(n)$ is limited because the side-rear-view camera cannot capture the rear-view if $\left|\Delta\theta(n)\right|$ exceeds 0.5π. Table 2 shows the average and the RMSE of each parameter calculated from 100 repeated experiments in different environments, and Figure 16 shows the parameter, precision, and recall values calculated over 100 experiments. If the precision and recall are 1, the two images are identical. We can see that the averages of precision and recall were 0.9758 and 0.9239, and both values were close to 1. This means that the edge image of the 3D vehicle model and the calibrated image were almost identical. The RMSE of rotational similarity was less than 1°, and the RMSE values of x- and y-axes translational similarity were 4.9041 and 13.4763 px, respectively. An RMSE value close to zero indicates that the experimental results are not affected by changes in the driving environment. Since we experimented in various environments without changing the camera orientation and the edge image of the 3D vehicle model, these quantitative results verified that the proposed method could perform online-calibration in most environments with RVA.

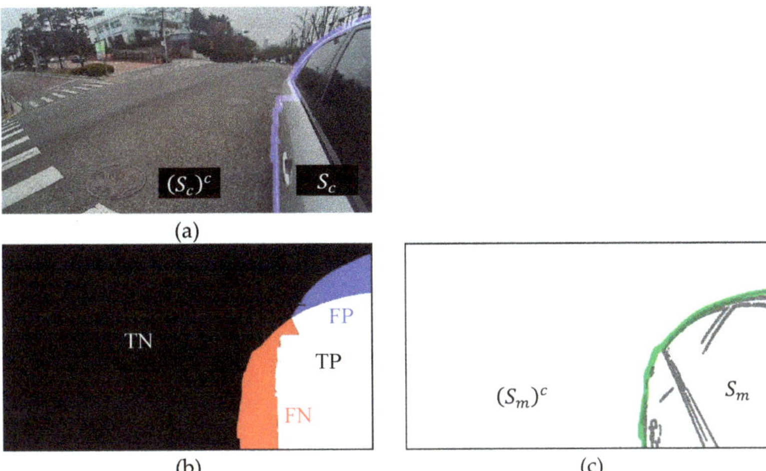

Figure 15. Visualization and parameterization of TP, FP, FN, and TN. (**a**) Captured image where the blue line is the boundary of RVA of captured image and S_c is RVA of calibrated image; (**b**) TP $= S_m \cap S_c$, TN $= (S_m)^c \cap (S_c)^c$, FN $= S_m \cap (S_c)^c$, and FP $= (S_m)^c \cap S_c$; and (**c**) edge image of the 3D vehicle model where the green line is the boundary of 3D vehicle model and S_m is the RVA of the edge image of the 3D vehicle model.

Table 2. Average and Root Mean Squared Error (RMSE) of the quantitative results of 100 repeated experiments in different environments.

	$\Delta\theta$ (Degree)	Δx (Pixel)	Δy (Pixel)	Precision with Calibration	Recall with Calibration	Precision without Calibration	Recall without Calibration
Average	−1.4000	−124.6400	−55.3000	0.9758	0.9239	0.6715	0.5929
RMSE	0.6164	4.9041	13.4763	-	-		

4.2.2. Experiments with Various Cameras

FOV of cameras used in ADAS depends on its purpose. For example, forward collision warning systems and parking assistance systems commonly use a narrow-angle camera and a wide-angle camera, respectively. In some cases, lens distortion may occur. In order to verify that the proposed algorithm can work in these various conditions, we experimented with three types of cameras: a 150-degree FOV camera with lens distortion, a 115-degree FOV camera with lens distortion, and a 115-degree FOV camera without lens distortion. Camera orientation was manually regulated by 5-degree per axis.

As shown in Figures 17 and 18, changes in FOV did not significantly affect the experimental results. Additionally, Figure 19 shows that the proposed method could calibrate both the lens distorted-image and lens distortion corrected-image. These qualitative results indicate that the proposed method could perform online calibration even if cameras' FOV and lens distortion were changed. Therefore, our method has the potential to be applied to various ADAS cameras.

4.3. Comparison with Previous Methods

As aforementioned in Section 2, camera calibration can be categorized according to which features and devices were used: offline calibration, online calibration with additional devices, and online calibration without additional devices. Table 3 shows the comparison of the related works and the proposed method from the viewpoint of the side-rear-view monitoring system calibration. Offline calibration is an inconvenient and restrictive method because the driver has to visit a large service center equipped with special facilities, and it cannot be conducted in natural driving environments. In addition, these facilities increase the price of offline calibration-based products. Likewise, additional devices of online calibration also increase the price.

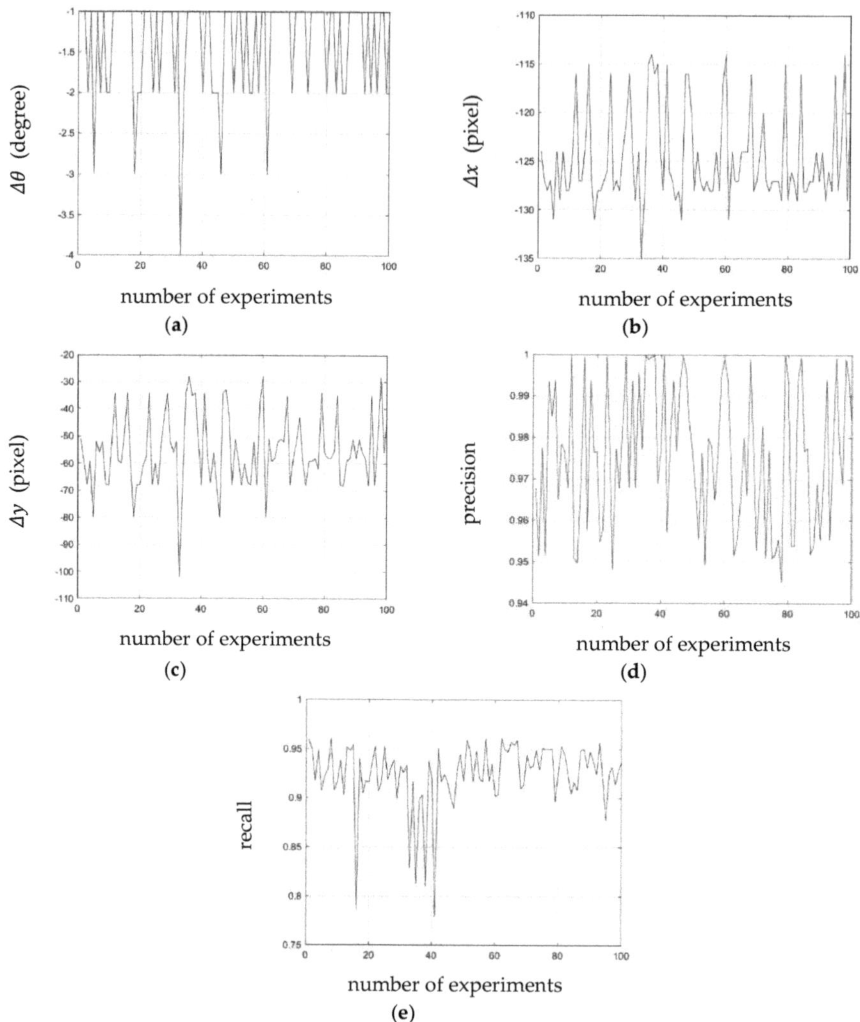

Figure 16. The result of 100 repeated experiments. (**a**) Estimated rotational similarity; (**b**) x-axis translational similarity; (**c**) y-axis translational similarity; (**d**) precision values; and (**e**) recall values.

Online calibration is convenient because it can automatically calibrate cameras in a natural driving situation. However, traditional online calibration can hardly calibrate side-rear-view cameras due to constraints. They must extract features such as lane from captured images, but the side-rear-view camera looking at the horizon behind the vehicle does not capture traffic lanes around the vehicle. In contrast to those methods, the proposed method can calibrate the side-rear-view camera using RVA that is being photographed at all times in natural driving environments.

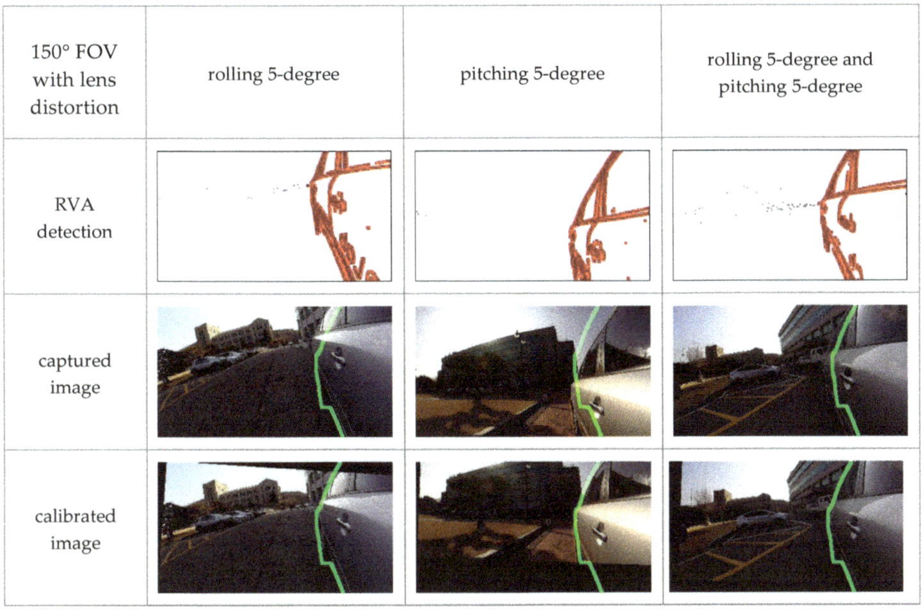

150° FOV with lens distortion	rolling 5-degree	pitching 5-degree	rolling 5-degree and pitching 5-degree
RVA detection			
captured image			
calibrated image			

Figure 17. Experimental results with a 150-degree Field Of View (FOV) camera with lens distortion in different orientation conditions, where the green line denotes the boundary of the 3D vehicle model.

115° FOV with lens distortion	rolling 5-degree	pitching 5-degree	rolling 5-degree and pitching 5-degree
RVA detection			
captured image			
calibrated image			

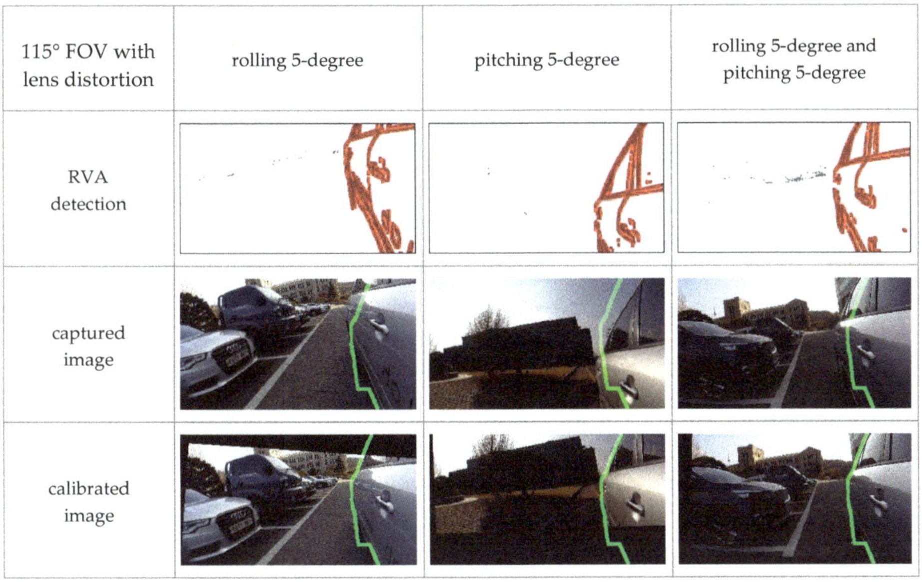

Figure 18. Experimental results with a 115 degree FOV camera with lens distortion in different orientation conditions, where the green line denotes the boundary of the 3D vehicle model.

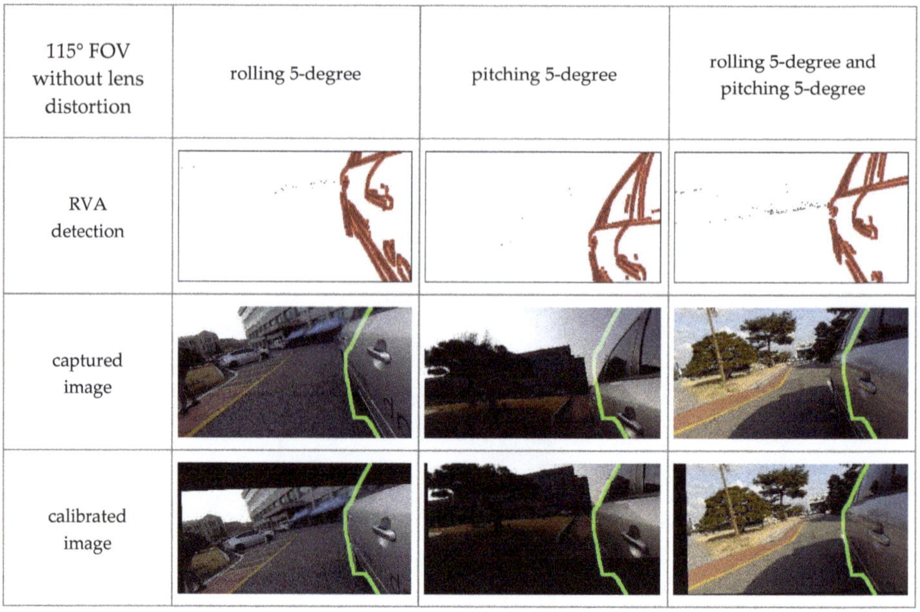

115° FOV without lens distortion	rolling 5-degree	pitching 5-degree	rolling 5-degree and pitching 5-degree
RVA detection			
captured image			
calibrated image			

Figure 19. Experimental results with 115-degree FOV camera and lens distortion correction in different orientation conditions, where the green line denotes the boundary of the 3D vehicle model.

Table 3. Comparison of the related works and the proposed method.

Method		Driver's Convenience	Product Cost	Calibration Constraint
Offline calibration		Poor	Poor	Fair
Online calibration with additional devices		Good	Fair	Poor
Online calibration without additional devices	Previous methods	Good	Good	Poor
	Proposed method	Good	Good	Good

Unfortunately, since there is no previous method that can calibrate side-rear-view monitoring system in natural driving environments, it is impossible to conduct quantitative performance comparison of the previous and proposed methods with the same dataset. However, the comparison summarized in Table 3 clearly explains that the proposed method is superior to the other previous methods in terms of side-rear-view camera calibration. Moreover, we can utilize the RVA information instead of the calibration patterns to implement offline calibration for calculating the similarity score and aligning images.

The similarity matrix consisting of image rotation and translation parameters can be estimated by minimizing an algebraic distance, called reprojection error, between matched feature points.

$$\hat{H}_S = \underset{H_S}{\operatorname{argmin}} \sum_i \|\breve{m}_i - H_S m_i\|^2 \tag{15}$$

where H_S is the similarity matrix, m_i is i-th feature points inside the RVA in a captured image, $\breve{m}_i$ is i-th feature points of the 3D vehicle model corresponding to m_i, and $\hat{H}_S$ is the estimated similarity matrix. We solved Equation (15) using the Levenberg–Marquardt method, one of the maximum likelihood estimation methods [45]. Experiments were performed using three types of cameras.

The proposed method can estimate similar parameter values to the RVA-based offline calibration method, as seen in Table 4. Furthermore, the experimental results with the 150-degree camera in Figure 20 show that the RVA boundary of the calibrated image by our method resembles the green line more than the previous methods. Additionally, other calibrated images in Figure 20 show that all RVA boundary locations of both methods almost fit the green line. These experimental results indicate that our method can provide similar results as the RVA-based offline calibration method even under conditions where the previous method cannot operate due to a lack of feature points.

Table 4. Quantitative performance comparison.

Camera Condition	Method	$\Delta\theta$ (Degree)	Δx (Pixel)	Δy (Pixel)
150° FOV with lens distortion	proposed	0	89	-25
	offline	1	75	0
115° FOV with lens distortion	proposed	−1	96	−71
	offline	1	103	−6
115° FOV without lens distortion	proposed	2	93	−33
	offline	0	117	−38

4.4. Limitation of Calibration

The proposed method compared the static edge points inside RVA of a captured image and a 3D vehicle model image to calculate the similarity between the two images. Therefore, the static edge points of RVA represent an essential factor, but RVA can be altered by various factors. To investigate the effect of RVA range, we repeated the experiment by gradually decreasing the RVA range. Since the goniometer has a limitation in changing the camera orientation, we decreased the RVA of the 3D vehicle model image instead of changing the camera orientation, as shown in Figure 21. We could predict that the calibrated images corresponding to Figure 21 will display the translated images along the x- and y-axis directions. Therefore, if the rotation parameter is changed or if the translation parameter is different from the decreased RVA value, the calibration fails.

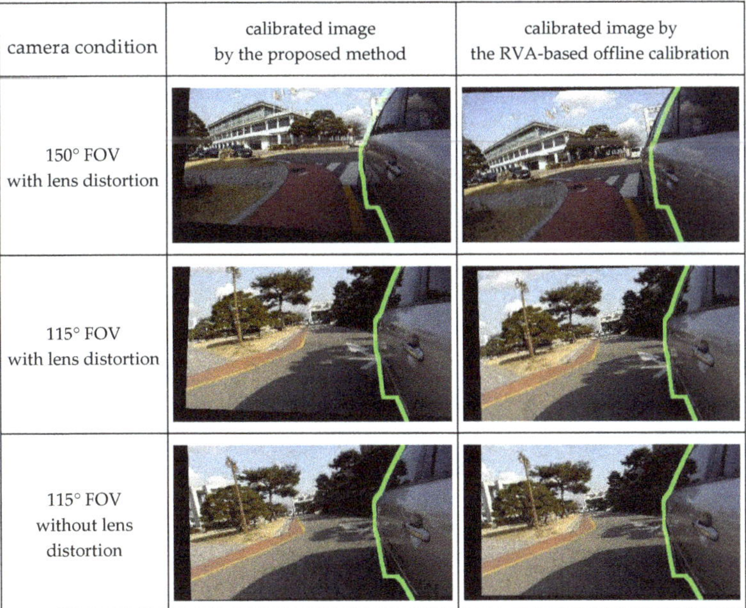

camera condition	calibrated image by the proposed method	calibrated image by the RVA-based offline calibration
150° FOV with lens distortion		
115° FOV with lens distortion		
115° FOV without lens distortion		

Figure 20. Qualitative performance comparison.

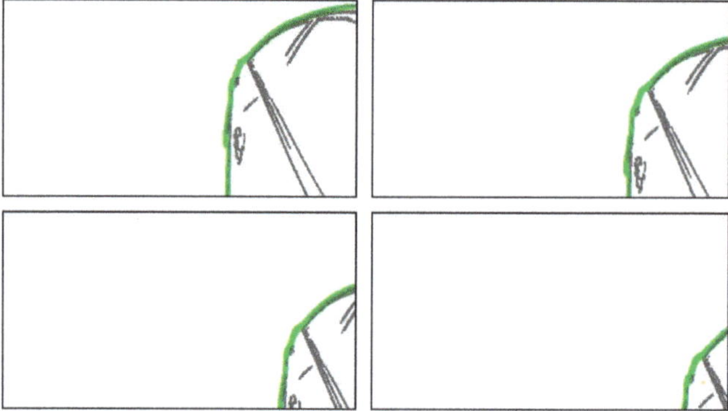

Figure 21. Images with gradually decreasing RVA of a 3D vehicle model image to verify the minimum required RVA, where the green lines are the boundaries of a 3D vehicle model image.

Figure 22 shows the calibrated images corresponding to the images in Figure 21. We can see that the rotation parameter of the bottom-right image in Figure 22 differed from those of the other calibrated images, and the vehicle boundary in all calibrated images in Figure 22 almost matched the green curves representing the boundaries of the 3D vehicle model. Accordingly, the calibration failed only in the bottom-right case. RVA in case of the failure had no static edge points inside it, unlike the other cases. This means that the calibration can fail when it uses only RVA boundary data. Through this experiment, we confirmed that the elements that serve as static edge points inside RVA (i.e., door handle or pillar) must be photographed for automatic calibration of the side-rear-view monitoring system.

Figure 22. Calibrated images corresponding to Figure 21, where the green lines denote the boundaries of a 3D vehicle model image.

5. Conclusions

We proposed an automatic online calibration method for the monitoring system of a vehicle equipped with a side-rear-view camera. The proposed method has the following advantages. The first advantage is that it can be used to automatically calibrate the camera while driving without using additional sensors or artificial markers. Therefore, no specialized facilities are required for calibration. In addition, there is no constraint that offline calibration must be performed before automatic calibration, which is true of conventional methods. The next advantage is that it provides consistent results, even when the driving environment changes. This is possible because we eliminate irrelevant data before utilizing RANSAC to provide consistent results in various driving environments. The third advantage is that the proposed method facilitates large-scale template matching by using information about edge points instead of color information because the method uses the Hough space. This advantage solves the problem of traditional large-scale template-matching methods that use color information, as well as the problem that the RVA color changes depending on the vehicle color and the driving environment. The last advantage is that the calibration requires only RVA information. Therefore, the proposed method can potentially be used to calibrate most cameras mounted on a vehicle.

Based on this potential, we expect the proposed automatic online calibration method to be applied not only to side-rear-view monitoring systems but also to various vision-based ADAS. These advantages indicate that the proposed method can provide convenience to motorists who require recalibration, and it can increase profits for vehicle manufacturers by reducing the usage of special facilities. As a disadvantage, the proposed method estimates the similarity instead of camera orientation. This disadvantage sometimes induces affine transformation errors. These errors can be solved by using a planar vehicle model, but it is difficult to overcome this disadvantage with the proposed method because it employs a 3D vehicle model. The results of experiments conducted in various driving environments indicate that the proposed automatic calibration method is suitable for use in real-world applications.

Author Contributions: J.H.L. developed the algorithm, and performed the experiments. D.-W.L. developed the system architecture and analyzed the experimental results. All authors wrote the paper together. All authors have read and agreed to the published version of the manuscript.

Funding: This research received no external funding.

Conflicts of Interest: The authors declare no conflicts of interest.

References

1. Zhang, B.; Appia, V.; Pekkucuksen, I.; Liu, Y.; Umit Batur, A.; Shastry, P.; Liu, S.; Sivasankaran, S.; Chitnis, K. A surround view camera solution for embedded systems. In Proceedings of the IEEE Conference on Computer Vision and Pattern Recognition Workshops, Columbus, OH, USA, 24–27 June 2014; pp. 662–667.
2. Yu, M.; Ma, G. 360 surround view system with parking guidance. *SAE Int. J. Commer. Veh.* **2014**, *7*, 19–24. [CrossRef]
3. Lee, J.H.; Han, J.Y.; You, Y.J.; Lee, D.W. Apparatus and Method for Matching Images. KR Patent 101,781,172, 14 September 2017.
4. Pan, J.; Appia, V.; Villarreal, J.; Weaver, L.; Kwon, D.K. Rear-stitched view panorama: A low-power embedded implementation for smart rear-view mirrors on vehicles. In Proceedings of the IEEE Conference on Computer Vision and Pattern Recognition Workshops, Honolulu, HI, USA, 21–26 July 2017; pp. 20–29.
5. Mammeri, A.; Lu, G.; Boukerche, A. Design of lane keeping assist system for autonomous vehicles. In Proceedings of the International Conference on New Technologies, Mobility and Security, Paris, France, 26–29 July 2015; pp. 1–5.
6. Eum, S.; Jung, H.G. Enhancing light blob detection for intelligent headlight control using lane detection. *IEEE Trans. Intell. Transp. Syst.* **2013**, *14*, 1003–1011. [CrossRef]
7. Cardarelli, E. Vision-based blind spot monitoring. In *Handbook of Intelligent Vehicles*; Azim, E., Ed.; Springer: London, UK, 2012; pp. 1071–1087, ISBN 978-0-85729-084-7.
8. Suzuki, S.; Raksincharoensak, P.; Shimizu, I.; Nagai, M.; Adomat, R. Sensor fusion-based pedestrian collision warning system with crosswalk detection. In Proceedings of the IEEE Intelligent Vehicles Symposium, San Diego, CA, USA, 21–24 June 2010; pp. 355–360.
9. Baró, X.; Escalera, S.; Vitrià, J.; Pujol, O.; Radeva, P. Traffic sign recognition using evolutionary adaboost detection and forest-ECOC classification. *IEEE Trans. Intell. Transp. Syst.* **2009**, *10*, 113–126. [CrossRef]
10. Liu, J.F.; Su, Y.F.; Ko, M.K.; Yu, P.N. Development of a vision-based driver assistance system with lane departure warning and forward collision warning functions. In Proceedings of the Digital Image Computing: Techniques and Applications, Canberra, Australia, 1–3 December 2008; pp. 480–485.
11. Chen, M.; Jochem, T.; Pomerleau, D. AURORA: A vision-based roadway departure warning system. In Proceedings of the IEEE/RSJ International Conference on Intelligent Robots and Systems, Pittsburgh, PA, USA, 5–9 August 1995; pp. 243–248.

12. Zhang, Z. A flexible new technique for camera calibration. *IEEE Trans. Pattern Anal. Mach. Intell.* **2000**, *22*, 1330–1334. [CrossRef]

13. Camera Calibration Facility. Available online: http://www.angtec.com/ (accessed on 12 May 2020).

14. Xia, R.; Hu, M.; Zhao, J.; Chen, S.; Chen, Y. Global calibration of multi-cameras with non-overlapping fields of view based on photogrammetry and reconfigurable target. *Meas. Sci. Technol.* **2018**, *29*, 065005. [CrossRef]

15. Mazzei, L.; Medici, P.; Panciroli, M. A lasers and cameras calibration procedure for VIAC multi-sensorized vehicles. In Proceedings of the IEEE Intelligent Vehicles Symposium, Alcala de Henares, Spain, 3–7 June 2012.

16. Hold, S.; Nunn, C.; Kummert, A.; Muller-Schneiders, S. Efficient and robust extrinsic camera calibration procedure for lane departure warning. In Proceedings of the IEEE Intelligent Vehicles Symposium, Xi'an, China, 3–5 June 2009.

17. Tan, J.; Li, J.; An, X.; He, H. An interactive method for extrinsic parameter calibration of onboard camera. In Proceedings of the IEEE Intelligent Vehicles Symposium, Baden-Baden, Germany, 5–9 June 2011.

18. Li, S.; Ying, H. Estimating camera pose from H-pattern of parking lot. In Proceedings of the IEEE International Conference on Robotics and Automation, Anchorage, AK, USA, 3–7 May 2010.

19. Wang, X.; Chen, H.; Li, Y.; Huang, H. Online Extrinsic Parameter Calibration for Robotic Camera-Encoder System. *IEEE Trans. Ind. Inform.* **2019**, *15*, 4646–4655. [CrossRef]

20. Schneider, S.; Luettel, T.; Wuensche, H.J. Odometry-based online extrinsic sensor calibration. In Proceedings of the IEEE/RSJ International Conference on Intelligent Robots and Systems, Tokyo, Japan, 3–7 November 2013; pp. 1287–1292.

21. Chien, H.J.; Klette, R.; Schneider, N.; Franke, U. Visual odometry driven online calibration for monocular lidar-camera systems. In Proceedings of the International Conference on Pattern Recognition, Cancún, México, 4–8 December 2016; pp. 2848–2853.

22. Li, M.; Mourikis, A.I. 3-D motion estimation and online temporal calibration for camera-IMU systems. In Proceedings of the IEEE International Conference on Robotics and Automation, Karlsruhe, Germany, 6–10 May 2013; pp. 5709–5716.

23. Hemayed, E.E. A survey of camera self-calibration. In Proceedings of the IEEE Conference on Advanced Video and Signal Based Surveillance, Miami, FL, USA, 22 July 2003; pp. 351–357.

24. Maybank, S.J.; Faugeras, O.D. A theory of self-calibration of a moving camera. *Int. J. Comput. Vis.* **1992**, *8*, 123–151. [CrossRef]

25. Catalá-Prat, Á.; Rataj, J.; Reulke, R. Self-calibration system for the orientation of a vehicle camera. In Proceedings of the ISPRS Commission V Symposium: Image Engineering and Vision Metrology, Dresden, Germany, 25–27 September 2006; pp. 68–73.

26. Xu, H.; Wang, X. Camera calibration based on perspective geometry and its application in LDWS. *Phys. Procedia* **2012**, *33*, 1626–1633. [CrossRef]

27. De Paula, M.B.; Jung, C.R.; da Silveira, L.G., Jr. Automatic on-the-fly extrinsic camera calibration of onboard vehicular cameras. *Expert Syst. Appl.* **2014**, *41*, 1997–2007. [CrossRef]

28. Wang, H.; Cai, Y.; Lin, G.; Zhang, W. A novel method for camera external parameters online calibration using dotted road line. *Adv. Robot.* **2014**, *28*, 1033–1042. [CrossRef]

29. Choi, K.; Jung, H.; Suhr, J. Automatic calibration of an around view monitor system exploiting lane markings. *Sensors* **2018**, *18*, 2956. [CrossRef] [PubMed]

30. Yang, K.; Hu, X.; Bergasa, L.M.; Romera, E.; Wang, K. PASS: Panoramic Annular Semantic Segmentation. *IEEE Trans. Intell. Transp. Syst.* **2019**, 1–15. [CrossRef]

31. Lowe, D.G. Distinctive image features from scale-invariant keypoints. *Int. J. Comput. Vis.* **2004**, *60*, 91–110. [CrossRef]

32. Szeliski, R. Image alignment and stitching: A tutorial. *Found. Trends® Comput. Graph. Vis.* **2007**, *37*, 1–104. [CrossRef]

33. Ullah, F.; Kaneko, S.I. Using orientation codes for rotation-invariant template matching. *Pattern Recognit.* **2004**, *37*, 201–209. [CrossRef]

34. Kim, H.Y.; de Araújo, S.A. Grayscale template-matching invariant to rotation, scale, translation, brightness and contrast. In Proceedings of the Pacific-Rim Symposium on Image and Video Technology, Santiago, Chile, 17–19 December 2007; pp. 100–113.

35. Yang, H.; Huang, C.; Wang, F.; Song, K.; Zheng, S.; Yin, Z. Large-scale and rotation-invariant template matching using adaptive radial ring code histograms. *Pattern Recognit.* **2019**, *91*, 345–356. [CrossRef]

36. Ruland, T.; Pajdla, T.; Krüger, L. Extrinsic autocalibration of vehicle mounted cameras for maneuvering assistance. In Proceedings of the Computer Vision Winter Workshop, Nove Hrady, Czech Republic, 3–5 February 2010; pp. 44–51.
37. Fischler, M.A.; Bolles, R.C. Random sample consensus: A paradigm for model fitting with applications to image analysis and automated cartography. *Commun. ACM* **1981**, *24*, 381–395. [CrossRef]
38. Unity Technologies. What's New in Unity 5.0. Available online: https://unity3d.com/unity/whats-new/unity-5.0 (accessed on 26 March 2020).
39. Duda, R.O.; Hart, P.E. Use of the Hough transformation to detect lines and curves in pictures. *Commun. ACM* **1972**, *15*, 11–15. [CrossRef]
40. Lewis, J.P. Fast template matching. In Proceedings of the European Conference on Computer Vision, Copenhagen, Denmark, 28–31 May 2002; pp. 120–123.
41. Antonelli, G.; Caccavale, F.; Grossi, F.; Marino, A. Simultaneous calibration of odometry and camera for a differential drive mobile robot. In Proceedings of the IEEE International Conference on Robotics and Automation, Anchorage, AK, USA, 3–7 May 2010; pp. 5417–5422.
42. Tang, H.; Liu, Y. A fully automatic calibration algorithm for a camera odometry system. *IEEE Sens. J.* **2017**, *17*, 4208–4216. [CrossRef]
43. Guo, C.X.; Mirzaei, F.M.; Roumeliotis, S.I. An analytical least-squares solution to the odometer-camera extrinsic calibration problem. In Proceedings of the IEEE International Conference on Robotics and Automation, Saint Paul, MN, USA, 14–18 May 2012; pp. 3962–3968.
44. Olson, D.L.; Delen, D. *Advanced Data Mining Techniques*; Springer: Berlin/Heidelberg, Germany, 2008; ISBN 978-3-540-76916-3.
45. Kanzow, C.; Yamashita, N.; Fukushima, M. Withdrawn: Levenberg–marquardt methods with strong local convergence properties for solving nonlinear equations with convex constraints. *J. Comput. Appl. Math.* **2005**, *173*, 321–343. [CrossRef]

Article

SD-VIS: A Fast and Accurate Semi-Direct Monocular Visual-Inertial Simultaneous Localization and Mapping (SLAM)

Quanpan Liu, Zhengjie Wang * and Huan Wang

School of Mechatronical Engineering, Beijing Institute of Technology, Beijing 100081, China;
liuquanpan@126.com (Q.L.); WH978992767@163.com (H.W.)
* Correspondence: wangzhengjie@bit.edu.cn; Tel.: +86-186-1164-6723

Received: 17 February 2020; Accepted: 7 March 2020; Published: 9 March 2020

Abstract: In practical applications, how to achieve a perfect balance between high accuracy and computational efficiency can be the main challenge for simultaneous localization and mapping (SLAM). To solve this challenge, we propose SD-VIS, a novel fast and accurate semi-direct visual-inertial SLAM framework, which can estimate camera motion and structure of surrounding sparse scenes. In the initialization procedure, we align the pre-integrated IMU measurements and visual images and calibrate out the metric scale, initial velocity, gravity vector, and gyroscope bias by using multiple view geometry (MVG) theory based on the feature-based method. At the front-end, keyframes are tracked by feature-based method and used for back-end optimization and loop closure detection, while non-keyframes are utilized for fast-tracking by direct method. This strategy makes the system not only have the better real-time performance of direct method, but also have high accuracy and loop closing detection ability based on feature-based method. At the back-end, we propose a sliding window-based tightly-coupled optimization framework, which can get more accurate state estimation by minimizing the visual and IMU measurement errors. In order to limit the computational complexity, we adopt the marginalization strategy to fix the number of keyframes in the sliding window. Experimental evaluation on EuRoC dataset demonstrates the feasibility and superior real-time performance of SD-VIS. Compared with state-of-the-art SLAM systems, we can achieve a better balance between accuracy and speed.

Keywords: visual-inertial; semi-direct SLAM; multi-sensor fusion

---

1. Introduction

Simultaneous localization and mapping (SLAM) plays an important role in self-driving cars, virtual reality, unmanned aerial vehicles (UAV), augmented reality and artificial intelligence [1,2]. This technology can provide reliable state estimation for UAV and self-driving cars in GPS-denied environments by relying on its sensors. Various types of sensors can be utilized in SLAM, such as stereo camera, lidar, inertial measurement units (IMU), and monocular camera. However, they have significant disadvantages when used individually: the metric scale of stereo camera can be obtained directly by using fixed baseline length, but it can only be estimated accurately in a limited depth range [3]; lidar has high precision in indoor, but it will encounter the reflection problem of glass surface in outdoor [4]; cheap IMUs are extremely susceptible to bias and noise [5]; monocular camera cannot estimate the absolute metric scale [6]. This paper mainly studies the monocular vision-inertial navigation system (VINS) based on multi-sensor fusion [7], which has the advantages of small size, lightweight, observable scale, roll, and pitch angle, etc.

According to the different methods of image information processing, there are two categories of SLAM: feature-based method and direct method. The standard process of feature-based method proceeds in three steps [8–10]. Firstly, a group of sparse point or line features is extracted from each image, and feature matching between adjacent frames is performed by using invariant feature descriptors. Secondly, estimate the camera motion and the 3D position of sparse feature points by using multiple view geometry theory. Finally, optimize the camera motion and the 3D position of sparse feature points by minimizing visual re-projection errors. However, the feature-based method has the following disadvantages: feature extraction and matching are very time-consuming; fewer features extractable in low-texture environments; repeated texture environments will cause incorrect feature matching. Therefore, the accuracy and robustness of the feature-based method depend on the correct feature extraction and matching.

The direct method considers the entire image or some pixels with a large gradient and directly estimates the camera motion and scene structure by minimizing the photometric error [11–14]. Therefore, in the low-texture environments and repeated texture environments, the performance of the direct method is better than the feature-based method. In addition, without feature extraction and matching, the calculation speed of the direct method is faster than the feature-based method. However, the photometric error function is highly non-convex, it is difficult for the direct method to converge when large baseline motion and image blur occurs. Due to the inability to perform effective loop closure detection, the cumulative drift generated during long-term operation is still an unresolved problem [15].

Academic research on SLAM is very extensive. Well known methods include PTAM, SVO, VINS-Mono, LSD-SLAM, ORB-SLAM, DSO. PTAM was one of the most representative systems in the early stage of the feature-based SLAM algorithm. This algorithm first proposed to divide tracking and mapping into two parallel threads [16]. After that, most feature-based SLAM algorithms have adopted this idea, including ORB-SLAM. ORB-SLAM is the most successful feature-based SLAM, which uses ORB features in tracking, mapping, re-location, and loop closure detection [17]. VINS- Mono is a compelling multi-sensor monocular vision-inertial SLAM based on tight coupling and non-linear optimization [18]. Pl-VIO is an extension of the line feature of VINS-Mono, which can optimize the re-projection error of point and line features in a sliding window [19]. VINS-Fusion is a multi-sensor fusion platform based on non-linear optimization, which is the continuation of VINS-Mono in the direction of multi-sensor fusion [20]. LSD-SLAM is a novel method of real-time monocular SLAM based on direct method and can create large-scale semi-dense maps in real-time on a laptop without GPU acceleration [21]. Based on LSD-SLAM, DSO adds complete photometric calibration and uniformly samples pixels with large gradients throughout the image, thereby improving tracking accuracy and robustness [11]. On the basis of DSO, LDSO adds loop closure detection, which makes up for the shortcomings of DSO incapable of loop closure detection and eliminates the cumulative error generated during long navigation [22]. Stereo DSO is an improved version of DSO, which can realize the real-time estimation of motion and 3D structure with high accuracy and robustness on the moving stereo camera [3]. Lee authors in [7] presents a new implementation method for efficient simultaneous localization and mapping using a forward-viewing monocular vision sensor. The method is developed to be applicable in real time on a low-cost embedded system for indoor service robots.

In recent years, SLAM systems that combine the feature-based method and direct method have become popular. Semi-direct visual odometry (SVO) is an effective hybrid method that combines the advantages of feature-based method and direct method [23]. However, since the algorithm was originally designed for the drone's look-down camera, it is easy to lose camera tracking in other settings. Lee [24] proposed a loose-coupled method by combining ORB-SLAM and DSO to improve positioning accuracy. However, its frontend and backend are almost independent, which cannot share estimation information to further improve the pose precision. In [25,26], different semi-direct approaches were proposed for stereo odometry. Both methods use feature-based tracking to obtain a motion prior, and then perform direct semi-dense or sparse alignment to refine the camera pose. SVL [27] can be considered a combination of ORB-SLAM and SVO. The method for ORB-SLAM is adopted in keyframes, and SVO is adopted in non-keyframes. Therefore SVL can achieve good balance between speed and accuracy according to camera motion and environment. Our method is motivated by SVL, but we go one step further in real-time performance. Specifically, thanks to the keyframe selection strategy and sliding window-based back-end, we only need to extract new feature points on the keyframes and track them with KLT sparse optical flow algorithm, which can further reduce the calculation complexity while ensuring accuracy.

In this paper, we present SD-VIS, a novel fast and accurate semi-direct visual-inertial SLAM framework, which combines the exactness of feature-based method and quickness of direct method. The keyframes in SD-VIS are tracked by feature-based method, which is used for sliding window-based non-linear optimization and loop closure detection. This strategy solves the problem of drift in the long-term operation and ensures the robustness of the algorithm in case of large baseline motion and image blur. Non-keyframes are tracked by the direct method, and the distance between adjacent non-keyframes is minimal, which ensures the convergence of error function. Compared with the direct method, SD-VIS exhibits the function of loop closure detection and solves the problem of drift in long-term operation. Compared with the feature-based method, SD-VIS can achieve the same accuracy while maintaining a faster speed.

2. System Framework Overview

Figure 1 demonstrates the framework of the semi-direct vision-inertial SLAM system. Sensor data comes from a monocular camera and IMU. IMU measurements between two consecutive images are pre-integrated, and the pre-integration is used as the constraint of IMU between two images (Section 3.2). In the initialization procedure, we detect the feature points on each image and track them with KLT sparse optical flow algorithm [15].

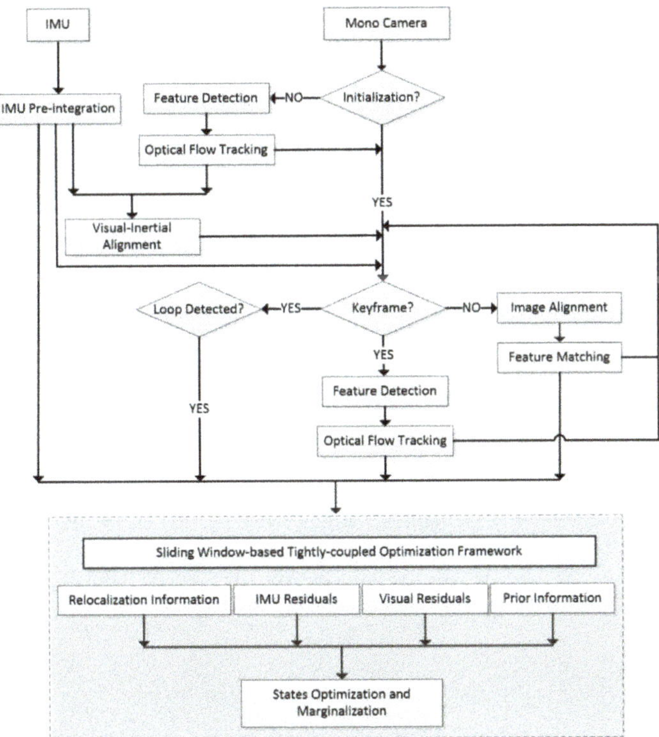

Figure 1. The semi-direct visual-inertial SLAM system framework

In the following visual-inertial alignment, we align the pre-integrated IMU measurements and visual images and calibrate out the metric scale, initial velocity, gravity vector, and gyroscope bias by using multiple view geometry (MVG) theory based on the feature-based method. (Section 3.3). After initialization, keyframe selection will be performed based on the IMU pre-integration and previous feature matching results. The previous feature matching refers to the feature matching between the last frame and the penultimate frame in the sliding window, and the matching is completed before the current frame arrives. Non-keyframes are used for fast-tracking and localization by direct method [11], and keyframes are tracked by feature-based method [18] and used for non-linear optimization and loop closure detection (Section 4). In the following tight coupling optimization framework, we can get more accurate state estimation by minimizing visual re-projection error, IMU residual, prior information from marginalization, and re-location information from loop closure detection (Section 5).

3. IMU Measurements and Visual-Inertial Alignment

3.1. Definition of Symbols

Figure 2 shows the definition of symbols in the semi-direct visual-inertial SLAM framework. C, B, and W are the camera coordinate system, the IMU body coordinate system, and the world coordinate system, respectively. We define $T_{WB} = \left(R_B^W, P_B^W\right)$ as the motion of B relative to W. $T_{BC} = \left(R_C^B, P_C^B\right)$ represents the extrinsic parameters between C and B, which can be calibrated in advance. $T_{t,t+1} \in SE(3)$ represents the motion from time t to time t + 1 in the coordinate system C, and $Z_{t,t+1}$ represents a pre-integrated IMU measurement between the camera coordinate system C_t and C_{t+1}.

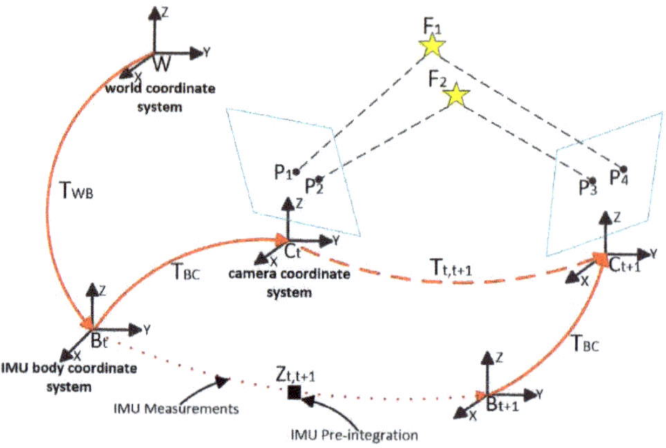

Figure 2. Symbol definition of algorithm

A 3D point F_1, $F_2 \in R^3$ represents the spatial feature points observed simultaneously by the camera coordinate system C_t and C_{t+1}. P_1, P_2, P_3, $P_4 \in R^2$ are the projections of feature points on the image coordinate system. We adopt the traditional pinhole camera model to map the F_1 in the camera coordinate system to the image coordinate system by the projection function $\pi: R^3 \to R^2$:

$$P_1 = \pi F_1 = \begin{bmatrix} f_u \frac{x_c}{z_c} + c_u \\ f_v \frac{y_c}{z_c} + c_v \end{bmatrix} F_1 = \begin{bmatrix} x_c & y_c & z_c \end{bmatrix}^T \tag{1}$$

where $\begin{bmatrix} f_u & f_v \end{bmatrix}^T$ and $\begin{bmatrix} c_u & c_v \end{bmatrix}^T$ is camera internal parameters.

3.2. IMU Pre-Integration

In the back-end optimization and visual-inertial alignment, the constraints of vision and IMU need to be optimized in the same frame, so the IMU measurements between two adjacent frames need to be integrated into one constraint.

IMU can output 3-axis angular velocity $\widetilde{\omega}$ and 3-axis acceleration $\widetilde{\alpha}$ including bias and Gaussian white noise:

$$\widetilde{\omega} = \omega^B + b_\omega^B + n_\omega^B \tag{2}$$

$$\widetilde{\alpha} = R_W^B\left(\alpha^W + g^W\right) + b_\alpha^B + n_\alpha^B \tag{3}$$

where $n_\omega^B \sim N\left(0, \sigma_\omega^2\right)$, $n_\alpha^B \sim N\left(0, \sigma_\alpha^2\right)$ represents the Gaussian white noise. $g^W = [0, 0, g]^T$ is the gravity vector. R_w^B represents the rotation matrix from W to B. b_ω^B, b_α^B represents the biases of gyroscope and accelerometer.

We define $P_{B_i}^W$, $V_{B_i}^W$, $q_{B_i}^W$ as the translation, velocity, and rotation quaternions in the IMU body coordinate system B_i. If we know the measurements value of IMU during $t \in [i, j]$, we can calculate $P_{B_j}^W$, $V_{B_j}^W$, $q_{B_j}^W$ in the coordinate system B_j:

$$P_{B_j}^W = P_{B_i}^W + V_{B_i}^W \Delta t + \iint_{t\in[i,j]}\left[R_{B_t}^W\left(\widetilde{\alpha}_t - b_{\alpha_t}^B\right) - g^W\right]\delta t^2 \tag{4}$$

$$V_{B_j}^W = V_{B_i}^W + \int_{t\in[i,j]}\left[R_{B_t}^W\left(\widetilde{\alpha}_t - b_{\alpha_t}^B\right) - g^W\right]\delta t \tag{5}$$

$$q_{B_j}^W = q_{B_i}^W \otimes \int_{t\in[i,j]}\frac{1}{2}\Omega\left(\widetilde{\omega}_t - b_{\omega_t}^B\right)q_{B_t}^W \delta t \tag{6}$$

where:

$$\Omega(\omega) = \begin{bmatrix} -\omega_\times & \omega \\ -\omega^T & 0 \end{bmatrix},\ \omega_\times = \begin{bmatrix} 0 & -\omega_z & \omega_y \\ \omega_z & 0 & -\omega_x \\ -\omega_y & \omega_x & 0 \end{bmatrix}$$

From formulas (4)–(6), we can know that the IMU body state $P_{B_j}^W$, $V_{B_j}^W$, $q_{B_j}^W$ is related to the IMU body coordinate system B_i. In the back-end tightly coupling optimization, we will continuously iteratively update the IMU state variables in the sliding window. When the IMU body state at time $t = i$ is iteratively updated, we need to recalculate the state at time $t = j$, which is very time-consuming. We adopt IMU pre-integration technology to avoid unnecessary time consumption. formulas (4)–(6) can be written as:

$$R_W^{B_i}P_{B_j}^W = R_W^{B_i}\left(P_{B_i}^W + V_{B_i}^W \Delta t - \frac{1}{2}g^W \Delta t^2\right) + \theta_{B_j}^{B_i} \tag{7}$$

$$R_W^{B_i}V_{B_j}^W = R_W^{B_i}\left(V_{B_i}^W \Delta t - g^W \Delta t\right) + \beta_{B_j}^{B_i} \tag{8}$$

$$q_W^{B_i} \otimes q_{B_j}^W = \gamma_{B_j}^{B_i} \tag{9}$$

where:

$$\theta_{B_j}^{B_i} = \iint_{t\in[i,j]}\left[R_{B_t}^{B_i}\left(\widetilde{\alpha}_t - b_{\alpha_t}^B\right)\right]\delta t^2 \tag{10}$$

$$\beta_{B_j}^{B_i} = \int_{t\in[i,j]}\left[R_{B_t}^{B_i}\left(\widetilde{\alpha}_t - b_{\alpha_t}^B\right)\right]\delta t \tag{11}$$

$$\gamma_{B_j}^{B_i} = \int_{t\in[i,j]}\frac{1}{2}\Omega\left(\widetilde{\omega}_t - b_{\omega_t}^B\right)q_{B_t}^{B_i}\delta t \tag{12}$$

From formulas (10)–(12), the pre-integration measurements $\theta_{B_j}^{B_i}$, $\beta_{B_j}^{B_i}$, $\gamma_{B_j}^{B_i}$, and the IMU body coordinate system B_i are independent of each other. This means that when the states in the B_i coordinate system are iteratively updated, there is no need to recalculate the states in the B_j coordinate system. Since the IMU pre-integrated measurements $\theta_{B_j}^{B_i}$, $\beta_{B_j}^{B_i}$, $\gamma_{B_j}^{B_i}$ is affected by the bias, when the bias is updated iteratively, we will update the pre-integrated measurement by the first-order approximation method:

$$\theta_{B_j}^{B_i} \approx \widetilde{\theta}_{B_j}^{B_i} + J_{b_\alpha}^{\theta} \delta b_\alpha^B + J_{b_\omega}^{\theta} \delta b_\omega^B \tag{13}$$

$$\beta_{B_j}^{B_i} \approx \widetilde{\beta}_{B_j}^{B_i} + J_{b_\alpha}^{\beta} \delta b_\alpha^B + J_{b_\omega}^{\beta} \delta b_\omega^B \tag{14}$$

$$\gamma_{B_j}^{B_i} \approx \widetilde{\gamma}_{B_j}^{B_i} \otimes \begin{bmatrix} 0 \\ \frac{1}{2} J_{b_\omega}^{\gamma} \delta b_\omega^B \end{bmatrix} \tag{15}$$

where $J_{b_\alpha}^{\theta}, J_{b_\omega}^{\theta}, J_{b_\alpha}^{\beta}, J_{b_\omega}^{\beta}, J_{b_\omega}^{\gamma}$ are the Jacobian matrices of pre-integrated measurements with respect to bias.

3.3. Visual-Inertial Alignment

The convergence speed and effect of nonlinear visual-inertial SLAM systems depend heavily on reliable initial values. Therefore, in the initialization procedure of SD-VIS, we align the pre-integrated IMU measurements with the visual image to complete the system initialization.

3.3.1. Gyroscope Bias Correction

We regard the camera coordinate system C_0 as the world coordinate system. We detect the feature points on each image and track them with KLT sparse optical flow algorithm, and then the rotation $R_{C_t}^{C_0}$ and $R_{C_{t+1}}^{C_0}$ of the two adjacent frames C_t and C_{t+1} can be estimated by using visual structure from motion (SfM). Rotation increment $\widetilde{\gamma}_{B_{t+1}}^{B_t}$ between the IMU body coordinate system B_t and B_{t+1} can be estimated by IMU pre-integration. We can get the following formula:

$$\min_{\delta b_{\omega_t}^B} \sum \| R_{C_{t+1}}^{C_0}{}^{-1} \otimes R_{C_t}^{C_0} \otimes \gamma_{B_{t+1}}^{B_t} \|^2 \tag{16}$$

where:

$$\gamma_{B_{t+1}}^{B_t} \approx \widetilde{\gamma}_{B_{t+1}}^{B_t} \otimes \begin{bmatrix} 0 \\ \frac{1}{2} J_{b_\omega}^{\gamma} \delta b_{\omega_t}^B \end{bmatrix} \tag{17}$$

We solve the above least squares problem to get the initial calibration of the gyroscope bias and use it to update $\theta_{B_j}^{B_i}$, $\beta_{B_j}^{B_i}$, $\gamma_{B_j}^{B_i}$.

3.3.2. Gravity Vector, Initial Velocity, and Metric Scale Correction

We define the variables that need to be calibrated as:

$$X_I = \begin{bmatrix} V_{b_0}^{b_0} & V_{b_1}^{b_1} & \cdots & V_{b_n}^{b_n} & g^{C_0} & s \end{bmatrix} \tag{18}$$

where $V_{b_k}^{b_k}$ is the initial velocity in the IMU body coordinate system, g^{C_0} is the gravity vector in the camera coordinate system, s represents the metric scale of semi-direct visual-inertial SLAM framework.

Suppose we have obtained pre-calibrated external parameter $T_{BC} = \left(R_C^B, P_C^B \right)$, we can transform the pose $T_{C_0 C_t} = \left(R_{C_t}^{C_0}, P_{C_t}^{C_0} \right)$ from the camera coordinate system to the IMU body coordinate system:

$$q_{B_t}^{C_0} = q_{C_t}^{C_0} \otimes q_C^{B-1} \tag{19}$$

$$sP^{C_0}_{B_t} = sP^{C_0}_{C_t} - R^{C_0}_{B_t} P^B_C \tag{20}$$

Considering two adjacent keyframes B_t and B_{t+1}, then formulas (7) and (8) can be rewritten as:

$$\theta^{B_t}_{B_{t+1}} = R^{B_t}_{C_0}\left(s\left(P^{C_0}_{B_{t+1}} - P^{C_0}_{B_t}\right) - R^{C_0}_{B_t} V^{B_t}_{B_t} \Delta t + \frac{1}{2}g^{C_0}\Delta t^2\right) \tag{21}$$

$$\beta^{B_t}_{B_{t+1}} = R^{B_t}_{C_0}\left(R^{C_0}_{B_{t+1}} V^{B_{t+1}}_{B_{t+1}} - R^{C_0}_{B_t} V^{B_t}_{B_t} + g^{C_0}\Delta t\right) \tag{22}$$

We combine formulas (19)–(22) to get the following formula:

$$\widetilde{Z}^{B_t}_{B_{t+1}} = \begin{bmatrix} \theta^{B_t}_{B_{t+1}} - P^B_C + R^{B_t}_{C_0} R^C_{B_{t+1}} P^B_C \\ \beta^{B_t}_{B_{t+1}} \end{bmatrix} = H^{B_t}_{B_{t+1}} X_I + n^{B_t}_{B_{t+1}} \tag{23}$$

where:

$$H^{B_t}_{B_{t+1}} = \begin{bmatrix} -I\Delta t & 0 & \frac{1}{2}R^{B_t}_C \Delta t^2 & R^{B_t}_{C_0}\left(P^{C_0}_{C_{t+1}} - P^{C_0}_{C_t}\right) \\ -I & R^{B_t}_{C_0} R^{C_0}_{B_{t+1}} & R^{B_t}_{C_0}\Delta t & 0 \end{bmatrix} \tag{24}$$

In the above formula $R^{C_0}_{B_t}, R^{C_0}_{B_{t+1}}, P^{C_0}_{B_t}, P^{C_0}_{B_{t+1}}$ can be obtained through visual SfM:

$$\min_{X_I} \sum \|\widetilde{Z}^{B_t}_{B_{t+1}} - H^{B_t}_{B_{t+1}} X_I\|^2 \tag{25}$$

Solving the above formula, we can calibrate the initial velocity for each keyframe, gravity vector, and absolute metric scale. After estimating the scale, we will adjust the translation vector of the vision SfM to make the system have an observable scale.

3.3.3. Gravity Vector Refinement

We can know the magnitude of the gravity vector in advance, so we refer to the VINS-Mono [18] method to re-parameterize the gravity vector obtained in Section 3.3.2 with two variables in tangent space, and perform further optimization.

After obtaining the accurate gravity vector, we can rotate the coordinate system C_0, which is temporarily the world coordinate system, to the real world coordinate system W. However, since the yaw angle in the visual-inertial SLAM system is unobservable, the yaw angle of the C_0 coordinate system remains unchanged during the rotation process. At this time, the initialization procedure of the semi-direct visual-inertial SLAM system is completed.

4. Visual Measurements

4.1. Keyframe Selection

We have three different keyframe selection strategies. Satisfying one of these three strategies makes the current frame a keyframe. The first and third strategies of keyframe selection are based on the feature matching results of the last frame and the penultimate frame in the sliding window, which has been matched before the current frame arrives. The first selection strategy is the tracking number of feature points. No new feature points will be extracted when tracking non-keyframes. The translational motion of the camera will lead to the decrease of tracking feature points. If the number of tracking points in the last frame in the sliding window is less than 70% of the minimum tracking point threshold, the current frame will be treated as a keyframe. The second selection strategy is related to IMU pre-integration. If the translation distance between the last two adjacent frames in the sliding window calculated by the IMU pre-integration exceeds a preset threshold, the current frame is also considered as a new keyframe. The third selection strategy is the average parallax of the feature points tracked on the the last frame and the penultimate frame in the sliding window. The translation or rotation of camera will cause parallax. When the average parallax exceeds the threshold, the current frame will also be regarded as a keyframe.

4.2. Keyframes Tracking

If the current frame is treated as a keyframe, we first use the fast feature detector [28] to add new feature points in the last frame in the sliding window and then use the KLT sparse optical flow algorithm to track them in the current frame (Figure 3). At least 200 feature points will be maintained in each frame. Since there is no need to calculate the feature point descriptor, the optical flow method can save more time. In addition, we also use RANSAC [29] with the fundamental matrix model to eliminate outliers generated during tracking.

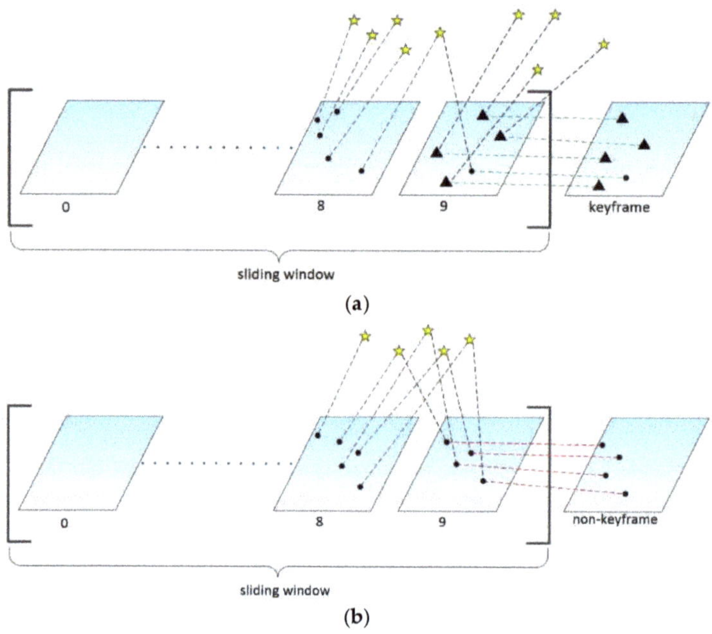

Figure 3. The top chart (**a**) shows how to track a keyframe. The black triangle represents the newly extracted feature points on the 10th frame (last frame) of the sliding window, while the green dotted line represents tracking them in the keyframe by KLT sparse optical flow algorithm. The bottom chart (**b**) shows how to track a non-keyframe. The red dotted line indicates that the feature points on the 10th frame of the sliding window are tracked on the non-keyframe by the direct method.

4.3. Non-Keyframes Tracking

If the current frame is considered a non-keyframe, we use direct image alignment to estimate the relative pose $T_{t,t+1}$ between the current frame and the last frame in the sliding window. The initial value of the relative pose can be obtained directly by IMU pre-integration. The feature points observed in the last frame are projected into the current frame according to the estimated pose $T_{t,t+1}$. Due to the hypothesis of photometric invariance, if the same feature point is observed by two adjacent frames, the photometric values of the projection points on the two adjacent frames are equal (Figure 4). Therefore, we can optimize the relative pose $T_{t,t+1}$ by minimizing the photometric error between image blocks $(4 \times 4$ pixels):

$$T_{t,t+1} = \underset{T}{\arg\min} \iint_R \rho[\delta I(T, u)]du \tag{26}$$

where $\rho[\cdot] = \frac{1}{2}\|\cdot\|^2$, u is the position of the feature point of the last frame in the sliding window. R is the image region where the depth is known in the last frame, and the back-projected points are visible in the current frame domain.

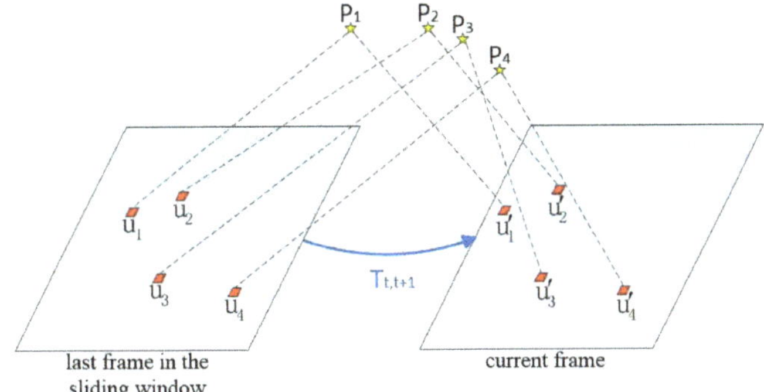

Figure 4. Adjusting the relative pose $T_{t,t+1}$ between the current frame and last frame in the sliding window means moving the re-projection position u' of feature points on the current frame.

The photometric error δI is:

$$\delta I(T_{t,t+1}, u) = I_{t+1}\left(\pi T_{t,t+1}\cdot\pi^{-1}\left(u, d_p\right)\right) - I_t(u) \quad \forall u \in R \tag{27}$$

where d_p is the depth of the feature point in the last frame in the sliding window. I_k represents the intensity image in the k-th frame.

We use the inverse compositional formulation [30] of the photometric error, which can avoid unnecessary Jacobian derivation. The update step $T(\xi)$ for the last frame in the sliding window is:

$$\delta I(\xi, u) = I_{t+1}\left(\pi T_{t,t+1}\cdot\pi^{-1}\left(u, d_p\right)\right) - I_t\left(\pi\left(T(\xi)\cdot\pi^{-1}\left(u, d_p\right)\right)\right) \forall u \in R \tag{28}$$

We solve it in an iterative Gauss Newton method and update $T_{t,t+1}$ in the following way:

$$T_{t,t+1} \leftarrow T_{t,t+1}\cdot T(\xi)^{-1} \tag{29}$$

After image alignment, we can get the optimized relative pose $T_{t,t+1}$ between the current frame and the last frame in the sliding window. We define all 3D points observed in all frames in the sliding window as the local map, and project the local map to the current frame to find the visible 3D points of the current frame. Due to the inaccuracy of the visible 3D point position and the camera pose, there will be errors in the projection position of the current frame. To make the projection position more accurate, the current frame needs to be aligned with the local map. The feature matching step optimizes the positions of all the projection points in the current frame by minimizing the photometric errors of the projection blocks (5 × 5 pixels) in the current frame and the reference frame (Figure 5):

$$u'_i = \underset{u'_i}{\arg\min}\frac{1}{2}\|I_{t+1}\left(u'_i\right) - A_i\cdot I_i\left(u_i\right)\|^2 \quad \forall i \tag{30}$$

Solving the above formula in an iterative Gauss Newton method, we can get the update of the projection block position. The reference frame is usually far away from the current frame, so we apply an affine warping A_i to the reference patch.

Through image alignment and feature matching, we get the implicit results of direct motion estimation — feature correspondence with sub-pixel accuracy. Note that when tracking non-keyframes with the direct method, no new feature points are extracted. In the back-end optimization, we will combine IMU residual, visual re-projection error, prior information, and re-localization information to optimize the camera pose and 3D point position again.

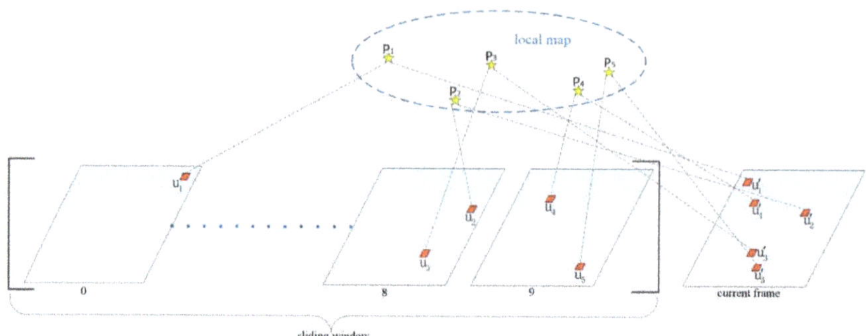

Figure 5. Adjust the position of the projection block u_i' on the current frame to minimize the photometric error of the projection block in the current frame and the reference frame in the sliding window.

5. Sliding Window-based Tightly-coupled Optimization Framework

After tracking non-keyframes and keyframes, we proceed with a sliding window-based tightly-coupled optimization framework for high accuracy and robust state estimation. In the optimization framework, we combined IMU residual, visual re-projection error, prior information, and re-localization information to optimize the camera pose and 3D point position again.

5.1. Formulation

The state variables to be estimated by SD-VIS are defined as:

$$
\begin{aligned}
X &= \begin{bmatrix} X_0 & X_1 & \dots & X_n & X_C^B \end{bmatrix} \\
X_k &= \begin{bmatrix} P_{B_k}^W & V_{B_k}^W & q_{B_k}^W & b_\alpha^B & b_\omega^B \end{bmatrix} k \in [0,n] \\
X_C^B &= \begin{bmatrix} P_C^B & q_C^B \end{bmatrix}
\end{aligned}
\tag{31}
$$

where X_k includes the translation, velocity, and rotation quaternions of the kth IMU body coordinate system concerning the world coordinate system, as well as the bias of gyroscope and accelerometer. n represents the size of the sliding window. By minimizing the sum of IMU residuals, visual re-projection errors, prior information, and relocation information in the sliding window, we can obtain a robust and accurate semi-direct visual-inertial SLAM system:

$$
\min_X \left\{ \sum \left\| r_B\!\left(\widetilde{Z}_{B_{k+1}}^{B_k}, X\right) \right\|_{P_{B_{k+1}}^{B_k}}^2 + \sum \left\| r_C\!\left(\widetilde{Z}_F^{C_j}, X\right) \right\|_{P_F^{C_j}}^2 + \left\| r_p - H_p X \right\|^2 + \sum \left\| r_C\!\left(\widetilde{Z}_F^{C_L}, X\right) \right\|_{P_F^{C_L}}^2 \right\}
\tag{32}
$$

where $r_B\!\left(\widetilde{Z}_{B_{k+1}}^{B_k}, X\right)$, $r_C\!\left(\widetilde{Z}_F^{C_j}, X\right)$, $\{r_p, H_p\}$ and $r_C\!\left(\widetilde{Z}_F^{C_L}, X\right)$ are IMU residuals, visual re-projection errors, prior information and re-localization information respectively.

5.2. IMU Residuals

According to the formulas (4)–(6) in Section 3.2, we can get the IMU measurement residual:

$$
r_B\left(\widetilde{Z}_{B_j}^{B_i}, X\right) =
\begin{bmatrix}
\delta\theta_{B_j}^{B_i} \\
\delta\beta_{B_j}^{B_i} \\
\delta\gamma_{B_j}^{B_i} \\
\delta b_\alpha^B \\
\delta b_\omega^B
\end{bmatrix}
=
\begin{bmatrix}
R_W^{B_i}\left(P_{B_j}^W - P_{B_i}^W - V_{B_i}^W \Delta t + \frac{1}{2}g^W \Delta t^2\right) - \widetilde{\theta}_{B_j}^{B_i} \\
R_W^{B_i}\left(V_{B_j}^W - V_{B_i}^W + g^W \Delta t\right) - \widetilde{\beta}_{B_j}^{B_i} \\
2\left[q_{B_i}^{W-1} \otimes q_{B_j}^W \otimes \left(\widetilde{\gamma}_{B_j}^{B_i}\right)^{-1}\right]_{xyz} \\
b_\alpha^{B_i} - b_\alpha^{B_j} \\
b_\omega^{B_i} - b_\omega^{B_j}
\end{bmatrix}_{15*1}
\tag{33}
$$

where $[\cdot]_{xyz}$ represents the real part of the quaternion. $\left[\widetilde{\theta}_{B_j}^{B_i}, \widetilde{\beta}_{B_j}^{B_i}, \widetilde{\gamma}_{B_j}^{B_i}\right]^T$ are the IMU pre-integration between two adjacent keyframes B_i and B_j.

5.3. Visual Re-Projection Errors

When the feature point F_1 is first observed in the ith image, the visual re-projection error in the jth image can be defined by the pinhole camera model as:

$$
r_C\left(\widetilde{Z}_{F_1}^{C_j}, X\right) = \pi^{-1}\left(\begin{bmatrix} \widetilde{u}_{F_1}^{C_i} \\ \widetilde{v}_{F_1}^{C_i} \end{bmatrix} - R_B^C\left(R_W^{B_j}\left(R_{B_i}^W\left(R_C^B \pi^{-1}\begin{bmatrix} \widetilde{u}_{F_1}^{C_j} \\ \widetilde{v}_{F_1}^{C_j} \end{bmatrix} + P_C^B\right) + P_{B_i}^W - P_{B_j}^W\right) - P_C^B\right)
\tag{34}
$$

where $\left[\widetilde{u}_{F_1}^{C_i}, \widetilde{v}_{F_1}^{C_i}\right]$ and $\left[\widetilde{u}_{F_1}^{C_j}, \widetilde{v}_{F_1}^{C_j}\right]$ represent the coordinates of the pixels projected from the feature point F_1 to the ith and j^{th} frame image, respectively. π^{-1} is the back-projection function of the pinhole camera model.

5.4. Marginalization Strategy

In order to limit the computational complexity of SD-VIS, the back-end adopts a sliding window-based tightly-coupled optimization framework, so we use the marginalization strategy [31] to make the correct operation of sliding windows. As shown in Figure 6, if the current frame is determined as a keyframe, the frame will remain in the sliding window, and the oldest frame is marginalized out When the oldest frame is marginalized, the feature points that can only be observed by the oldest frame will be discarded directly, and other visual and inertial measurements associated with the frame will be removed from the sliding window by Schur complement. The new prior information constructed by Schur complement will be added to the existing prior information. If the current frame is not a keyframe, the last frame in the sliding window will be marginalized, and all visual measurements related to that frame will be removed directly from the sliding window.

5.5. Re-Localization

Due to the global 3D position and yaw angle are unobservable, there will be inevitable accumulative errors in the vision-inertial SLAM system. To eliminate the accumulated error, we introduce the re-localization module. After the keyframe is traced successfully, it can be judged whether the SLAM system has been here before by loop closure detection. We utilize DBoW2, a state-of-the-art bag-of-word place recognition approach, for loop closure detection. When a loop is detected, the re-localization module can effectively align the current sliding window, thus eliminating the accumulated error. For a detailed description of re-location, readers may refer to [18].

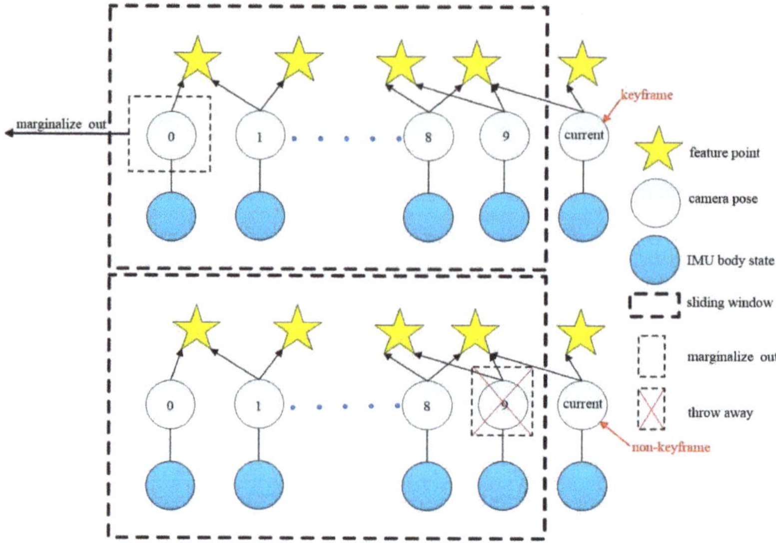

Figure 6. Marginalization strategy of SD-VIS.

6. Experiment

We evaluate the accuracy, robustness, and real-time performance of SD-VIS on the EuRoC dataset [32]. The SD-VIS method is compared with the state-of-the-art vision SLAM methods, such as VINS-mono [18] and VINS-Fusion [20]. In Section 6.1, the accuracy and robustness of SD-VIS are evaluated, and the experimental results show that the accuracy and robustness of the proposed method reach the same level as the state-of-the-art method. Section 6.2 evaluates real-time performance. The experimental results show that the proposed method achieves a good balance between accuracy and real-time performance. Section 6.3 evaluates the loop closure detection capability and verifies the overall feasibility of the SLAM system.

6.1. Accuracy and Robustness Evaluate

In the experiments on the EuRoC dataset, we adopt the open source tool EVO [33] to evaluate the performance of SD-VIS. By comparing the estimated value with the actual value, we calculate the absolute pose error (APE) as an index of the evaluation algorithm [34]. Table 1 shows the root mean square error (RMSE) of the translation on the EuRoC dataset. For fairness, the following algorithms do not use the loop closure detection module.

Table 1. The root mean square error (RMSE) results of the translation (m).

Dataset	VINS-Mono	VINS-Fusion	SD-VIS
MH_01_easy	**0.254651**	0.364247	0.260793
MH_02_easy	**0.263258**	0.339122	0.289663
MH_03_medium	0.547901	**0.483257**	0.577422
MH_04_medium	0.590191	0.614950	**0.497288**
MH_05_difficult	**0.512011**	0.524107	0.512458
V1_01_easy	**0.217083**	0.247467	0.245990
V1_02_medium	0.492645	**0.434756**	0.502134
V1_03_difficult	0.361521	**0.345895**	0.388959
V2_01_easy	**0.170790**	0.177467	0.202474
V2_02_medium	0.424259	**0.370081**	0.454785
V2_03_difficult	0.475561	0.521573	**0.444759**

As can be seen from Table 1, in terms of accuracy, SD-VIS and VINS-Mono and VINS-Fusion are at the same level. The accuracy of SD-VIS is slightly lower than that of VINS-Mono when moving at low and medium speed in the environment with abundant feature points (such as MH_01_easy and MH_03_medium). This is due to the susceptibility to various illumination changes when tracking non-keyframes using the direct method. We have observed that some datasets exhibit strong exposure changes between images and, therefore, the tracking effect of the direct method is reduced in these cases. In addition, in order to further improve the real-time performance of the algorithm, we only extract new feature points in keyframes, which results in that the number of feature points in non-keyframes will be less than keyframes, which will also bring some negative effects. Although the accuracy performance is not significantly better than the traditional method, it has achieved the same level of accuracy as VINS-Mono while greatly improving real-time performance. Therefore, our algorithm is very suitable for small-sized unmanned platform with limited computing resources. The accuracy of SD-VIS is higher than that of VINS-Mono and VINS-Fusion when moving fast in the low-texture environment (such as V2_03_difficult). This is due to the excellent performance of the direct method in the low-texture environment. In addition, the keyframe selection strategy will tend to generate more keyframes during fast motion, which will also improve the accuracy performance of the algorithm. Figure 7 shows more intuitively the trajectory heat map estimated by SD-VIS, VINS-Mono, and VINS-Fusion in MH_01_easy. Figure 8 shows the change of translation absolute pose error with time in MH_01_easy, MH_04_medium, and V2_03_difficult. Through Figures 7 and 8, we came to the conclusion that the accuracy and robustness of our algorithm have reached the level of the state-of-the-art algorithm. Especially in the initialization procedure and low-texture environment, our algorithm performs better.

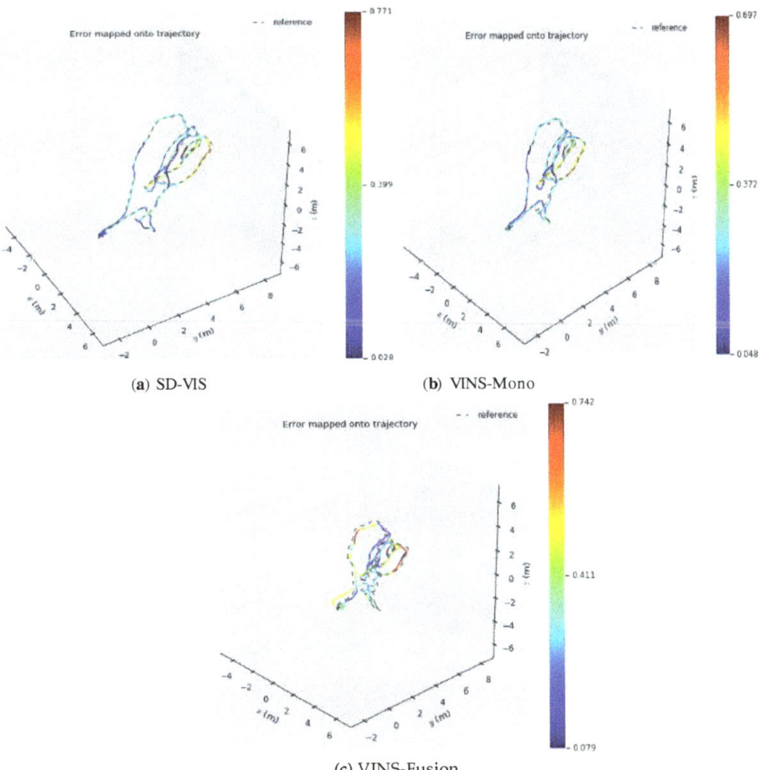

(a) SD-VIS

(b) VINS-Mono

(c) VINS-Fusion

Figure 7. The trajectory heat map estimated by SD-VIS, VINS-Mono, and VINS-Fusion in MH_01_easy.

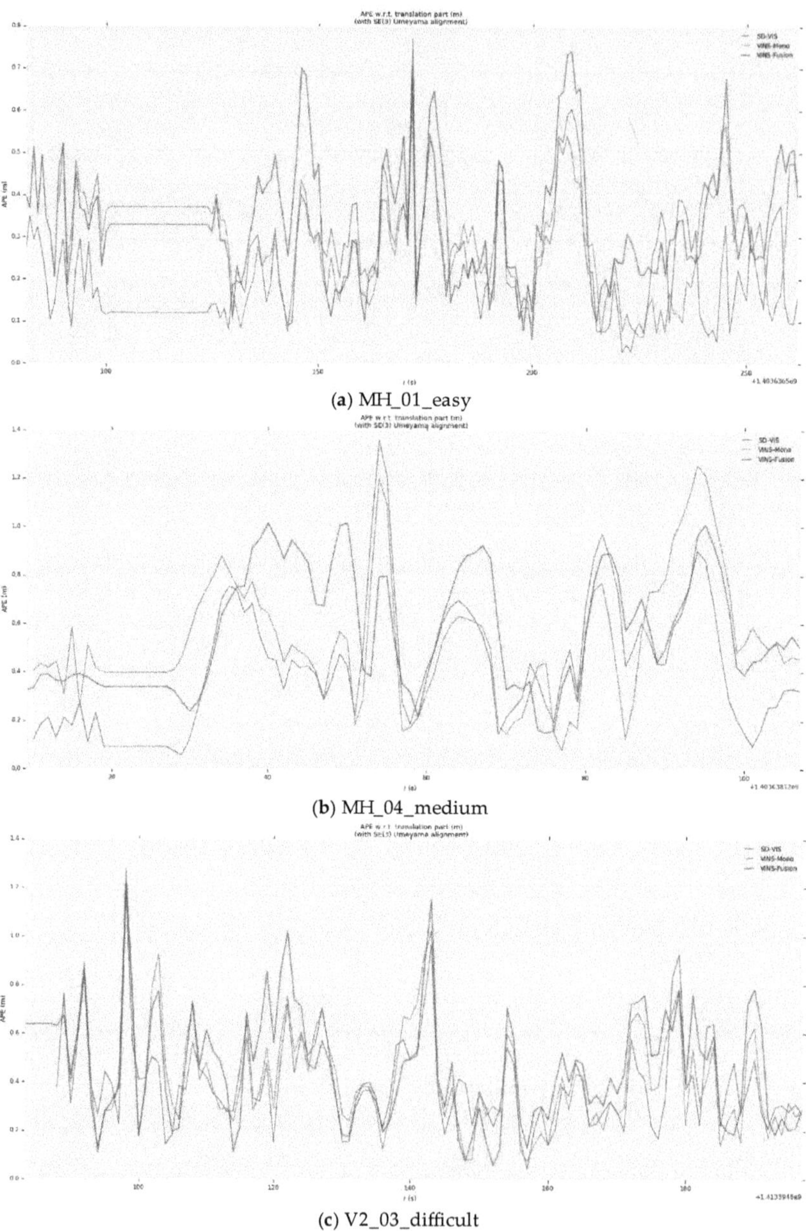

(a) MH_01_easy

(b) MH_04_medium

(c) V2_03_difficult

Figure 8. The change of translation absolute pose error with time in MH_01_easy, MH_04_medium and V2_03_difficult. Blue, green, and red represent SD-VIS, VINS-Mono, and VINS-Fusion respectively

6.2. Real-Time Performance Evaluate

In this section, we evaluate the real-time performance of SD-VIS. We compared the average time required to track an image (Table 2).

Table 2. Average time (ms) spent tracking an image.

Dataset	ORB-SLAM2	VINS-Mono	VINS-Fusion	SD-VIS
MH_01_easy	37.82	17.67	38.91	**6.72**
MH_02_easy	35.31	13.60	45.86	**6.58**
MH_03_medium	34.20	12.28	44.15	**7.87**
MH_04_medium	30.35	12.92	42.54	**6.93**
MH_05_difficult	30.43	12.77	42.65	**6.91**
V1_01_easy	37.15	13.38	45.87	**7.14**
V1_02_medium	28.46	13.89	45.36	**10.96**
V1_03_difficult	×	13.50	44.40	**7.81**
V2_01_easy	33.01	15.30	49.10	**6.28**
V2_02_medium	31.13	13.04	45.56	**10.22**
V2_03_difficult	×	17.90	45.07	**12.38**

As can be seen from Table 2, the image tracking of ORB-SLAM2 [11] uses the feature-based method to extract and match the ORB features of each frame, which takes a long time. However, VINS-Mono uses the optical flow method to track FAST features, which saves the calculation of feature descriptors, so the time consumption is less than ORB-SLAM2. VINS-Fusion is a stereo visual-inertial fusion SLAM algorithm, and image tracking also takes a long time. In SD-VIS, non-keyframes are used for fast-tracking and localization by direct method, and keyframes are tracked by feature-based method and used for back-end optimization and loop closure detection. This algorithm saves a lot of time and minimizes the average time of SD-VIS tracking images. Due to the keyframe selection strategy, in some low-speed motion scenes (such as MH_01_easy and V1_01_easy), the number of keyframes will be less, and the time to track a frame of the image will be reduced. In some fast-moving scenes (such as V1_02_medium and V2_03_difficult), as the number of keyframes increases, the time required to track an image will be increased.

In summary, the reason why we can obtain good real-time performance is due to the use of KLT sparse optical flow algorithm when tracking keyframes, which eliminates the calculation of descriptors and feature matching. In addition, for non-keyframes, only the direct method is used to track existing feature points, and new feature points are not extracted. Due to the close distance between two adjacent non-keyframes, the direct method of image alignment and feature matching can quickly converge.

Compared with the feature-based method, we use the direct method to track non-keyframes and accelerate the algorithm without reducing the accuracy and robustness. As shown in Figure 9, in MH_02_easy, 26% of the frames are determined to be keyframes, while 74% of the frames are determined to be non-keyframes. The time consumption of tracking keyframes is 65%, while that of non-keyframes are only 35%. Combined with Section 6.1, we can conclude that compared with the state-of-the-art SLAM systems, we can achieve a better balance between quickness and exactness.

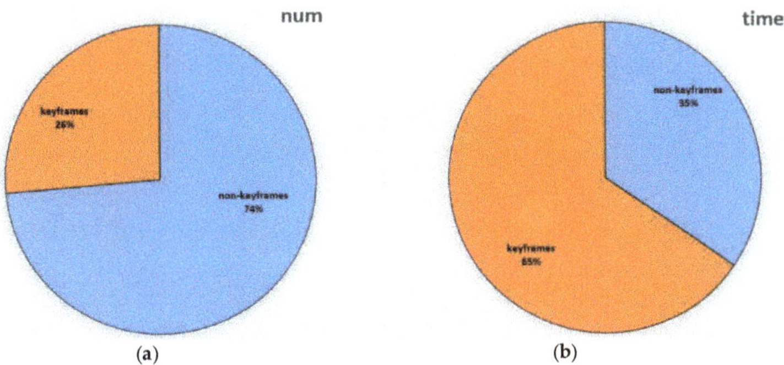

Figure 9. The left picture (**a**) shows a comparison of the number of keyframes and non-keyframes. The right picture (**b**) shows the comparison of time consumption between tracking keyframes and non-keyframes

6.3. Loop Closure Detection Evaluate

Finally, in order to verify the integrity and feasibility of the proposed algorithm, we evaluate the loop closure detection capability of SD-VIS. As can be seen from Figures 10 and 11, the accuracy of SD-VIS with loop detection is improved obviously. Compared with the direct method, SD-VIS exhibits the function of loop closure detection and solves the problem of drift in long-term operation.

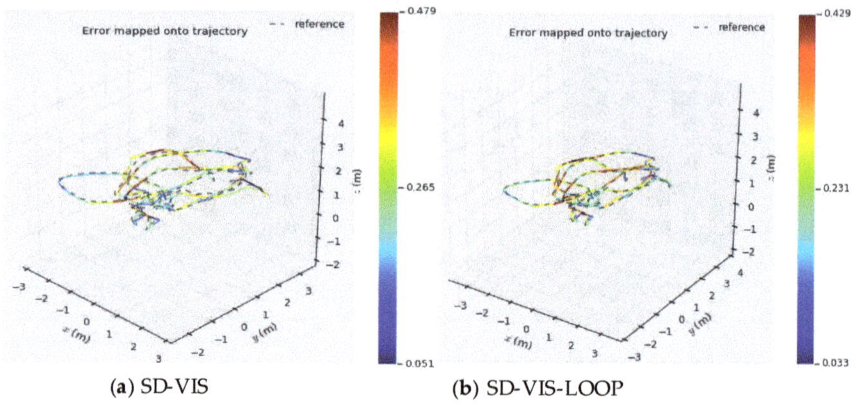

(**a**) SD-VIS   (**b**) SD-VIS-LOOP

Figure 10. The trajectory heat map estimated by SD-VIS and SD-VIS-LOOP in V1_01_easy.

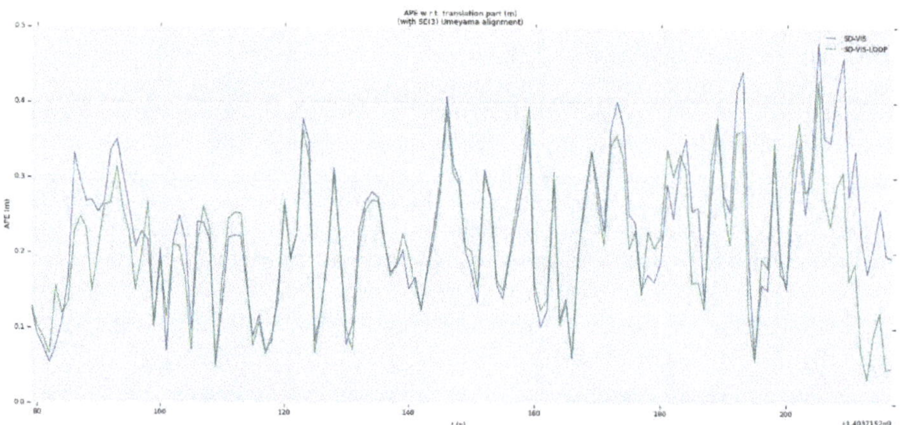

Figure 11. The change of translation absolute pose error with time in V1_01_easy. Blue and green represent SD-VIS and SD-VIS-LOOP respectively.

7. Conclusions

We present SD-VIS, a novel fast and accurate semi-direct visual-inertial SLAM framework, which combines the exactness of feature-based method and quickness of direct method. Compared with the state-of-the-art feature-based method, we use the direct method to track non-keyframes and accelerate the algorithm without reducing the accuracy and robustness. Compared with the direct method, SD-VIS exhibits the function of loop closure detection and solves the problem of drift in long-term operation. We get a better balance between accuracy and speed, so the algorithm is more suitable for the platform with limited computing resources. In the future, we will extend the algorithm to support more types of multi-sensor fusion to increase its robustness in complex environments.

Author Contributions: Conceptualization, Q.L. and H.W.; methodology, Q.L.; software, Q.L. and H.W.; validation, Q.L., Z.W. and H.W.; formal analysis, Q.L.; investigation, Q.L.; resources, Q.L. and H.W.; data curation, Q.L.; writing—original draft preparation, Q.L.; writing—review and editing, Q.L. and Z.W.; visualization, Q.L.; supervision, Z.W.; funding acquisition, Z.W. All authors have read and agreed to the published version of the manuscript.

Funding: This work was supported by the National defense Innovation Fund.

Acknowledgments: We express our high respect and gratitude to the editors and reviewers, and their help and suggestions are very helpful to our manuscript.

Conflicts of Interest: The authors declare no conflict of interest.

References

1. Fuentes-Pacheco, J.; Ruiz-Ascencio, J.; Manuel Rendon-Mancha, J. Visual simultaneous localization and mapping: A survey. *Artif. Intell. Rev.* **2012**, *43*, 55–81. [CrossRef]
2. Bailey, T.; Durrant-Whyte, H. Simultaneous localization and mapping (SLAM): Part II. *IEEE Robot. Autom. Mag.* **2006**, *13*, 108–117. [CrossRef]
3. Wang, R.; Schworer, M.; Cremers, D. Stereo DSO: Large-scale direct sparse visual odometry with stereo cameras. In Proceedings of the IEEE International Conference on Computer Vision, Venice, Italy, 22–29 October 2017; pp. 3903–3911.
4. Wang, X.; Wang, J. Detecting glass in simultaneous localization and mapping. *Robot. Auton. Syst.* **2017**, *88*, 97–103. [CrossRef]
5. Titterton, D.; Weston, J.L. *Strapdown Inertial Navigation Technology*; The Institution of Engineering and Technology: Stevenage, UK, 2004; Volume 17.
6. Davison, A.J.; Reid, I.D.; Molton, N.D.; Stasse, O. MonoSLAM: Real-Time Single Camera SLAM. *IEEE Trans. Pattern Anal. Mach. Intell.* **2007**, *29*, 1052–1067. [CrossRef]
7. Lee, T.; Kim, C.; Cho, D.D. A Monocular Vision Sensor-Based Efficient SLAM Method for Indoor Service Robots. *IEEE Trans. Ind. Electron.* **2019**, *66*, 318–328. [CrossRef]
8. Shen, S.J.; Michael, N.; Kumar, V. Tightly-Coupled Monocular Visual-Inertial Fusion for Autonomous Flight of Rotorcraft MAVs. In Proceedings of the 2015 IEEE International Conference on Robotics and Automation (ICRA), Seattle, WA, USA, 26–30 May 2015; pp. 5303–5310.
9. Leutenegger, S.; Lynen, S.; Bosse, M.; Siegwart, R.; Furgale, P. Keyframe-based visual-inertial odometry using nonlinear optimization. *Int. J. Robot. Res.* **2015**, *34*, 314–334. [CrossRef]
10. Mur-Artal, R.; Montiel, J.M.M.; Tardos, J.D. ORB-SLAM: A Versatile and Accurate Monocular SLAM System. *IEEE Trans. Robot.* **2015**, *31*, 1147–1163. [CrossRef]
11. Engel, J.; Koltun, V.; Cremers, D. Direct sparse odometry. *IEEE Trans. Pattern Anal. Mach. Intell.* **2018**, *40*, 611–625. [CrossRef] [PubMed]
12. Usenko, V.; Engel, J.; Stuckler, J.; Cremers, D. Direct Visual-Inertial Odometry with Stereo Cameras. In Proceedings of the 2016 IEEE International Conference on Robotics and Automation (ICRA), Stockholm, Sweden, 16–21 May 2016; pp. 1885–1892.
13. Falquez, J.M.; Kasper, M.; Sibley, G. Inertial Aided Dense & Semi-Dense Methods for Robust Direct Visual Odometry. In Proceedings of the IEEE/RSJ International Conference on Intelligent Robots & Systems, Daejeon, Korea, 9–14 October 2016; pp. 3601–3607.
14. Newcombe, R.A.; Lovegrove, S.J.; Davison, A.J. DTAM: Dense tracking and mapping in real-time. In Proceedings of the 2011 IEEE International Conference on Computer Vision (ICCV), Barcelona, Spain, 6–13 November 2011; pp. 2320–2327.
15. Lucas, B.D.; Kanade, T. An iterative image registration technique with an application to stereo vision. In Proceedings of the International Joint Conference on Artificial Intelligence, Vancouver, BC, Canada, August 1981; pp. 24–28.
16. Klein, G.; Murray, D. Parallel tracking and mapping for small AR workspaces (PTAM). In Proceedings of the 2007 6th IEEE and ACM International Symposium on Mixed and Augmented Reality, Washington, DC, USA, 13–16 November 2007; pp. 1–10.
17. Mur-Artal, R.; Tardos, J.D. ORB-SLAM2: An Open-Source SLAM System for Monocular, Stereo, and RGB-D Cameras. *IEEE Trans. Robot.* **2017**, *33*, 1255–1262. [CrossRef]

18. Qin, T.; Li, P.; Shen, S. VINS-Mono: A Robust and Versatile Monocular Visual-Inertial State Estimator. *IEEE Trans. Robot.* **2018**, *34*, 1004–1020. [CrossRef]
19. He, Y.; Zhao, J.; Guo, Y.; He, W.; Yuan, K. Pl-VIO: Tightly coupled monocular visual inertial odometry using point and line features. *Sensors* **2018**, *18*, 1159. [CrossRef] [PubMed]
20. Qin, T.; Pan, J.; Cao, S.; Shen, S.J. A General Optimization-based Framework for Local Odometry Estimation with Multiple Sensors. *arXiv Preprint* **2019**, arXiv:1901.03638.
21. Engel, J.; Schöps, T.; Cremers, D. LSD-SLAM: Large-scale direct monocular SLAM. In Proceedings of the European Conference on Computer Vision, Zurich, Switzerland, 6–12 September 2014; pp. 834–849.
22. Gao, X.; Wang, R.; Demmel, N.; Cremers, D. LDSO: Direct Sparse Odometry with Loop Closure. In Proceedings of the International Conference on Intelligent Robots and Systems (IROS), Madrid, Spain, 1–5 October 2018.
23. Forster, C.; Pizzoli, M.; Scaramuzza, D. SVO: Fast semi-direct monocular visual odometry. In Proceedings of the 2014 IEEE International Conference on Robotics and Automation (ICRA), Hong Kong, China, 31 May–5 June 2014; pp. 15–22.
24. Lee, S.H.; Civera, J. Loosely-coupled semi-direct monocular slam. *arXiv Preprint* **2018**, arXiv:1807.10073. [CrossRef]
25. Krombach, N.; Droeschel, D.; Houben, S.; Behnke, S. Feature based visual odometry prior for real-time semi-dense stereo slam. *Robot. Auton. Syst.* **2018**, *109*, 38–58. [CrossRef]
26. Kim, O.; Lee, H.; Kim, H.J. Autonomous flight with robust visual odometry under dynamic lighting conditions. *Auton. Robot.* **2019**, *43*, 1605–1622. [CrossRef]
27. Li, S.P.; Zhang, T.; Gao, X.; Wang, D.; Xian, Y. Semi-direct monocular visual and visual-inertial SLAM with loop closure detection. *Robot. Auton. Syst.* **2019**, *112*, 201–202. [CrossRef]
28. Shi, J. Good features to track. In Proceedings of the 1994 IEEE Computer Society Conference on Computer Vision and Pattern Recognition (CVPR'94), Seattle, WA, USA, 21–23 June 1994; pp. 593–600.
29. Hartley, R.; Zisserman, A. *Multiple View Geometry in Computer Vision*; Cambridge University Press: Cambridge, UK, 2003.
30. Baker, S.; Matthews, I. Lucas-Kanade 20 Years On: A Unifying Framework: Part 1. *Int. J. Comput. Vis.* **2002**, *56*, 221–255. [CrossRef]
31. Sibley, G.; Matthies, L.; Sukhatme, G. Sliding window filter with application to planetary landing. *J. Field Robot.* **2010**, *27*, 587–608. [CrossRef]
32. Burri, M.; Nikolic, J.; Gohl, P.; Schneider, T.; Rehder, J.; Omari, S.; Achtelik, M.W.; Siegwart, R. The EuRoC micro aerial vehicle datasets. *Int. J. Robot. Res.* **2016**, *35*, 1157–1163. [CrossRef]
33. Michael Grupp. EVO. Available online: https://github.com/MichaelGrupp/evo (accessed on 8 August 2019).
34. Sturm, J.; Engelhard, N.; Endres, F.; Burgard, W.; Cremers, D. A benchmark for the evaluation of RGB-D SLAM systems. In Proceedings of the 2012 IEEE/RSJ International Conference on Intelligent Robots and Systems (IROS), Vilamoura, Portugal, 7–12 October 2012; pp. 573–580.

 *sensors* 

Article

Camera Calibration with Weighted Direct Linear Transformation and Anisotropic Uncertainties of Image Control Points

Francesco Barone [1],* , Marco Marrazzo [2],* and Claudio J. Oton [1]

[1] Scuola Superiore Sant'Anna, TeCIP Institute, Via Giuseppe Moruzzi 1, 56127 Pisa, Italy; c.oton@santannapisa.it

[2] Baker Hughes, Via Felice Matteucci 2, 50127 Florence, Italy

* Correspondence: f.barone@santannapisa.it (F.B.); marco.marrazzo@bhge.com (M.M.)

Received: 12 November 2019; Accepted: 30 December 2019; Published: 20 February 2020

Abstract: Camera calibration is a crucial step for computer vision in many applications. For example, adequate calibration is required in infrared thermography inside gas turbines for blade temperature measurements, for associating each pixel with the corresponding point on the blade 3D model. The blade has to be used as the calibration frame, but it is always only partially visible, and thus, there are few control points. We propose and test a method that exploits the anisotropic uncertainty of the control points and improves the calibration in conditions where the number of control points is limited. Assuming a bivariate Gaussian 2D distribution of the position error of each control point, we set uncertainty areas of control points' position, which are ellipses (with specific axis lengths and rotations) within which the control points are supposed to be. We use these ellipses to set a weight matrix to be used in a weighted Direct Linear Transformation (wDLT). We present the mathematical formalism for this modified calibration algorithm, and we apply it to calibrate a camera from a picture of a well known object in different situations, comparing its performance to the standard DLT method, showing that the wDLT algorithm provides a more robust and precise solution. We finally discuss the quantitative improvements of the algorithm by varying the modules of random deviations in control points' positions and with partial occlusion of the object.

Keywords: camera calibration; DLT; PnP; weighted DLT; uncertainty; covariance; robustness

1. Introduction

Many computer vision applications, such as robotics, photogrammetry, or augmented reality, require camera calibration algorithms. The calibration is the estimation of the parameters of the camera model from given photos and videos, acquired with the camera. Camera parameters are both extrinsic and intrinsic: extrinsic parameters are pose dependent, while the intrinsic ones are related to the intrinsic properties of the camera itself.

A camera model is a mathematical description of the projection of a 3D point in the real world on the 2D image plane. The pinhole camera model assumes no distortion and a small aperture size; thus, the projection is linear, and it is fully described by a 3×4 projection matrix (or camera matrix). Control points are usually used for the parameters' estimation. They are points whose coordinates are known both in the 3D real world and in the 2D image plane. How the control points are chosen influences the estimation of parameters: a poor choice of control points requires a robust estimation algorithm. The most common procedures exploit a photo of an object whose geometry and pose in space are known. A chessboard pattern is usually used because corners in the chessboard pattern are very easy to identify and its geometry is simple.

When such a pattern is not available, auto-calibration algorithms could be used: they require multiple images to extract information to cope with the lack of a known object in the scene. In this case, features from multiple frames of a moving target (or moving camera) have to be extracted, then the correspondences between features of the different images have to be found [1,2]. Several video frames could be used for accurate auto-calibration procedures, especially offline when there are no constraints on execution times [3]. Plane surfaces are also commonly found in a scene, and some works use them to calibrate cameras [4,5].

Some calibration procedures aim for a one-off estimation of the intrinsic parameters of the camera, which will not vary in time. Once the intrinsic parameters are known, an online algorithm has to estimate only the camera pose. Given a set of n control points and given the intrinsic parameters, the estimation of the camera pose is called the Perspective-n-Points problem (PnP). Three is the smallest number of correspondences that yields a finite number of solutions [6,7]. PnP with three correspondences points is called P3P.

There are many different solutions to PnP [8,9], and some of them are very efficient with an $O(n)$ complexity [10,11]. Other algorithms also provide some internal parameters, such as the focal length [12,13]. The sensitivity to the control points is critical. In particular, the RANSAC algorithm [14] became very popular because of its robustness to outliers. A recent algorithm developed by Ferraz (EPPnP [15]) reaches even better performance and higher robustness. He also integrated the uncertainties of features point [16], given the internal parameters.

If intrinsic parameters are unknown, a complete calibration is required. Many analytical methods were developed in the past especially by photogrammetrists [17]. The most known is the Direct Linear Transformation (DLT [18]) algorithm, which can compute the projection matrix from at least six control points. If there are more than six points, DLT minimizes the least squared error (LSQ) of reprojected points. Then, from the projection matrix, it is possible to extract the parameters of the pinhole camera model [2]. Because of LSQ, DLT is not robust to outliers [19]. To reduce gross errors, Molnár [20] used the Huber estimator, based on a weight function that limits the accounting of errors for outliers. Uncertainties were also integrated in the EPPnP ([15]) algorithm to solve the PnP problem, but it assumed that the internal parameters were known [16].

Current research for camera calibration faces many different problems, such as: automatic calibration methods [21], multiple cameras' calibration [22], and distortion parameters' estimation. For non-linear cameras' calibration, Bouguet developed a general method, which included radial and tangential distortion [23], while Mei extended the algorithm also to omnidirectional cameras by means of multiple views of a planar pattern [24]. Further advances for omnidirectional camera calibration were done by Scaramuzza, who developed and implemented a fully automatic and iterative procedure for non-linear camera calibration [25,26].

Compared to DLT, non-linear camera models have the disadvantage of needing multiple images to estimate the distortion parameters; indeed, trying to estimate many parameters from only few points of a single image leads to an ill conditioned problem. A large set of control points is available especially when calibration patterns are used and corners are easy to detect. In particular applications, the number of control points is required to be low. For example, even if calibration patterns for an infrared camera exist [27], they cannot be used for an infrared camera inside a gas turbine.

Temperature mapping of gas turbine rotor blades is extremely important both for condition health monitoring and blade design optimization. Infrared thermography allows contactless measurements of the rotor blade surface temperature in gas turbines [28], but the mapping between the 2D image and the 3D blade requires adequate camera calibration. Thus, the blade itself has to be used as the calibration frame: indeed, its geometry is well known, and usually, an accurate 3D model is available. Nevertheless, the blade is always partially in the field of view.

Moreover, camera lenses are subject to high temperature variation and the intrinsic parameters, such as the focal length changes with temperature; thus, offline calibration of intrinsic parameters cannot be done.

Manual calibration is required because infrared images of blades usually have low quality and low resolution; thus, images have a limited number of features (such as corners and edges), and the number of control point is very limited.

To solve this problem, in this paper, we propose a method for camera calibration suitable for a low number (less than 10) of control points with anisotropic uncertainty. Assuming a different bivariate Gaussian distribution of the position error of each control point, its anisotropic uncertainty is determined by the uncertainty ellipse, within which the control point is estimated to be. The proposed method, based on direct linear transformation, consists of minimizing the weighted reprojection error of control points. The weight matrix is chosen according to the axis lengths and axis rotation of the uncertainty ellipses. Because of that, the presented weighted variation of the DLT (wDLT) includes the uncertainty information, improving the average accuracy, especially with few control points.

The wDLT algorithm we propose requires a photo of a well known 3D object (such as the rotor blade in a gas turbine).

Uncertainty ellipses could also be used to constrain a point along the direction of an edge, or a segment: in this case, the uncertainty is high along the edge, while it is low in the direction perpendicular to it. Exploiting the uncertainty becomes necessary when there are only a few control points, and some of them are almost exact, while the other ones are approximated.

In the following sections, we show the DLT algorithm for camera calibration and how to set the weight matrix for wDLT according to the uncertainty areas of control points. Then, we compare the performance of three algorithms (DLT, Bouguet's method [23], and wDLT) using the reprojection error on 21 real images with only seven control points each. We evaluate the robustness of the algorithms, introducing fictitious errors to control points. Infrared images of rotor blades inside gas turbine always have a partial view of the blade itself, because of small spaces and the design constraint of the camera system; thus, we also simulate different scenarios where the object is partially occluded, making it difficult to find the control points manually, leading to an accuracy error. We also evaluate the sensitivity of uncertainty values. Results show that wDLT reaches higher average accuracy when the control points are perturbed with fictitious errors, and it allows camera calibration in some evaluated scenarios where the standard DLT and Bouguet's method fail.

2. Materials and Methods

In this section, we first describe the weighted direct linear transformation algorithm and how we propose to choose the weight matrix; then, we describe the tests we performed on real images. The implementations of the DLT and wDLT algorithms were realized with MATLAB R2018b [29]; Bouguet's method was performed by using the functions and scripts of the Camera Calibration toolbox for MATLAB [23].

2.1. The Weighted Direct Linear Transformation Algorithm

Considering the pinhole approximation, the camera model maps the 3D world points to 2D image points through a linear projection operation, which is expressed by the following relation:

$$\lambda \hat{x}_i = \mathbf{P} X_i \tag{1}$$

where $\mathbf{X}_i = [X_i, Y_i, Z_i, 1]^T$ is the ith world point in homogeneous coordinates, $\hat{x}_i = [x_i, y_i, z_i]^T$ is a 3D representation of the ith projected point, $\mathbf{P} \in \Re^{3 \times 4}$ is the projection matrix, and $\lambda \in \Re$ is a free parameter. The parameter λ is usually set equal to the reciprocal of z_i, such as $\tilde{x}_i = \lambda \hat{x}_i = [u_i \; v_i \; 1]^T$

becomes the ith image point in the homogeneous coordinate. Applying the DLT algorithm (see Appendix A), for each correspondence between $\mathbf{X}_i$ and $\tilde{\mathbf{x}}_i$, two equations could be written:

$$\begin{bmatrix} -\mathbf{X}_i^T & \mathbf{0}^T & u_i\mathbf{X}_i^T \\ \mathbf{0}^T & -\mathbf{X}_i^T & v_i\mathbf{X}_i^T \end{bmatrix} \begin{bmatrix} \mathbf{p}_1^T \\ \mathbf{p}_2^T \\ \mathbf{p}_3^T \end{bmatrix} = \mathbf{M}_i\mathbf{p} = \begin{bmatrix} 0 \\ 0 \end{bmatrix} \tag{2}$$

where $\mathbf{p}_1$, $\mathbf{p}_2$, and $\mathbf{p}_3$ are the rows of the projection matrix $\mathbf{P}$, which have to be determined. Writing the equations for Ncorrespondences, the following equation system is obtained:

$$\mathbf{Mp} = \begin{bmatrix} \mathbf{M}_1 \\ \vdots \\ \mathbf{M}_N \end{bmatrix} \begin{bmatrix} \mathbf{p}_1^T \\ \mathbf{p}_2^T \\ \mathbf{p}_3^T \end{bmatrix} = \begin{bmatrix} 0 \\ \vdots \\ 0 \end{bmatrix} \tag{3}$$

Considering that $\mathbf{p}$ is a 12 element array and each correspondence adds two equations, at least six equations are needed. Thus, for $N \geq 6$, an LSQ solution of the homogeneous linear system $\mathbf{Mp} = 0$ is the eigenvector of $\mathbf{M}^T\mathbf{M}$ corresponding to the minimum eigenvalue.

2.1.1. Weights of the Equations

The proposed approach is based on modeling the error on the image plane as a bivariate Gaussian distribution, such as Ferraz et al. [16] did for the PnP problem. The covariance matrix $\mathbf{C}_i$ of the error distribution could be written as follows:

$$\mathbf{C}_i = \begin{bmatrix} \sigma_{x_i}^2 & 0 \\ 0 & \sigma_{y_i}^2 \end{bmatrix} \mathbf{R}(\theta_i) \tag{4}$$

where $\mathbf{R}(\theta_i) \in \Re^{2\times2}$ is the rotation matrix that rotates an angle θ_i and σ_{x_i} and σ_{y_i} are the standard deviations along the principal axes.

The weight matrix is defined by the following equation:

$$\mathbf{W}_i = \begin{bmatrix} 1/\sigma_{x_i} & 0 \\ 0 & 1/\sigma_{y_i} \end{bmatrix} \mathbf{R}(\theta_i) \tag{5}$$

Each equation of (3) is left multiplied by a weight matrix $\mathbf{W}_i$ that depends on the covariance matrix of the expected error distribution for the ith point. This yields the following equation:

$$\hat{\mathbf{M}}\mathbf{p} = \mathbf{WMp} = \begin{bmatrix} \mathbf{W}_1\mathbf{M}_1 \\ \vdots \\ \mathbf{W}_N\mathbf{M}_N \end{bmatrix} \begin{bmatrix} \mathbf{p}_1^T \\ \mathbf{p}_2^T \\ \mathbf{p}_3^T \end{bmatrix} = \begin{bmatrix} 0 \\ \vdots \\ 0 \end{bmatrix} \tag{6}$$

where $\mathbf{W}$ is a diagonal block matrix whose blocks on the diagonal are $\mathbf{W}_i$.

The LSQ solution of the homogeneous linear system in Equation (6) is the eigenvector of $\hat{\mathbf{M}}^T\hat{\mathbf{M}}$ corresponding to the minimum eigenvalue. The error to minimize is weighted by the matrix $\mathbf{W}$, so that for the ith point, the reprojection error along the axis x_i (or y_i), which is tilted by an angle θ_i with respect to the horizontal (or vertical) image axis, is accounted for less if the σ_{x_i} (or is σ_{y_i}) is high and vice versa.

2.1.2. Covariance Matrix Estimation

The error is considered to be a bivariate Gaussian distribution, whose average is the ith control point and covariance matrix is $\mathbf{C}_i$. Here, the estimation of $\mathbf{C}_i$ is done by the manual assessment of the confidence region, the region within which each point is almost surely. For a bivariate distribution, this region is an ellipse. This ellipse is fully determined by the length l_{x_i} and l_{y_i} of the two orthogonal principal axis and the direction θ_i of x_i axes. The standard deviation of the bivariate Gaussian distribution is set to a third of the length of the axis: $\sigma_{x_i} = l_{x_i}/3$ and $\sigma_{y_i} = l_{y_i}/3$. Thus, the covariance matrix is computed using Equation (4). The rotation matrix $\mathbf{R}(\theta_i)$ aligns the image axis to the axis of the confidence ellipse of the ith point. For our purposes, it is not the absolute value of σ_{x_i} and σ_{y_i} that is relevant, but their ratio.

2.2. Tests on Real Images

The following tests aimed to compare the wDLT performance with the standard DLT and Bouguet's method (BOU), highlighting the advantages of using uncertainty information when the number of control points is low. The approach would reach similar results if the uncertainties of each point were equal (i.e., if the covariance matrix $\mathbf{C}_i$ was the same scalar matrix for each i) or if the number of points was equal to six, that is the least needed for both algorithms.

For the test, we used 21 images[1] of a metal parallelepiped (see Figure 1a) whose size was $1 \times 3 \times 5$ cm. The sizes of the blocks were accurate up to ± 0.01 mm, but the edges and the vertices were blunt. The block stood on a table, lying on one of the two 3×5 cm faces.

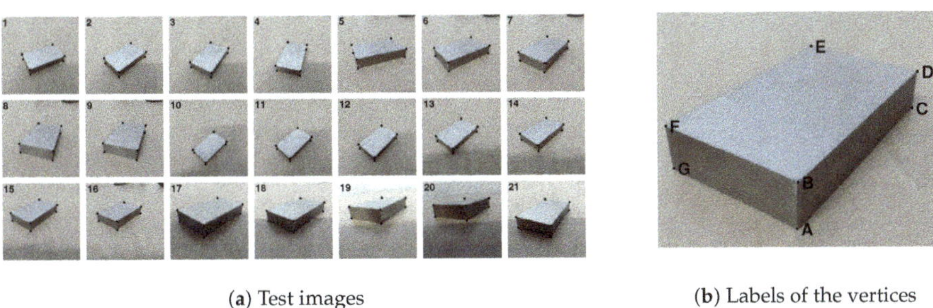

(a) Test images   (b) Labels of the vertices

Figure 1. The 21 images[1] of the metal block used for testing the DLT, Bouguet's method (BOU), and weighted DLR (wDLT) algorithms. The control points (black dots) are manually selected on the images. They correspond to the seven visible vertices of the parallelepiped. The reprojection error, using DLT to calibrate the camera, is 2.5 ± 1.3 pixels.

The images were recorded varying the point of view with respect to the object. Each image had 3000×3000 pixels, and the same seven vertices of the block were visible, while the eighth was always hidden. Vertices were labeled with capital letters, as shown in Figure 1b.

The geometry of the object in the scene was known; thus, a 3D model of the block was generated. The seven visible vertices of the block were chosen as control points, because they were easy to recognize: from the 3D model of the object, the positions of the vertices were exactly known, while on the images, they were set manually. The DLT was applied: the reprojection error was 2.5 ± 1.3 pixels (average $\pm$ STD), and the maximum error was 6.9 pixels for a single point. The average and the standard deviation of the reprojection error were first computed among points of the same image, and the results were averaged across different images. The average reprojection error for BOU was 2.5 ± 1 pixels without distortion and 2.2 ± 1.2 with second order radial distortion approximation. Adding more distortion parameters led to an ill conditioned estimation problem and numerical errors. The error of a few pixels on a high resolution image could be considered negligible, and the additional improvement obtained by including radial distortion was very slight. The two tests we performed are described below.

2.2.1. Robustness to Random Error

In order to evaluate robustness to random errors, a first test was performed: the calibration was performed with DLT, BOU, and wDLT using perturbed control points. For each image n_E, control points of the image were randomly chosen, and an error, whose module was E, was added to these points in a random direction. Because of the randomness, the test was performed 10 times for each image, and $n_E = \{1,2,3\}$ and $E = \{10,20,30,40\}$. For each point, the standard deviation was $\sigma_{x_k} = \sigma_{y_k} = 1$, except for perturbed points, where $\sigma_{x_p} = \sigma_{y_p} = \{5,8,12\}$. In this test, the axes were equal, their rotation had no influence, and the confidence ellipses were circles. An example with $n_E = 1$ and $\sigma_{x_p} = \sigma_{y_p} = 8$ is shown in Figure 2: the real position of the vertices (unperturbed control points) is in black circles, and the perturbed ones are in blue. For wDLT, the confidence regions are shown as well. The green crosses are the reprojected points for DLT (Figure 2a), BOU (Figure 2b), and wDLT (Figure 2c). The reprojected points of the BOU algorithm, if no distortion was estimated, were very close to the DLT ones.

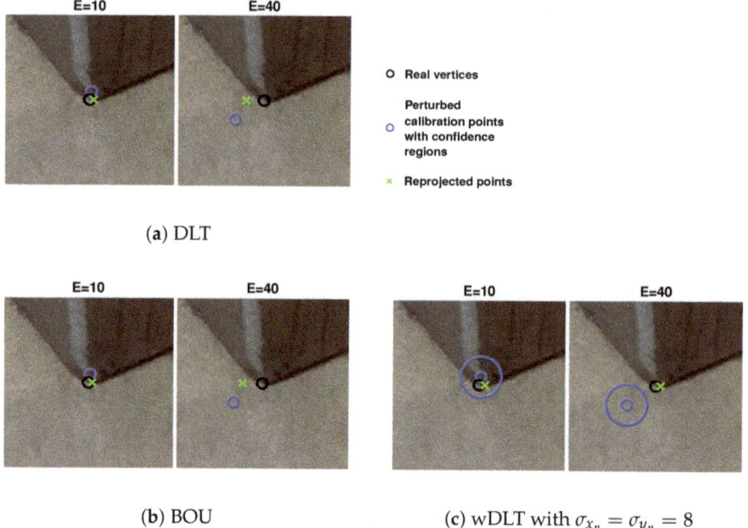

Figure 2. Example: Here, the calibration is performed on the seventh image[1] with an error E on a control point of 10 and 40 pixels. The black circles are the unperturbed position of the control point, while in blue, we show the perturbed ones. In Figure 2c, the confidence region is shown as well: the radius of the circles is 24 pixels, corresponding to σ=8. Here, wDLT can recover an error of 40 pixels on a single perturbed point, while DLT and BOU cannot.

For each image, the average, the standard deviation, and the maximum of the reprojection error were computed. The reprojection error was the Euclidean distance on the image plane between the seven visible vertices and the control points without error.

2.2.2. Occluded Vertices Scenario

The objective of the second test was to simulate a manual choice of the control points when reference points of the objects in the scene (such as corners) are occluded in the image. We performed the calibration without using some of the vertices' positions in the images. Instead of using control

---

[1]   Photo Credits: *Copyright 2020 Baker Hughes Company. All rights reserved.*

points corresponding to the occluded vertices, the median points of the block edges were used in addition. The world points corresponding to the median were exactly known, while the corresponding points on the image were not. In order to have comparable results, each additional control point on the images was chosen systematically, using the following equation:

$$\begin{bmatrix} u_{XY} \\ v_{XY} \end{bmatrix} = \begin{bmatrix} u_X \\ v_X \end{bmatrix} + m \begin{bmatrix} u_Y - u_X \\ v_Y - v_X \end{bmatrix} \tag{7}$$

where $[u_{XY}, v_{XY}]$ are the coordinates of the point along the edge XY, $[u_X, v_X]$ and $[u_Y, v_Y]$ are the coordinates of two vertices X and Y of the edge, and $m \in [0,1]$ is an adimensional parameter. The coordinates of the control point $[u_{XY}, v_{XY}]$ varied with m between the vertices along the edge XY.

Two scenarios were simulated:

1. Vertices C and D were occluded, and two additional control points were chosen on the edges AC and DE.
2. Vertices C, D, and E were occluded, and three additional control points were chosen on the edges AC, DE, and EF.

For each scenario and for each image, three sets of control points with $m = \{0.45, 0.50, 0.55\}$ were computed. Figure 3 shows, for both scenarios, the control points on the vertices in black, the occluded vertices in red, and in blue, the additional control points along the edge for $m = \{0.45, 0.50, 0.55\}$. The reprojection error was the Euclidean distance between the real vertices and the reprojected vertices (i.e., without considering the error on the additional control points).

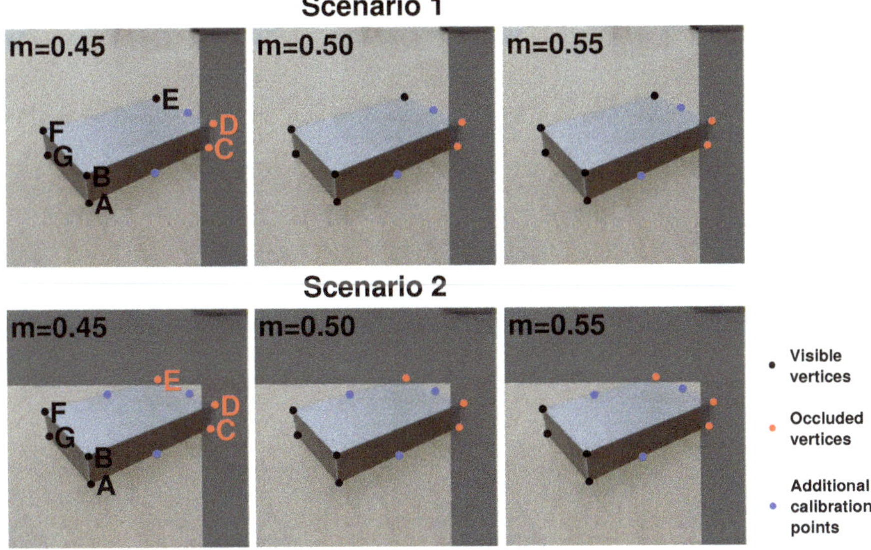

Figure 3. Scenarios with two and three occluded vertices: the black dots are the visible vertices, the red dots the occluded vertices (not used in calibration), and blue dots the additional points along the edges. The figures show the additional control points chosen along the edges with $m = \{0.45, 0.50, 0.55\}$.

wDLT was tested with different values of uncertainty. For control points that were on vertices, the axis lengths of the confidence ellipses were the same ($l_{x_i} = l_{y_i} = 1$). Instead, for control points that were not on vertices, but on edges, the axes of the confidence ellipses were set to be equal to one in the direction perpendicular to the edge ($l_{y_i} = 1$) and equal to 3, 15, and 10^6 in the direction parallel to the edge ($l_{x_i} = \{3, 15, 10^6\}$). We chose these values in order to evaluate the sensitivity of wDLT with different ellipse eccentricities to the estimation of the uncertainty σ.

Figure 4 shows for the second scenario and for different ratios l_{x_i}/l_{y_i} the control points, the occluded vertices, and the confidence region of the additional control points. As an example, the reprojected control points and reprojected vertices are also shown. For the sake of clear representation, all the confidence ellipses had sizes 50 times bigger than the real ones.

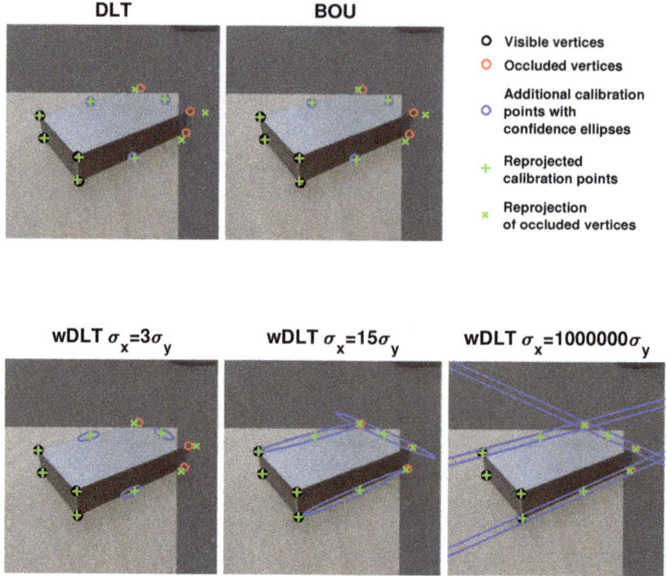

Figure 4. Scenario 2 with three occluded points and m = 0.50: black circles are the visible vertices, the red ones the occluded vertices (not used in calibration), and the blue ellipses the confidence ellipses of the additional control points on the edges. We represent each ellipse 50 times bigger in order to make them clearly visible. The green pluses and the green crosses are respectively the reprojected control points and the reprojected occluded vertices. While DLT and BOU reach good alignment of the control points (black and blue circles), the reprojection error on the occluded vertices is high. Instead, wDLT reaches a better calibration because the error on the occluded vertices' reprojection is low.

3. Results

3.1. Robustness to Error

The first test aimed to evaluate the robustness of the three algorithms (DLT, BOU, and wDLT); thus, the reprojection error was calculated with control points perturbed with random errors and with different values of uncertainty for wDLT. We varied both the module of the error E and the number of points n_E affected by the error E. BOU calibration was performed without distortion estimation, because not all the random configurations of control points led to a solution; this fact was due to the estimation problem, which became ill conditioned when data did not contain enough information.

The reprojection error for random perturbations of control points (see Section 2.2.1) is summarized in the charts in Figure 5: vertical bars show the mean absolute error (MAE); the standard deviation (STD) is indicated by the black vertical lines, which show the $\pm1\sigma$ range; and red squares are the maximum absolute errors (MaxErr). The whole distributions of the error are shown in Appendix B.

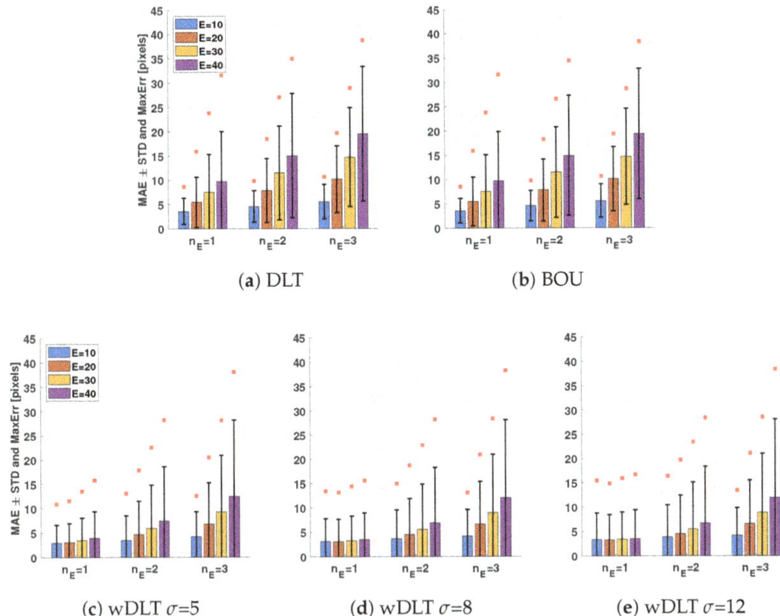

Figure 5. Reprojection error of the calibration with DLT, BOU, and wDLT with n_E perturbed control points and error E; bars represent the mean absolute error (MAE), the black lines the average STD of the errors, and the red squares the average of the maximum error.

As expected, the DLT and BOU performances decreases with high error E and number of errors n_E (see Figure 5a). While the DLT and BOU algorithms only relied on the position of the control points, wDLT had the uncertainty of the points' position. Because higher uncertainty was set for perturbed points, wDLT assigned less weight to the information about their position, while it focused more on the position of the others. In this way, the average error for each E and n_E was lower for wDLT. Moreover, with wDLT, the MAE did not vary significantly when the values of uncertainties σ_{x_p} and σ_{y_p} of perturbed control points changed.

With only one perturbed point of seven control points ($n_E = 1$), the error was almost fully recovered, and the performance was similar to the DLT with no perturbed points. Figure 2c shows an example where with wDLT, the reprojected point was almost unaffected by the position of the control point and was near the real vertex. For more than one of seven perturbed points, it was shown that the error was no longer recovered. The reason for that was the fact that the algorithms needed at least six control points, and with $n_E = 1$ and high uncertainty, wDLT would consider only the remaining six low uncertainty points. On the contrary, with $n_E \geq 1$, less than six points were reliable; thus, wDLT needed the information of the perturbed points. However, wDLT would assign more weight to points with low uncertainty, and the average error would decrease as well.

It is worth noting that when the perturbation was small (E = 10), the maximum error did not improve with the exploitation of the uncertainty by wDLT. The reason was that the value of E was comparable with the uncertainty of the unperturbed points; thus, the best performance was achieved by setting in wDLT an equal uncertainty for all the points, which was equivalent to using DLT or BOU.

Summarizing: If the estimated uncertainty σ was too low for a subset of control points, the wDLT would take high error points more into account, and the MAE would increase because of the wrong information in input; if the uncertainty was too high, the algorithm would not consider important information from those points, and the MAE would increase as well. Therefore, in order to have better

MAE than DLT and BOU, it was enough to set the uncertainty such that if a point was more uncertain than another one, the first had to have higher uncertainty than the second one, regardless of the exact ratio between the two uncertainties.

3.2. Occluded Vertices Scenario

The results of the second test are shown in the charts in Figure 6. As for the first test, the BOU estimation of distortion parameters was not feasible; thus, they were not estimated. In both scenarios (with two or three occluded points), the algorithms showed the same behaviors, but the errors with three occluded points were always higher than the ones with only two occluded points, as expected.

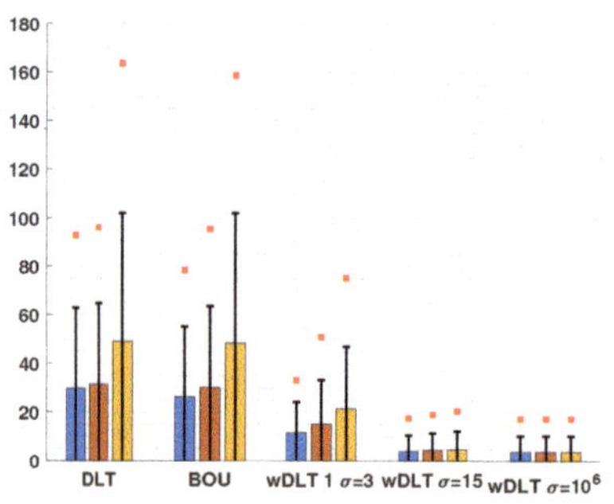

(**a**) Scenario 1: two occluded points

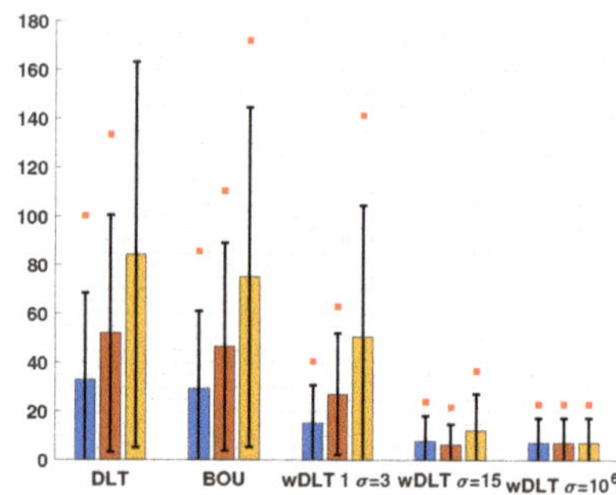

(**b**) Scenario 2: three occluded points

Figure 6. Reprojection error of the calibration with DLT, BOU, and wDLT with different settings of the uncertainty σ for points along the edges. The bars represent the average error, the black lines the average STD of the errors, and the red squares the average of the maximum error.

The results with $m = 0.45$ were better than the others because they corresponded to the points that were nearer to the real ones. Indeed, points with $m = 0.45$ were chosen along the edge BD and nearer to D than to B, along the edge DE and nearer to E than to D, and along EF and nearer to E than to F; due to the perspective of the image, these points were nearer to the median of the edges, which was the point that was considered in the 3D model (see Section 2.2.2). For the same reason, the worst results were obtained with $m = 0.55$. Figure 4 shows the reprojected vertices for the seventh image: both DLT and BOU minimized the error on control points (black and blue circles), but because blue circles had position errors, the alignment was wrong, which was clear by looking at the reprojection of occluded vertices. For wDLT, the confidence ellipses allowed the reprojected points to slide more along their major axes; thus, the results were better because the high variance along the edges produced a better estimation of the positions of the occluded vertices.

With a very high uncertainty set along the edge ($\sigma = 10^6$), wDLT reached a similar MAE without depending on the position of the point along the edge (i.e., the values of m, according to our parametrization of its position). The reason was that with a very high σ along the direction of the edge, the resulting system of Equation (6) constrained the median points of the edge to be along the lines of each edge, but made them free to move along that. Thus, wDLT did not take into account if the positions of those points slid along the edges.

4. Discussion

The proposed method was a weighted DLT based algorithm, suitable when the uncertainty areas of the control points on the image were known or could be estimated. It did not require the knowledge of the exact values of uncertainty, because a gross estimation still improved the mean absolute error. To the best of our knowledge, no one has yet presented a calibration algorithm that exploits the anisotropic uncertainty of the control points by setting a weight matrix according to the lengths and rotation of the axis of uncertainty ellipses.

wDLT was also useful when there was no calibration pattern in the scene and the choice of the control points was manual. In this kind of situation, there were only a few control points: some of them corresponded to corners, and their uncertainty was low; others could be on edges, and their uncertainty was low only in the direction perpendicular to the edge; other points could be assigned in correspondence to a wide area, and their uncertainty was higher. If there were only a few corners, it was necessary to exploit as much information as possible about the points.

For infrared thermography in gas turbines, the image of the rotor blade was only partial and had low resolution. Many defects and artifacts were present on the image, and manual calibration was necessary. The lack of corners, reference points (the shape of the blade was smooth and curvy), and clear edges made the choice of control points very hard. Allowing the use of uncertainty ellipses (instead of single points), this step became easier and calibration more accurate.

Some works presented different calibration algorithms that could cope with outliers, but they required many points to be extracted. Tens of points are usually given from automatic feature extraction algorithms, and usually, many outliers are present. The presented results showed the advantages of using uncertainty information in the calibration problem. In the presented cases, the obtained error was lower compared to the DLT algorithm and Bouguet's method.

The cost of this improvement was the knowledge of the uncertainty. In the first test, we supposed knowing the points with higher error. Huge improvement was reached only with $n_E = 1$, but almost the same result could be reached applying the DLT to the six points without errors (which we supposed were known). For $n_E \geq 1$, there were only minor improvements.

The second test showed that the wDLT algorithm was useful when there were partially occluded objects with few visible corners. Introducing anisotropy in the uncertainty area of the edge points strongly improved the estimation of the positions of the hidden vertices.

The main advantage of this algorithm was that it offered the possibility of including anisotropic uncertainty information in the calibration, improving the solution. Even though this method did not provide a precise way to estimate the absolute uncertainty of the individual points, when these values were manually set, a strong improvement was observed as soon as the relative differences between uncertainties were assigned.

Infrared camera calibration for gas turbine thermography by using single control points and standard DLT led to gross and unacceptable errors in camera parameters, whereas by setting uncertainty areas, instead of control points, the camera could be properly calibrated by using weighted DLT.

5. Conclusions

We presented a method to set the weight matrix in a weighted Direct Linear Transformation (wDLT) algorithm, in order to take into account the anisotropic uncertainties of the positions of individual control points. Instead of control points, we proposed to use uncertainty ellipses with different axis lengths and angles. This feature became important when few control points were available. To demonstrate the effectiveness of the algorithms, we performed two tests: the first, to evaluate the robustness to random error, and the second, to simulate a scenario with few control points. We showed that wDLT with the uncertainty performed better (lower MAE) in both tests. In particular, in the second test, the calibration failed both with DLT and Bouguet's method when several vertices were hidden, while wDLT still provided a successful calibration in these cases.

A suitable application of wDLT is in infrared thermography in gas turbines, where only a few control points can be chosen accurately, and then, manual selection of them is required. Accuracy in infrared camera calibration allows associating the correct temperature with the 3D model of the object in the field of view.

Author Contributions: Conceptualization, methodology, software, validation, formal analysis, investigation, and draft preparation, F.B.; manuscript review and editing, C.J.O. and M.M.; supervision, C.J.O. and M.M. All authors have read and agreed to the published version of the manuscript.

Funding: This research received no external funding.

Acknowledgments: This work was done in collaboration with Baker Hughes, through the Scuola Sant'Anna—Baker Hughes Joint Lab "Advanced Sensors for Turbomachinery".

Conflicts of Interest: The authors F.B. and C.J.O. declare no conflict of interest. The author M.M. declares that he works for Baker Hughes.

Appendix A. Direct Linear Transformation

The Direct Linear Transformation (DLT) algorithm aims to solve the problem of determining the pinhole camera parameters from at least six correspondences between 2D image points and 3D world points. A camera model maps each point of the 3D world to a point of the 2D image through a projection operation. The pinhole camera model makes the assumption that the aperture size of the camera is small, so that it can be considered as a point. Thus, the ray of light has to pass across a single point, the camera center, and there are no lenses, no distortion, and an infinite depth of field.

The pinhole camera model is fully represented by a linear projection, which is expressed by the following equation:

$$\lambda \hat{\mathbf{x}}_i = \mathbf{P} \mathbf{X}_i \tag{A1}$$

where $\mathbf{X}_i = [X_i, Y_i, Z_i, 1]^T$ is the ith world point in homogeneous coordinates, $\hat{\mathbf{x}}_i = [x_i, y_i, z_i]^T$ is a 3D representation of the ith projected point, $\mathbf{P} = [p_{ij}] \in \Re^{3 \times 4}$ is the projection matrix, or camera matrix, and $\lambda \in \Re$ is a free parameter. The parameter λ is usually set equal to the reciprocal of z, such as:

$$\tilde{\mathbf{x}}_i = \lambda \hat{\mathbf{x}}_i = \frac{1}{z} \hat{\mathbf{x}}_i = \begin{bmatrix} \frac{x_i}{z_i} & \frac{y_i}{z_i} & 1 \end{bmatrix}^T = \begin{bmatrix} u_i & v_i & 1 \end{bmatrix}^T = \begin{bmatrix} \mathbf{x}_i^T & 1 \end{bmatrix}^T \tag{A2}$$

where $\tilde{\mathbf{x}}_i$ is the ith projected point $\mathbf{x}_i$ in the image plane in homogeneous coordinates.

Combining Equations (A1) and (A2), the coordinates of the image point $\mathbf{x}_i$ are expressed by the following relationship:

$$\mathbf{x}_i = \begin{bmatrix} u_i \\ v_i \end{bmatrix} = \begin{bmatrix} x_i/z_i \\ y_i/z_i \end{bmatrix} = \begin{bmatrix} \mathbf{p}_1 \mathbf{X}_i / \mathbf{p}_3 \mathbf{X}_i \\ \mathbf{p}_2 \mathbf{X}_i / \mathbf{p}_3 \mathbf{X}_i \end{bmatrix} \tag{A3}$$

where $\mathbf{p}_1$, $\mathbf{p}_2$, and $\mathbf{p}_3$ are the rows of the projection matrix $\mathbf{P}$. This relationship is not linear in the unknown parameters $\mathbf{p}_1$, $\mathbf{p}_2$ and $\mathbf{p}_3$.

Through DLT, the nonlinear problem becomes linear in the unknown parameters, rearranging Equation (A3) as follows:

$$\begin{bmatrix} u\mathbf{p}_3\mathbf{X}_i \\ v\mathbf{p}_3\mathbf{X}_i \end{bmatrix} = \begin{bmatrix} \mathbf{p}_1\mathbf{X}_i \\ \mathbf{p}_2\mathbf{X}_i \end{bmatrix} \Rightarrow \begin{bmatrix} -\mathbf{X}_i^T\mathbf{p}_1^T + \mathbf{0}^T\mathbf{p}_2^T + u_i\mathbf{X}_i^T\mathbf{p}_3^T \\ \mathbf{0}^T\mathbf{p}_1^T - \mathbf{X}_i^T\mathbf{p}_2^T + v_i\mathbf{X}_i^T\mathbf{p}_3^T \end{bmatrix} = \begin{bmatrix} -\mathbf{X}_i^T & \mathbf{0}^T & u_i\mathbf{X}_i^T \\ \mathbf{0}^T & -\mathbf{X}_i^T & v_i\mathbf{X}_i^T \end{bmatrix} \begin{bmatrix} \mathbf{p}_1^T \\ \mathbf{p}_2^T \\ \mathbf{p}_3^T \end{bmatrix} = \begin{bmatrix} 0 \\ 0 \end{bmatrix} \tag{A4}$$

These homogeneous equations can be solved linearly. Combining the equations for each point i, the homogeneous linear systems shown in Equation (3) is obtained.

Appendix B. Error Distribution for Random Perturbation of Control Points

In this section, the error distributions for the test described in Section 2.2.1 are shown. Figure A1a,b show the error distribution using the DLT algorithm and Bouguet's method, while in Figure A1c–e, we show the distributions for the wDLT algorithm with different σ settings for the perturbed points. Each bar chart represents the error distribution for different n_E and E.

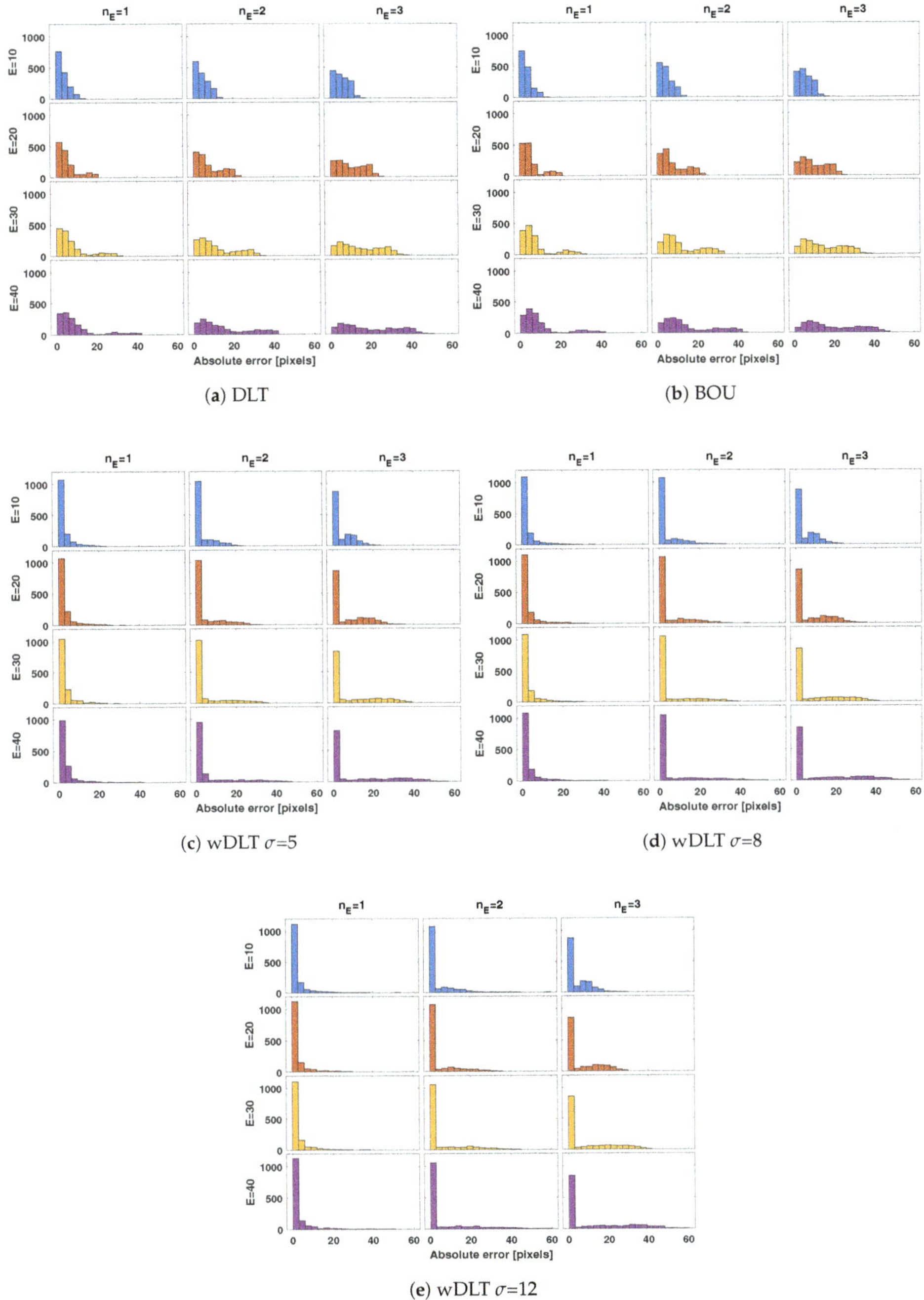

(**a**) DLT

(**b**) BOU

(**c**) wDLT σ=5

(**d**) wDLT σ=8

(**e**) wDLT σ=12

Figure A1. Distribution of the reprojection error of the calibration with DLT, BOU, and wDLT with n_E perturbed control points and error E.

References

1. Faugeras, O.D.; Luong, Q.T.; Maybank, S.J. Camera self-calibration: Theory and experiments. In *Computer Vision — ECCV'92, Proceedings of the European Conference on Computer Vision, Ligure, Italy, 19–22 May 1992*; Sandini. G.; Springer: Berlin/Heidelberg, 2005; pp. 321–334.
2. Hartley, R.; Zisserman, A. *Multiple View Geometry in Computer Vision*; Cambridge University Press: Cambridge, UK, 2003.
3. Kim, J.S.; Hong, K.S. A recursive camera resectioning technique for off-line video-based augmented reality. *Pattern Recognit. Lett.* **2007**, *28*, 842–853.
4. Zhang, Z. A flexible new technique for camera calibration. *IEEE Trans. Pattern Anal. Mach. Intell.* **2000**, *22*, 1330–1334.
5. Strum, P.F.; Maybank, S.J. On plane-based camera calibration: A general algorithm. In Proceedings of the 1999 IEEE Computer Society Conference on Computer Vision and Pattern Recognition, Fort Collins, CO, USA, 23–25 June 1999.
6. Gao, X.S.; Hou, X.R.; Tang, J.; Cheng, H.F. Complete solution classification for the perspective-three-point problem. *IEEE Trans. Pattern Anal. Mach. Intell.* **2003**, *25*, 930–943.
7. Persson, M.; Nordberg, K. Lambda Twist: An Accurate Fast Robust Perspective Three Point (P3P) Solver. In Proceedings of the European Conference on Computer Vision (ECCV), Munich, Germany, 8–14 September 2018; pp. 318–332.
8. Quan, L.; Lan, Z. Linear n-point camera pose determination. *IEEE Trans. Pattern Anal. Mach. Intell.* **1999**, *21*, 774–780.
9. Hesch, J.A.; Roumeliotis, S.I. A direct least-squares (DLS) method for PnP. In Proceedings of the 2011 International Conference on Computer Vision, Barcelona, Spain, 6–13 November 2011; pp. 383–390.
10. Lepetit, V.; Moreno-Noguer, F.; Fua, P. Epnp: An accurate O(n) solution to the pnp problem. *Int. J. Comput. Vis.* **2009**, *81*, 155.
11. Schweighofer, G.; Pinz, A. Globally Optimal O(n) Solution to the PnP Problem for General Camera Models. In Proceedings of the British Machine Vision Conference 2008, Leeds, UK, 1–4 September 2008; pp. 1–10. doi:10.5244/C.22.55.
12. Penate-Sanchez, A.; Andrade-Cetto, J.; Moreno-Noguer, F. Exhaustive linearization for robust camera pose and focal length estimation. *IEEE Trans. Pattern Anal. Mach. Intell.* **2013**, *35*, 2387–2400.
13. Zheng, Y.; Kneip, L. A direct least-squares solution to the PnP problem with unknown focal length. In Proceedings of the IEEE Conference on Computer Vision and Pattern Recognition, Las Vegas, NV, USA, 27–30 June 2016; pp. 1790–1798.
14. Fischler, M.A.; Bolles, R.C. Random sample consensus: A paradigm for model fitting with applications to image analysis and automated cartography. *Commun. ACM* **1981**, *24*, 381–395.
15. Ferraz, L.; Binefa, X.; Moreno-Noguer, F. Very fast solution to the PnP problem with algebraic outlier rejection. In Proceedings of the IEEE Conference on Computer Vision and Pattern Recognition, Columbus, OH, USA, 23–28 June 2014; pp. 501–508.
16. Ferraz, L.; Binefa, X.; Moreno-Noguer, F. Leveraging Feature Uncertainty in the PnP Problem. In Proceedings of the British Machine Vision Conference, Nottingham, UK, 1–5 September 2014. doi:10.5244/C.28.83.
17. Fraser, C.S. Photogrammetric camera component calibration: A review of analytical techniques. In *Calibration and Orientation of cameras in Computer Vision*; Springer: Berlin/Heidelberg, Germany, 2001; pp. 95–121.
18. Abdel-Aziz, Y.; Karara, H. Direct Linear Transformation from Comparator Coordinates into Object Space Coordinates in Close-Range Photogrammetry. *Photogramm. Eng. Remote Sens.* **2015**, *81*, 103–107.
19. Maronna, R.A.; Martin, R.D.; Yohai, V.J.; Salibián-Barrera, M. *Robust Statistics: Theory and Methods (With R)*; Wiley: Hoboken, NJ, USA, 2018.
20. Molnár, B. Direct linear transformation based photogrammetry software on the web. *Int. Arch. Photogramm. Remote Sens. Spatial Inf. Sci.* **2010**, *38*. Part 5, Commission V Symposium, Newcastle upon Tyne, UK.
21. Hillemann, M.; Weinmann, M.; Mueller, M.S.; Jutzi, B. Automatic Extrinsic Self-Calibration of Mobile Mapping Systems Based on Geometric 3D Features. *Remote Sens.* **2019**, *11*, 1955.
22. Khoramshahi, E.; Campos, M.B.; Tommaselli, A.M.G.; Vilijanen, N.; Mielonen, T.; Kaartinen, H.; Kukko, A.; Honkavaara, E. Accurate Calibration Scheme for a Multi-Camera Mobile Mapping System. *Remote Sens.* **2019**, *11*, 2778.

23. Bouguet, J.Y. Caltech Vision Group Offical Site. Available online: http://www.vision.caltech.edu (accessed on 7 January 2020)
24. Mei, C.; Rives, P. Single view point omnidirectional camera calibration from planar grids. In Proceedings of the 2007 IEEE International Conference on Robotics and Automation, Roma, Italy, 10–14 April 2007; pp. 3945–3950.
25. Scaramuzza, D.; Martinelli, A.; Siegwart, R. A flexible technique for accurate omnidirectional camera calibration and structure from motion. In Proceedings of the Fourth IEEE International Conference on Computer Vision Systems, New York, NY, USA, 4–7 January 2006; pp. 45–45.
26. Scaramuzza, D.; Martinelli, A.; Siegwart, R. A toolbox for easily calibrating omnidirectional cameras. In Proceedings of the 2006 IEEE/RSJ International Conference on Intelligent Robots and Systems, Beijing, China, 9–15 October 2006; pp. 5695–5701.
27. Kelly, J.; Kljun, N.; Olsson, P.O.; Mihai, L.; Liljeblad, B.; Weslien, P.; Klemedtsson, L.; Eklundh, L. Challenges and Best Practices for Deriving Temperature Data from an Uncalibrated UAV Thermal Infrared Camera. *Remote Sens.* **2019**, *11*, 567.
28. Mevissen, F.; Meo, M. A Review of NDT/Structural Health Monitoring Techniques for Hot Gas Components in Gas Turbines. *Sensors* **2019**, *19*, 711.
29. MATLAB. *Version 9.5 (R2018b)*; The MathWorks Inc.: Natick, MA, USA, 2018.

Article

A Self-Assembly Portable Mobile Mapping System for Archeological Reconstruction Based on VSLAM-Photogrammetric Algorithm

Pedro Ortiz-Coder *,† and Alonso Sánchez-Ríos *,† 

Department of Graphic Expression, University Centre of Mérida, University of Extremadura, 06800 Mérida, Spain

* Correspondence: peorco04@alumnos.unex.es (P.O.-C.); schezrio@unex.es (A.S.-R.)

† These authors contributed equally to this work.

Received: 19 July 2019; Accepted: 9 September 2019; Published: 12 September 2019

Abstract: Three Dimensional (3D) models are widely used in clinical applications, geosciences, cultural heritage preservation, and engineering; this, together with new emerging needs such as building information modeling (BIM) develop new data capture techniques and devices with a low cost and reduced learning curve that allow for non-specialized users to employ it. This paper presents a simple, self-assembly device for 3D point clouds data capture with an estimated base price under €2500; furthermore, a workflow for the calculations is described that includes a Visual SLAM-photogrammetric threaded algorithm that has been implemented in C++. Another purpose of this work is to validate the proposed system in BIM working environments. To achieve it, in outdoor tests, several 3D point clouds were obtained and the coordinates of 40 points were obtained by means of this device, with data capture distances ranging between 5 to 20 m. Subsequently, those were compared to the coordinates of the same targets measured by a total station. The Euclidean average distance errors and root mean square errors (RMSEs) ranging between 12–46 mm and 8–33 mm respectively, depending on the data capture distance (5–20 m). Furthermore, the proposed system was compared with a commonly used photogrammetric methodology based on Agisoft Metashape software. The results obtained demonstrate that the proposed system satisfies (in each case) the tolerances of 'level 1' (51 mm) and 'level 2' (13 mm) for point cloud acquisition in urban design and historic documentation, according to the BIM Guide for 3D Imaging (U.S. General Services).

Keywords: self-assembly device; 3D point clouds; accuracy analysis; VSLAM-photogrammetric algorithm; portable mobile mapping system; low-cost device; BIM

---

1. Introduction

The tridimensional modeling of an object starts with its original design or with the process of acquiring the data necessary for its geometric reconstruction. In both cases, the result is a 3D virtual model that can be visualized and analyzed interactively on a computer [1,2]. In many cases, the process continues with the materialization of the model in the form of a prototype, which serves as a sample of what will be the final product, allowing us to check if its design is correct, thus changing the traditional manufacturing or construction industry [3–5].

The applications of 3D models (virtual or prototype) are numerous and widely used; they are usually used in the scope of clinical applications [6,7], geosciences [8–13], cultural heritage preservation [14,15] and engineering [16].

In this context, to address this wide variety of application areas, both data capture techniques and devices, as well as the specific software for data processing and management tend to be simplified.

265

This is done in order to be accessible to the greatest number of users, even with limited knowledge in 3D measurement technologies.

In this sense, the classical methods of photogrammetry are combined with new techniques and procedures which are usually adopted for other areas [17], such as visual odometry (VO), the simultaneous localization and mapping (SLAM) and the visual slam (VSLAM) techniques. These are normally used to solve localization and mapping problems in the areas of robotics and autonomous systems [18–21], but also the combination of photogrammetry techniques with methodologies based on instruments like terrestrial or aerial laser scanners have obtained successful results [22,23].

These combined methods provide support and analytical robustness for the development of low/middle-cost capture systems, usually based on tablets or mobile devices that incorporate inertial sensors, absolute positioning and low cost cameras which can achieve medium 3D positional accuracy scanning, in compliance with technical requirements of a wide range of applications at a low cost and reduced learning curve [24–28]. As a result, handheld mobile mapping systems have appeared in recent years, using different technologies to perform 3D reconstructions that use fully automated processes [27,28]. Among which we can find systems based exclusively on images, requiring a fully automated process, taking into account the usual technical constraints in photogrammetry, and the free user movements in data capture [17]. In this field, different lines of research have been developed, depending on whether the final result is obtained in real time [29,30] or not. In the first case, the reduction in the time needed for data processing is the most important factor in the approach to research objectives (even at the expense of a metric accuracy reduction); in the second, however, metric accuracy is the most important factor, although the temporal cost is higher [17,31,32].

There are many commercial mobile mapping systems for urban, architectural or archaeological applications with high accuracy results [33]. Those systems are based on the integration on different sensors such as (Inertial Measurement Unit) IMU, line scanners [28], cameras, Global Navigation Satellite System (GNSS) [34], odometers and other sensors. The price and complexity of those systems are normally high [35].

The classical applications require a known level of data accuracy and quality, however, the emerging needs of Industry 4.0, building information modeling (BIM) or digital transformation, next to the appearance of new devices and information processing techniques pose new challenges and research opportunities in this field. Each capture method has its advantages and drawbacks, offering a particular level of quality in its results; in this sense, numerous investigations have linked these parameters, allowing people to choose the most cost-effective approach [35]. This can be achieved by way of evaluating the use of the laser scanner and the vision-based reconstruction among different solutions for progress monitoring inspection in construction and concluding (among other characteristics) that both of them are appropriate for spatial data capture. This could [36] include among 3D sensing technologies, photo/video-grammetry, laser scanning and the range of images that make a detailed assessment of the content (low, medium or high) into BIM working environments. It may also include [37] comparing photo/video-grammetry capture techniques with laser scanning, considering aspects such as accuracy, quality, time efficiency and cost needed for collecting data on site. The combination of data capture methods has also been traditionally analyzed; thus, [38] presently, there is a combined laser scanning/photogrammetry approach to optimize data collection, cutting around 75% of the time required to scan the construction site.

A common aspect, taken into account in most of the research, is the point cloud accuracy evaluation, that has been addressed in three different ways [39]: (a) By defining levels of quality parameters defined in national standards and guidelines that come from countries like the United States, Canada, the United Kingdom and Scandinavian countries that lead BIM implementation in the world [40], such as the U.S. General Services Administration (GSA) BIM Guide for 3D Imaging [41] that sets the quality requirements of point clouds in terms of their level of accuracy (LOA) and level of detail (LOD); (b) by evaluating quality parameters of a point cloud, in terms of accuracy and completeness [37] or (c) following three quality criteria: Reference system accuracy, positional accuracy and completeness [42].

Furthermore, in the specific environment of 3D indoor models, [43] we propose a method that provides suitable criteria for the quantitative evaluation of geometric quality in terms of completeness, correctness, and accuracy by defining parameters to optimize a scanning plan in order to minimize data collection time while ensuring that the desired level of quality is satisfied, in some cases, with the implementation of an analytical sensor model that uses a "divide and conquer" strategy based on segmentation of the scene [44], or one that captures the relationships between parameters like data collection parameters and data quality metrics [45]. In other cases, the influence of scan geometry is considered in order to optimize measurement setups [46], or are compared to different known methods for obtaining accurate 3D modeling applications, like in the work of [47], in the context of cultural heritage documentation.

This paper extends on past surveys of classical photogrammetry solutions, adopting an extended solution approach for outdoor environments based on the use of a simple and hand-held self-assembly device for data capture, based on images, that consist on two cameras: One, which data will be used to calculate in real time, the path followed by the device using a VSALM algorithm, while with other one; a high-resolution video recorded and used to achieve the scene reconstruction using photogrammetric techniques. Finally, after following simple data collection and fully automated processing, a 3D point cloud with associated color is obtained.

To determine the effectiveness of the proposed system, we evaluate it in one study site performed outdoors in the facades of the Roman Aqueduct of Miracles, in terms of the requirements laid down in the GSA BIM Guide for 3D Imaging. In this experiment, we obtain 3D point clouds from different data capture conditions, that vary according to the distance from the device and the monument; the measurements acquired by a total station serve to compare the coordinates of fixed points in both systems, and therefore, determining the LOA of each point cloud. The results obtained, with root mean square errors (RMSEs) between eight and 33 mm, stress the feasibility of the proposed system for urban design and historic documentation projects, in the context of allowable dimensional deviations in BIM and CAD deliverables.

This paper is divided into four sections. Following the Introduction, the portable mobile mapping system is described, including the proposed algorithm schema for the computations; therefore, a case study in which the system is applied is described in Section 2. The results are presented in Section 3 and finally, the conclusions are presented in Section 4.

2. Materials and Methods

This study was conducted with a simple and self-assembly prototype specifically built for data capture (Figure 1), that consists of two cameras from the Imaging Source Europe GmbH company (Bremen, Germany): Camera A (model DFK 42AUC03) and camera B (model DFK 33UX264) were fixed to a platform with the condition of its optical axes being parallel; each camera incorporated a lens; for camera A, the model was TIS-TBL 2.1 C, from the Imaging Source Europe GmbH company and for camera B, the model was the Fujinon HF6XA–5M, from FUJIFILM Corporation (Tokyo, Japan). The technical characteristics of cameras and lenses appear in Tables 1 and 2, respectively. Both cameras were connected to a laptop (with an Intel core i7 7700 HQ CPU processor and RAM of 16 Gb, running under Windows 10 Home), via USB 2.0 (camera A) and 3.0 (camera B). This beta version of the prototype had an estimated base price under €2500.

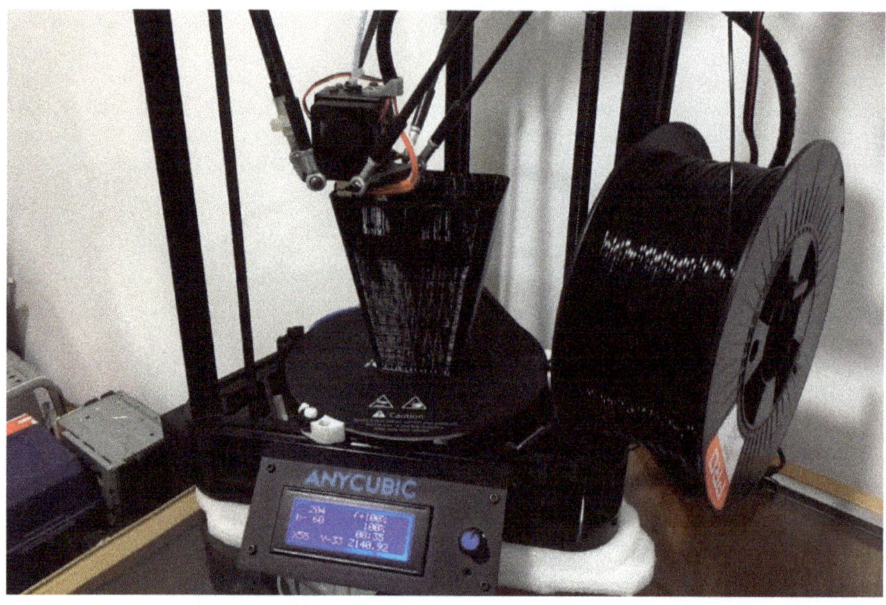

Figure 1. (**a**) 3D printing process of the prototype case; (**b**) cameras A and B with their placement inside the case; and (**c**) the final portable mobile mapping system prototype.

Table 1. Main technical characteristics of cameras used in the prototype (from the Imaging Source Europe GmbH company).

Model	Resolution (Pixels)	Megapixels	Pixel Size (μm)	Frame Rate (fps)	Sensor	Sensor Size	A/D (bit)
DFK 42AUC03	1280 × 960	1.2	3.75	25	Aptina MT9M021 C	1/3″CMOS	8
DFK 33UX264	2448 × 2048	5	3.45	″8	Sony IMX264	2/3″ CMOS	8/12

Table 2. Main technical data of lenses used in the prototype (from the Imaging Source Europe GmbH company and FUJIFILM Corporation).

Model	Focal Length (mm)	Iris Range	Angle of view (H × V)
TIS-TBL 2.1 C	2.1	2	97° × 81.2°
Fujinon HF6XA–5M	6	1.9–16	74.7° × 58.1°

The calibration process of the cameras was carried out with a checkerboard target (60 cm × 60 cm) using a complete single camera calibration method [48], that provided the main internal calibration parameters: The focal length, radial and tangential distortions, optical center coordinates and camera axe skews. In addition, to know the parameters that related to the position of one camera compared to the other, we designed the following, practical test: To use as ground control points we placed 15 targets on two perpendicular walls and measured the coordinates of each target with a TOPCON Robotic total station, with an accuracy of 1" measuring angles (ISO 17123-3:2001) and 1.5 mm + 2 ppm measuring distances (ISO 17123-4:2001). After running the observations with the prototype, we used a seven-parameter transformation, using the 15 targets, to determine the relative position of one camera in relation to the other [17].

Camera A and camera B had different configuration parameters which defined image properties such as brightness, gain or exposure between others. In order to automate the capture procedure, automatic parameters options had been chosen. In such a way, the data collection was automatic and the user didn't need to follow specials rules since the system accepted convergent or divergent turns of the camera, stops or changes in speed. The algorithm processed all this data properly using the proposed methodology.

During the capture (Figure 2), the user needed to see the VSLAM tracking in the screen of the computer in real time. In this way, the user was sure he didn´t make a fast movement or if an item appeared that interrupted camera visualization and, therefore, the tracking could not continue. In this case, the user must return again to a known place and continue the tracking from this point.

Figure 2. In-field data capture. The white arrow indicates the user direction movement, parallel to the monument façade, followed during this test.

Workflow of the Proposed Algorithm for the Computation

The application of the VSLAM technique on a low weight device, normally with limited calculation capabilities, needed the implementation of a low computational cost VSLAM algorithm to achieve effective results. The technical literature provided a framework that consisted of the following basic modules: The initialization module; to define a global coordinate system, and the tracking and mapping modules; to continuously estimate camera poses. In addition, two additional modules were used for a more reliable and accurate result: The re-localization module, that has to be used when, due to a fast device motion or some disruptions in data capture, the camera pose must be computed again and the global map optimization, which is performed to estimate and remove accumulative errors in the map, produced during camera movements.

The characteristics of the VSLAM-photogrammetric algorithm, including identified strong and weak points, depend on the methodology used for each module which sets its advantages and limitations. In our case, we proposed the following sequential workflow (Figure 3) divided into four threaded processes, which have been implemented in C++.

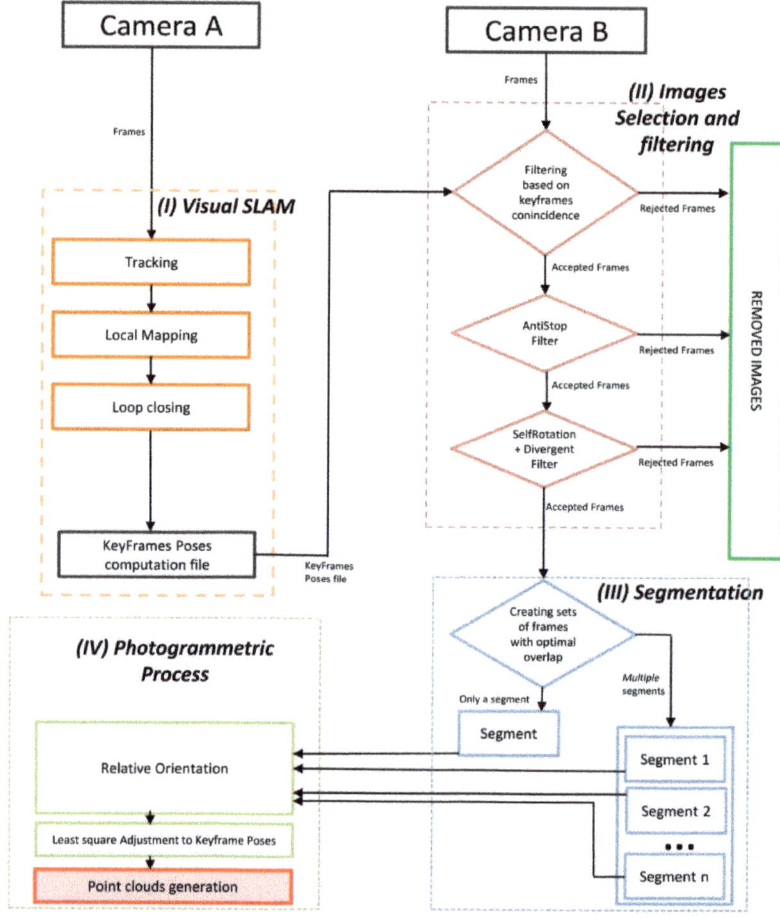

Figure 3. General workflow of the algorithm implemented in C++ for the computations.

Basically, the four processes consisted in the following: (I) A VSLAM algorithm to estimate both motion and structure, that is applied in frames obtained from camera A, (II) an image selection and filtering process of frames obtained with camera B, (III) the application of an image segmentation algorithm and finally, (IV) a classical photogrammetric process applied to obtain the 3D point cloud. Each process is explained in more detail below.

The first process (I) started with the simultaneous acquisition of videos with cameras A and B, with speeds of 25 FPS and 4 FPS, respectively. With the frames from camera A, used as an ORB descriptor [49] for object recognition, detection and matching was used. This descriptor built on the FAST key-point detector and the BRIEF descriptor, with good performance and low cost, and therefore, was appropriate for our case. An ORB-SLAM algorithm was then applied to estimate camera positioning and trajectory calculation [50]; this was an accurate monocular SLAM system that worked in real time and can be applied in indoor/outdoor scenarios, and has modules to loop closure detection, re-localization (to recover from situations where the system becomes lost) and to a totally automatic initialization, taking into account the calibration parameters of the camera. From these remarks, our process was carried out in three steps as follows [50]. The first step was the tracking, which calculated the positioning of the camera for each frame and selected keyframes and decided which frames were added to the list; the second one was local mapping, which performed keyframes optimization, incorporating those that were being taken and removing the redundant keyframes. With these data, through a local bundle adjustment, in addition to increasing the quality of the final map, it was possible to reduce the computational complexity of the processes that were just running, and equally for the subsequent steps. The third one was loop closing, which looked for redundant areas where the camera had already passed before, which could be found in each new keyframe; the transformation of similarity on the accumulated drift in the loop was calculated, the two ends of the loop were aligned [50], the duplicate points were merged and the trajectory was recalculated and optimized to achieve overall consistency. The result of this process was a text file with UNIX time parameters and camera poses of the selected keyframes.

The above information together with the frames recorded by camera B, was used to start the second process (II), in which a selection and filtering of the images obtained by camera B was carried out, which consisted in the direct deletion of images whose baseline was very small, and therefore, which made it difficult to compute an optimum relative orientation [17,51,52]. The filtering process was performed in three consecutive steps: The first, a filtering based on keyframes coincidence, which consisted of incorporating a β number of frames (in our case $\beta = 2$) from camera B between each two consecutive frames from camera A and, at the same time, the remaining frames were removed. To run this filter, it was necessary that the cameras were synchronized by UNIX time. The second process, applied the so-called AntiStop filter, which removed those frames obtained in the event that the camera had been in a static position, or with a very small movement, recorded images of the same zone, which we described as redundant and which should, therefore, be eliminated. To determine the redundant frames, it was assumed that cameras A and B were synchronized and that we knew the coordinates of the projection center of each frame, computed in (I). We continued with the calculation of the distances between the projection centers of every two consecutive keyframes i and j (D_{ij}) as well as the mean value of all the distances between consecutive frames (D_m) and the definition of the minimum distance (D_{min}) from which the device was either stopped or was in motion, by the expression:

$$D_{min} = D_m * p,$$

where p is a parameter that depends on the data capture conditions (in our case, after performing several tests, we defined a value of $p = 0.7$). Finally, the keyframes took by camera B in which the distance between the projection centers of each two consecutive keyframes was less than the minimum distance ($D_{ij} < D_{min}$) were removed.

The third, called the divergent self-rotation filter, was able to remove those keyframes captured by camera B when they met two conditions: The rotation angles of the camera ω_i (X axis) and κ_i (Z axis) (Figure 4) increase or decrease their value permanently during data capture for of at least three consecutive frames at a value of $\pm 9°$ (in our case), and besides, their projection centers were very close to each other; for the calculation of the same procedure is the same one used as the one used for the AntiStop filter, but with a different value of p (in our case, we considered a value of $p = 0.9$).

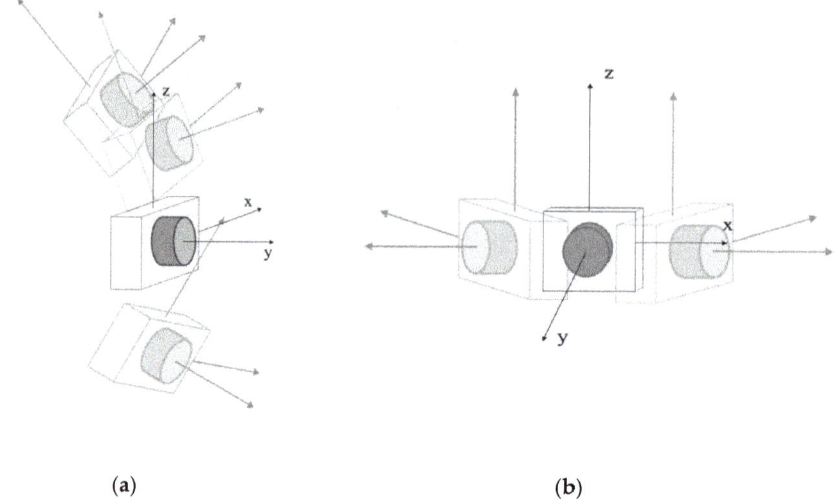

(a)   (b)

Figure 4. Divergent self-rotation in the (**a**) *X*-axis; and (**b**) *Z*-axis.

The next process (III) was segmentation, which aimed to obtain more significant and easy to analyze images in the subsequent photogrammetric process. It started searching for the homologous points belonging to the keyframes resulting from the filtering process carried out in (II) [53–55], which was performed between an image, the earlier one and the later one. The resulting images were stored in a set, called a "segment". The result of this process generated one or more independent segments among themselves, which had a number of homologous points and an appropriate distribution to be properly oriented (in our case, 200 points and 10% of these points were in each quadrant of the image; in addition, if the segment did not have at least three images, it was discarded and its images were removed).

The last process (IV) was called the photogrammetric process, which was structured in three steps: The first was to compute a relative image orientation [53] setting the first image as the origin of the relative reference system and used the homologous points of each segment and algorithms leading to direct solutions [17,51,53]; then, a bundle adjustment) was used on the oriented images to avoid divergences [56], obtaining the coordinates of the camera poses and computed tie points. The second step consisted of an adjustment of the camera poses in each segment to adapt them to the overall trajectory, computed in (I). This procedure was performed using minimum square techniques [57] in each segment, and a three-dimensional transformation [10] to correct the positions of camera B with respect to camera A.

In the third step, the scene was reconstructed using MICMAC software [54], in order to obtain dense cloud points with color. MICMAC is a free open-source photogrammetry software developed by the French National Mapping Agency (IGN) and the National School of Geographic Sciences (ENSG) [58]. This software generates a depth map from the main image and a series of secondary images to obtain parallax values. The calculation was carried out having taken into account that the scene could be described by a single function $Z = f(X; Y)$ (with $X; Y; Z$ using Euclidean coordinates) with several parameters of MICMAC to calculate the density correlation and obtain the cloud of dense points with color [54,55,59] which was the final result of the process.

3. Accuracy Assessment and Results

This work determined the accuracy of a set of point clouds obtained with the prototype in order to validate the device for BIM work environments. Additionally, the results were compared with a usual photogrammetric procedure, using a reflex camera and photogrammetric software (Agisoft Metashape [60]), in order to compare the advantages and disadvantages of the proposed prototype in respect to this known methodology. For this, an experimental test was carried out in the Roman aqueduct of "The Miracles" in the city of Mérida (Spain). This monument, built in the first-century A.C, has a total dimension of 12 km in length between underground and aerial sections with arches. The test was carried out on an archery stretch which was 23 m high and 60 m wide, performing a set of three data capture scenarios at different observation distances (5, 12, and 20 m) from the prototype to the base of the monument (Figure 5).

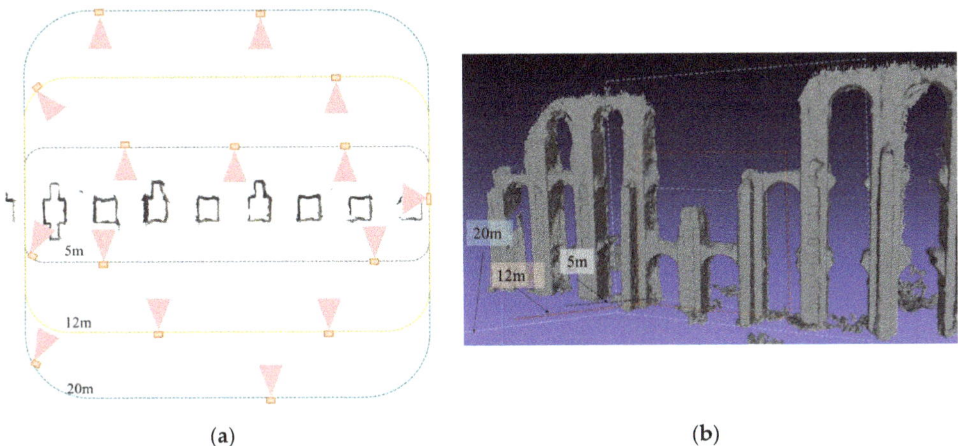

(a)   (b)

Figure 5. (**a**) Scheme with the data capture trajectories and (**b**) the areas covered by a frame, for 5, 12 and 20 m of distance prototype-monument. The figure that appears in (**b**), is a 3D model (mesh) generated by the software Meshlab [61] from the 20m points cloud made only for visualization purposes.

In this test, the data collection was carried out in such a way that the movement of the user followed a perpendicular direction to the camera optical axe (Figure 2), avoiding divergent turns since this kind of movement was not necessary in this case. In this way, this prevented the algorithm from using the divergent self-rotation filter in an unnecessary situation.

In order to evaluate the metric quality of the measures obtained with the prototype and the Agisoft Metashape photogrammetric procedure, a control network was performed to be used in the dimensional control study, following the procedures carried out by [62] and [63]. The network was used as reference points and consisted on a set of targets and natural targets whose three-dimensional coordinates in a local coordinate system were obtained by a second measuring instrument (more precise than the device we want to evaluate). In this case, a total station Pentax V-227N (Pentax Ricoh Imaging Company, Ltd, Tokyo, Japan) was used, with an accuracy of 7′ for angular measurements (ISO 17123-3:2001) and 3 mm ± 2 ppm for distance measurements (ISO 17123-3:2001) with which a total of 40 uniformly distributed targets have been measured (Figure 6).

Then, the method proposed by [62] was used, in which the accuracy of the 3D point cloud was quantified according to the Euclidean average distance error (δ_{avg}) as:

$$\delta_{avg} = \frac{1}{n} \sum_{i=1}^{n} |Ra_i - T - b_i| \tag{1}$$

where a_i is the *i*th checkpoint measured by the prototype, b_i is the corresponding reference point acquired by the total station, R and T are the rotation and translation parameters for 3D Helmert transformation.

And the quality of the 3D point cloud was also evaluated by the root mean square error (RMSE) as:

$$RMSE = \sqrt{\frac{1}{n}\sum_{i=1}^{n}\left(a_i^t - b_i\right)^2} \tag{2}$$

where a_i^t indicates the a_i point after the 3D conformal transformation to bring the model coordinates in the same system of the reference points.

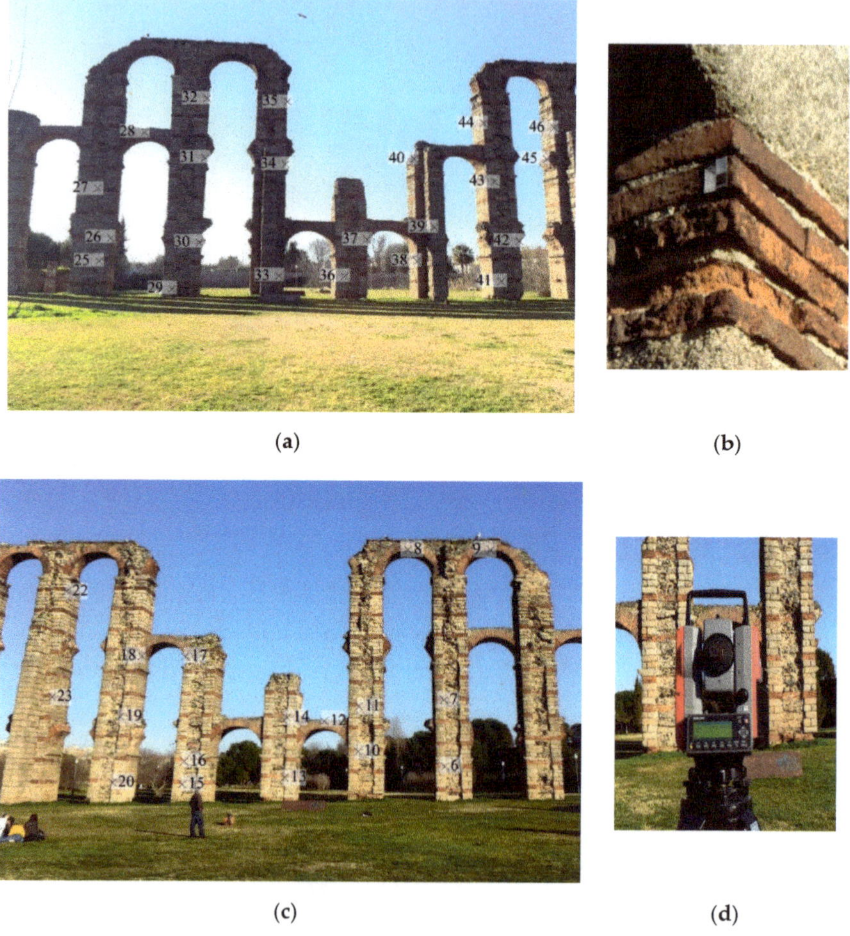

(a)   (b)

(c)   (d)

Figure 6. (**a**,**c**) Reference points spread on the two fronts of the monument. (**b**) Target model used in the test and (**d**) total station Pentax V-227N used to measure the network coordinates.

As mentioned previously in Section 1 of this paper, the point cloud accuracy evaluation can be done according to different criteria. In our case, we have used the GSA BIM Guide for 3D Imaging criteria, that defines four levels of detail (LOD) with dimensions of the smallest recognizable feature ranging between 13 mm × 13 mm to 152 mm × 152 mm; and also defines the level of accuracy (LOA) associated to each LOD, ranging between three and 51 mm of tolerance, considering it as the allowable dimensional deviation in the deliverable from truth (that has been obtained by some more precise other means). In the case of a point cloud, the guide specifies that the distance between two points from the model must be compared to the true distance between the same two points, and be less than or equal to the specified tolerance; the guide also defines the area of interest as a hierarchical system of scale in which each scan is registered, depending on the LOD. In Table 3, we summarize the data quality parameters defined by the GSA for registering point clouds.

Table 3. Data quality parameters defined by U.S. General Services Administration (GSA) for registering point clouds. (Unit: Millimeters).

Level of Detail (LOD)	Level of Accuracy (LOA, Tolerance)	Resolution	Areas of Interest (Coordinate Frame, c. f.)
Level 1	±51	152 × 152	Total Project area (Local or State c. f.)
Level 2	±13	25 × 25	e.g., building (local or project c. f.)
Level 3	±6	13 × 13	e.g., floor level (project or instrument c. f.)
Level 4	±3	13 × 13	e.g., room or artifact (instrument c. f.)

In order to complete the study, other photogrammetric system was analyzed under conditions similar to the prototype (Figure 7). The camera used was a Canon EOS 1300D and the lens was an EFS 18–55 mm, but we only used the focal length of 18mm for this experiment. Multiple images were taken in this experiment for each distance (35 images for 5 m, 41 images for 12 m and 43 images for 20 m) and the camera was configured with a resolution of 2592 pixels × 1728 pixels with the aim of comparing the results in an equitable way with the proposed approach which have a similar image resolution. The reflex camera´s parameters (shutter, diaphragm, ISO, etc.) were chosen in automatic mode during the test to match the conditions to the prototype test. The pictures were taken standing on the same trajectories previously followed by the prototype, at the same distances from the aqueduct: 5, 12 and 20 m. These circumstances increase the time consumed in the field during the data capture, as can be seen in the Table 4, because the user must focus each image and ensure that the picture has been taken with enough overlap and quality. On the other hand, the prototype cameras also have a configuration with automatic parameters which allowed the user, along with the methodology used, to make a continuous capture, without stopping to take the images. The images of the Canon camera were processed using the software Agisoft Metashape 1.5.4 [60] which is commercialized by the company Agisoft LLC, sited in St. Petersburg, Russia (Figure 8).

Point cloud density for each system was measured. Two points clouds, one for each system, were processed using the same 10 images of the aqueduct at a distance of 5 m. The density [37] of the point cloud was 328 points/dm^2 for the proposed prototype system and 332 points/dm^2 for the Agisoft Metashape photogrammetric software.

Table 4. Comparison between the proposed approach and the camera with Agisoft Metashape software in regards to the time spent in the field for data capture and processing time using the same laptop (Intel core i7 7700 HQ CPU processor, 16Gb RAM, Operative System Windows 10 Home). Distance values are measured from the camera to the monument.

System	Data Capture Distance (m)	Data Capture Time (min)	Processing Time (min)
Prototype and Visual Slam (VSLAM)-Photogrammetric Algorithm	5	4.25	80
	12	4.53	85
	20	4.65	99
Canon Camera and Agisoft Metashape Software	5	7.83	72
	12	9.08	80
	20	9.50	89

With the prototype and VSLAM-Photogrammetric algorithm we have computed the average error and the RMSE in each direction (x, y, and z) of each data capture distance, that are listed in Table 5, with overall accuracies of 12, 26 and 46 mm for 5, 12 and 20 m respectively and the RMSEs on each axis ranging between 5 to 8 mm (5 m), 10 to 21 mm (12 m) and 30 to 38 mm (20 m) (Figure 7) which satisfied the error tolerance of 'level 1' (51 mm) for data capture distances from 12–20 m and 'level 2' (13 mm) for data capture distances about 5 m.

Table 5. This table compares the accuracy assessment results with the root mean square errors (RMSEs) and average errors for data capture distances from 5 to 20 m from the camera to the monument, between the prototype and VSLAM-photogrammetric algorithm and the Canon camera with Agisoft Metashape software. The RMSE error values have been computed in the three vector components: X, Y and Z.

	Methodology							
	Prototype and VSLAM-Photogrammetric Algorithm				Canon Camera and Agisoft Metashape Software			
	Distance 5 m							
	Error Vector X (mm)	Error Vector Y (mm)	Error Vector Z (mm)	Error (mm)	Error Vector X (mm)	Error Vector Y (mm)	Error Vector Z (mm)	Error (mm)
Average Error				12				11
RMSE	5	8	8	8	4	9	8	12
	Distance 12 m							
AVERAGE Error				26				23
RMSE	21	16	10	16	12	18	17	28
	Distance 20 m							
Average Error				46				35
RMSE	32	30	38	33	18	24	24	39

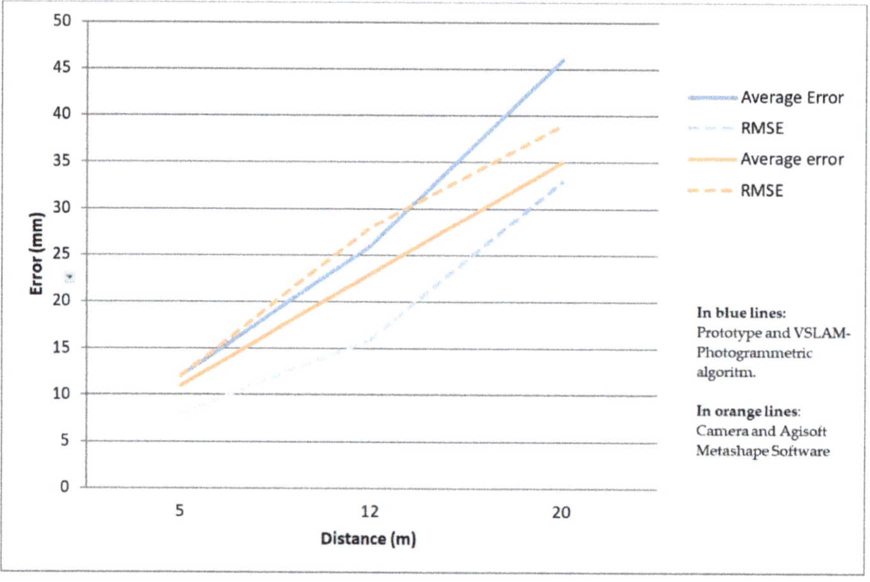

Figure 7. Graphic on the evolution of the average errors and RMSEs for the distances of 5, 12 and 20 m from the camera to the monument. The results are shown for both systems: Prototype and VSLAM-photogrammetric algorithm and Canon camera with Agisoft Metashape software.

(a)   (b)

Figure 8. Comparison between points clouds resulting from both systems (with a data capture distance of 12 m): (**a**) Prototype and VSLAM-photogrammetric algorithm; and (**b**) Canon camera EOS 1300D with Agisoft Metashape software.

The point clouds obtained at the different distances of observation shown in Figure 9. Small holes or missing parts can be seen in those points clouds. This occurs due to the camera's trajectory, since it needs to focus directly on all the desired areas and capture a minimum number of images to perform optimal triangulation. No filter has been applied in the results shown in Figure 9.

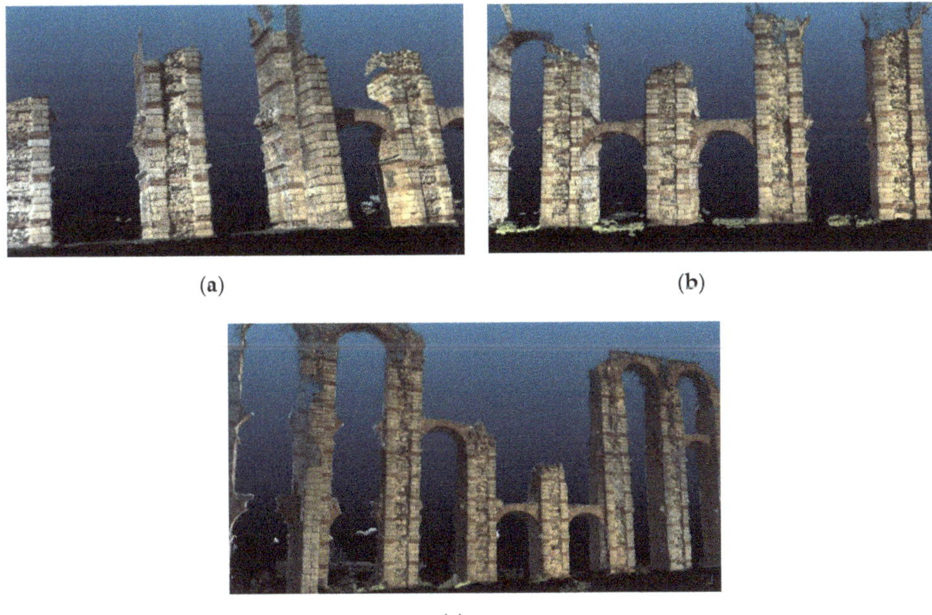

(a)   (b)

(c)

Figure 9. Points clouds resulted at the distances established in the experimental test: (**a**) 5 m; (**b**) 12 m; and (**c**) 20 m. The images show the central part of the color points clouds that resulted from the test. The points clouds have not been filtered or edited.

4. Conclusions

The major innovations of this study are as follows: First, the proposed approach for the 3D data capture and the implementation of the VSLAM-photogrammetric algorithm has been materialized in a functional and low-cost prototype, which has been checked in an experimental test, the results of which have been presented in the context of the BIM work environment.

Second, the results obtained in the experimental test comply with the precision requirements of the GSA BIM Guide for 3D Imaging for point cloud capture work with a resolution (minimum artifact size) of 152 mm × 152 mm, for observation distances of approximately 20 m. For distances between 5 and 12 m, we saw that better accuracies and resolution results were achieved.

Third, the possibility of using the instrument at different distances facilitates the data capture in shaded areas or areas with difficult access. This, together with the fact that the device has been designed for outdoor data collection, makes it suitable for urban design and historic documentation, which are usually carried out in outdoor environments, registering information for plans, sections, elevations and details and 3D point cloud in PLY format (positioning: x, y, z and color: R, G, B), following the GSA PBS(Public Building Service) CAD standards (2012) and the GSA BIM Guide for 3D Imaging Standards.

In order to increase the knowledge of the proposed approach, it has been compared with a well-known photogrammetric methodology consisting of a Reflex Canon 1300D camera and the software Agisoft Metashape. The results of the comparison test have provided interesting conclusions:

1. The accuracy results of both methods are similar as can be seen in Table 5. Although the average error is slightly higher in the proposed approach, the RMSE is a bit lower than with the Agisoft Metashape methodology. This indicates a small, but greater dispersion of the points of the proposed approach in respect to the Agisoft software. But as can be seen in the results, this factor does not imply an increase of RMSE, but this error is slightly less in the proposed approach in relation to Agisoft software.

2. The processing time was a bit higher in the proposed approach for the distances of 5 and 12 m but not for 20 m, for which the time was slightly less. The differences are not significant, in our opinion, and indicate that the proposed method optimizes the number of images extracted and the photogrammetric process, thus equating well-known procedures such as the use of a Reflex camera and the Agisoft Metashape software.

3. In our opinion, the greatest improvement occurred in the data capture field. The user does not worry about how to use the camera or where to take the picture, because in the proposed approach, the capture is continuous and the system chooses the images automatically, as is explained in Section 2. In this way, the learning curve changes significantly, provided the user doesn´t need to have previous knowledge about photography or photogrammetry. For this reason, the proposed approach here described, reduces significantly the time spent in the field, as can be seen in Table 4.

A new handheld mobile mapping system based on images have been presented in this paper. This proposed methodology does not adversely affect the known photogrammetric process (accuracy, processing time, point cloud density) but it proposes a new, easier and faster way to capture the data in the field, based on continuous data capture and fully automatic processing, without human intervention in any phase.

Author Contributions: Conceptualization, P.O.-C.; data curation, P.O.-C. and A.S.-R.; formal analysis, A.S.-R.; investigation, P.O.-C.; methodology, P.O.-C.; resources, P.O.-C.; software, P.O.-C.; supervision, A.S.-R.; validation, A.S.-R.; writing—original draft, P.O.-C.; writing—review, A.S.-R.

Funding: This research received no external funding.

Acknowledgments: We are grateful to the "Consorcio de la Ciudad Monumental de Mérida" for allowing the work in this monument.

Conflicts of Interest: The authors declare no conflict of interest.

References

1. Remondino, F.; El-Hakim, S. Image-Based 3D Modelling: A Review. *Photogramm. Rec.* **2006**, *21*, 269–291. [CrossRef]
2. Raza, K.; Khan, T.A.; Abbas, N. Kinematic Analysis and Geometrical Improvement of an Industrial Robotic Arm. *J. King Saud Univ. Eng. Sci.* **2018**, *30*, 218–223. [CrossRef]
3. Rayna, T.; Striukova, L. From Rapid Prototyping to Home Fabrication: How 3D Printing Is Changing Business Model Innovation. *Technol. Soc. Chang.* **2016**, *102*, 214–224. [CrossRef]
4. Wu, P.; Wang, J.; Wang, X. A Critical Review of the Use of 3-D Printing in the Construction Industry. *Autom. Constr.* **2016**, *68*, 21–31. [CrossRef]
5. Tay, Y.W.D.; Panda, B.; Paul, S.C.; Mohamed, N.A.N.; Tan, M.J.; Leong, K.F. 3D Printing Trends in Building and Construction Industry: A Review. *Virtual Phys. Prototyp.* **2017**, *12*, 261–276. [CrossRef]
6. Yan, Q.; Dong, H.; Su, J.; Han, J.; Song, B.; Wei, Q.; Shi, Y. A Review of 3D Printing Technology for Medical Applications. *Engineering* **2018**, *4*, 729–742. [CrossRef]
7. Moruno, L.; Rodríguez Salgado, D.; Sánchez-Ríos, A.; González, A.G. An Ergonomic Customized-Tool Handle Design for Precision Tools Using Additive Manufacturing: A Case Study. *Appl. Sci.* **2018**, *8*, 1200. [CrossRef]
8. Liang, X.; Wang, Y.; Jaakkola, A.; Kukko, A.; Kaartinen, H.; Hyyppä, J.; Honkavaara, E.; Liu, J. Forest Data Collection Using Terrestrial Image-Based Point Clouds from a Handheld Camera Compared to Terrestrial and Personal Laser Scanning. *IEEE Trans. Geosci. Remote Sens.* **2015**, *53*. [CrossRef]
9. Behmann, J.; Mahlein, A.-K.; Paulus, S.; Kuhlmann, H.; Oerke, E.-C.; Plümer, L. Calibration of Hyperspectral Close-Range Pushbroom Cameras for Plant Phenotyping. *ISPRS J. Photogramm. Remote Sens.* **2015**, *106*, 172–182. [CrossRef]
10. Abellán, A.; Oppikofer, T.; Jaboyedoff, M.; Rosser, N.J.; Lim, M.; Lato, M.J. Terrestrial Laser Scanning of Rock Slope Instabilities. *Earth Surf. Process. Landf.* **2014**, *39*, 80–97. [CrossRef]
11. Ghuffar, S.; Székely, B.; Roncat, A.; Pfeifer, N. Landslide Displacement Monitoring Using 3D Range Flow on Airborne and Terrestrial LiDAR Data. *Remote Sens.* **2013**, *5*, 2720–2745. [CrossRef]
12. Lotsari, E.; Wang, Y.; Kaartinen, H.; Jaakkola, A.; Kukko, A.; Vaaja, M.; Hyyppä, H.; Hyyppä, J.; Alho, P. Gravel Transport by Ice in a Subarctic River from Accurate Laser Scanning. *Geomorphology* **2015**, *246*, 113–122. [CrossRef]
13. Harpold, A.; Marshall, J.; Lyon, S.; Barnhart, T.; Fisher, A.B.; Donovan, M.; Brubaker, K.; Crosby, C.; Glenn, F.N.; Glennie, C.; et al. Laser Vision: Lidar as a Transformative Tool to Advance Critical Zone Science. *Hydrol. Earth Syst. Sci.* **2015**, *19*, 2881–2897. [CrossRef]
14. Cacciari, I.; Nieri, P.; Siano, S. 3D Digital Microscopy for Characterizing Punchworks on Medieval Panel Paintings. *J. Comput. Cult. Herit.* **2014**, *7*, 19. [CrossRef]
15. Jaklič, A.; Erič, M.; Mihajlović, I.; Stopinšek, Ž.; Solina, F. Volumetric Models from 3D Point Clouds: The Case Study of Sarcophagi Cargo from a 2nd/3rd Century AD Roman Shipwreck near Sutivan on Island Brač, Croatia. *J. Archaeol. Sci.* **2015**, *62*, 143–152. [CrossRef]
16. Camburn, B.; Viswanathan, V.; Linsey, J.; Anderson, D.; Jensen, D.; Crawford, R.; Otto, K.; Wood, K. Design Prototyping Methods: State of the Art in Strategies, Techniques, and Guidelines. *Des. Sci.* **2017**, *3*, 1–33. [CrossRef]
17. Luhmann, T.; Robson, S.; Kyle, S.; Harley, I. *Close Range Photogrammetry: Principles, Techniques and Applications*; Whittles Publishing: Dunbeath, UK, 2006.
18. Ciarfuglia, T.A.; Costante, G.; Valigi, P.; Ricci, E. Evaluation of Non-Geometric Methods for Visual Odometry. *Robot. Auton. Syst.* **2014**, *62*, 1717–1730. [CrossRef]
19. Yousif, K.; Bab-Hadiashar, A.; Hoseinnezhad, R. An Overview to Visual Odometry and Visual SLAM: Applications to Mobile Robotics. *Intell. Ind. Syst.* **2015**, *1*, 289–311. [CrossRef]
20. Strobl, K.H.; Mair, E.; Bodenmüller, T.; Kielhöfer, S.; Wüsthoff, T.; Suppa, M. Portable 3-D Modeling Using Visual Pose Tracking. *Comput. Ind.* **2018**, *99*, 53–68. [CrossRef]
21. Kim, P.; Chen, J.; Cho, Y. SLAM-Driven Robotic Mapping and Registration of 3D Point Clouds. *Autom. Constr.* **2018**, *89*, 38–48. [CrossRef]

22. Balsa-Barreiro, J.; Fritsch, D. Generation of Visually Aesthetic and Detailed 3D Models of Historical Cities by Using Laser Scanning and Digital Photogrammetry. *Digit. Appl. Archaeol. Cult. Herit.* **2018**, *8*, 57–64. [CrossRef]

23. Balsa-Barreiro, J.; Fritsch, D. Generation of 3D/4D Photorealistic Building Models. The Testbed Area for 4D Cultural Heritage World Project: The Historical Center of Calw (Germany). In Proceedings of the International Symposium on Visual Computing, Las Vegas, NV, USA, 14–16 December 2015; pp. 361–372. [CrossRef]

24. Dupuis, J.; Paulus, S.; Behmann, J.; Plümer, L.; Kuhlmann, H. A Multi-Resolution Approach for an Automated Fusion of Different Low-Cost 3D Sensors. *Sensors* **2014**, *14*, 7563–7579. [CrossRef]

25. Sirmacek, B.; Lindenbergh, R. Accuracy Assessment of Building Point Clouds Automatically Generated from Iphone Images. *ISPRS Int. Arch. Photogramm. Remote Sens. Spat. Inf. Sci.* **2014**, *45*, 547–552. [CrossRef]

26. Lachat, E.; Macher, H.; Landes, T.; Grussenmeyer, P. Assessment and Calibration of a RGB-D Camera (Kinect v2 Sensor) Towards a Potential Use for Close-Range 3D Modeling. *Remote Sens.* **2015**, *7*, 13070–13097. [CrossRef]

27. Sánchez, A.; Gómez, J.M.; Jiménez, A.; González, A.G. Analysis of Uncertainty in a Middle-Cost Device for 3D Measurements in BIM Perspective. *Sensors* **2016**, *16*, 1557–1574. [CrossRef]

28. Zlot, R.; Bosse, M.; Greenop, K.; Jarzab, Z.; Juckes, E.; Roberts, J. Efficiently Capturing Large, Complex Cultural Heritage Sites with a Handheld Mobile 3D Laser Mapping System. *J. Cult. Herit.* **2014**, *15*, 670–678. [CrossRef]

29. Pollefeys, M.; Nistér, D.; Frahm, J.-M.; Akbarzadeh, A.; Mordohai, P.; Clipp, B.; Engels, C.; Gallup, D.; Kim, S.-J.; Merrell, P.; et al. Detailed Real-Time Urban 3D Reconstruction from Video. *Int. J. Comput. Vis.* **2008**, *78*, 143–167. [CrossRef]

30. Zingoni, A.; Diani, M.; Corsini, G.; Masini, A. Real-Time 3D Reconstruction from Images Taken from an UAV. *ISPRS Int. Arch. Photogramm. Remote Sens. Spat. Inf. Sci.* **2015**, *40*, 313–319. [CrossRef]

31. Sapirstein, P. Accurate Measurement with Photogrammetry at Large Sites. *J. Archaeol. Sci.* **2016**, *66*, 137–145. [CrossRef]

32. O'Driscoll, J. Landscape Applications of Photogrammetry Using Unmanned Aerial Vehicles. *J. Archaeol. Sci. Rep.* **2018**, *22*, 32–44. [CrossRef]

33. Campi, M.; di Luggo, A.; Monaco, S.; Siconolfi, M.; Palomba, D. Indoor and Outdoor Mobile Mapping Systems for Architectural Surveys. *ISPRS Int. Arch. Photogramm. Remote Sens. Spat. Inf. Sci.* **2018**, *42*, 201–208. [CrossRef]

34. Petrie, G. Mobile Mapping Systems: An Introduction to the Technology. *Geoinformatics* **2010**, *13*, 32–43.

35. Kopsida, M.; Brilakis, I.; Antonio Vela, P. A Review of Automated Construction Progress Monitoring and Inspection Methods. In Proceedings of the 32nd CIB W78 Conference, Eindhoven, The Netherlands, 27–29 October 2015.

36. Omar, T.; Nehdi, M.L. Data Acquisition Technologies for Construction Progress Tracking. *Autom. Constr.* **2016**, *70*, 143–155. [CrossRef]

37. Dai, F.; Rashidi, A.; Brilakis, I.; Vela, P. Comparison of Image-Based and Time-of-Flight-Based Technologies for Three-Dimensional Reconstruction of Infrastructure. *J. Constr. Eng. Manag.* **2013**, *139*, 69–79. [CrossRef]

38. El-Omari, S.; Moselhi, O. Integrating 3D Laser Scanning and Photogrammetry for Progress Measurement of Construction Work. *Autom. Constr.* **2008**, *18*, 1–9. [CrossRef]

39. Rebolj, D.; Pučko, Z.; Babič, N.Č.; Bizjak, M.; Mongus, D. Point Cloud Quality Requirements for Scan-vs-BIM Based Automated Construction Progress Monitoring. *Autom. Constr.* **2017**, *84*, 323–334. [CrossRef]

40. Wu, P. *Integrated Building Information Modelling*; Li, H., Wang, X., Eds.; Bentham Science Publishers: Sharjah, UAE, 2017. [CrossRef]

41. U.S. General Services Administration, Public Buildings Service. *GSA Building Information Modeling Guide Series: 03—GSA BIM Guide for 3DImaging*; General Services Administration: Washington, DC, USA, 2019.

42. Akca, D.; Freeman, M.; Sargent, I.; Gruen, A. Quality Assessment of 3D Building Data: Quality Assessment of 3D Building Data. *Photogramm. Rec.* **2010**, *25*, 339–355. [CrossRef]

43. Tran, H.; Khoshelham, K.; Kealy, A. Geometric Comparison and Quality Evaluation of 3D Models of Indoor Environments. *ISPRS J. Photogramm. Remote Sens.* **2019**, *149*, 29–39. [CrossRef]

44. Zhang, C.; Kalasapudi, V.S.; Tang, P. Rapid Data Quality Oriented Laser Scan Planning for Dynamic Construction Environments. *Adv. Eng. Inf.* **2016**, *30*, 218–232. [CrossRef]

45. Tang, P.; Alaswad, F.S. Sensor Modeling of Laser Scanners for Automated Scan Planning on Construction Jobsites. In *Construction Research Congress 2012*; American Society of Civil Engineers: West Lafayette, IN, USA, 2012; pp. 1021–1031. [CrossRef]
46. Soudarissanane, S.; Lindenbergh, R.; Menenti, M.; Teunissen, P. Scanning Geometry: Influencing Factor on the Quality of Terrestrial Laser Scanning Points. *ISPRS J. Photogramm. Remote Sens.* **2011**, *66*, 389–399. [CrossRef]
47. Shanoer, M.M.; Abed, F.M. Evaluate 3D Laser Point Clouds Registration for Cultural Heritage Documentation. *Egypt. J. Remote Sens. Space Sci.* **2018**, *21*, 295–304. [CrossRef]
48. Zhang, Z. A Flexible New Technique for Camera Calibration. *IEEE Trans. Pattern Anal. Mach. Intell.* **2000**, *22*, 1330–1334. [CrossRef]
49. Rublee, E.; Rabaud, V.; Konolige, K.; Bradski, G. ORB: An Efficient Alternative to SIFT or SURF. In Proceedings of the 2011 International Conference on Computer Vision, Barcelona, Spain, 6–13 November 2011; pp. 2564–2571. [CrossRef]
50. Mur-Artal, R.; Montiel, M.M.J.; Tardós, J.D. ORB-SLAM: A Versatile and Accurate Monocular SLAM System. *IEEE Trans. Robot.* **2017**, *31*, 1255–1262. [CrossRef]
51. Hartley, R.; Zisserman, A. *Multiple View Geometry in Computer Vision*, 2nd ed.; Cambridge University Press: Cambridge, UK, 2004. [CrossRef]
52. Stewénius, H.; Engels, C.; Nistér, D. Recent Developments on Direct Relative Orientation. *ISPRS J. Photogramm. Remote Sens.* **2006**, *60*, 284–294. [CrossRef]
53. Pierrot Deseilligny, M.; Clery, I. Apero, an Open Source Bundle Adjusment Software for Automatic Calibration and Orientation of Set of Images. *ISPRS Int. Arch. Photogramm. Remote Sens. Spat. Inf. Sci.* **2012**, *269277*. [CrossRef]
54. Georgantas, A.; Brédif, M.; Pierrot-Desseilligny, M. An Accuracy Assessment of Automated Photogrammetric Techniques for 3D Modeling of Complex Interiors. *ISPRS Int. Arch. Photogramm. Remote Sens. Spat. Inf. Sci.* **2012**, *39*, 23–28. [CrossRef]
55. Cerrillo-Cuenca, E.; Ortiz-Coder, P.; Martínez-del-Pozo, J.-Á. Computer Vision Methods and Rock Art: Towards a Digital Detection of Pigments. *Archaeol. Anthropol. Sci.* **2014**, *6*, 227–239. [CrossRef]
56. Triggs, B.; Mclauchlan, P.; Hartley, R.; Fitzgibbon, A. Bundle Adjustment—A Modern Synthesis. In Proceedings of the International Workshop on Vision Algorithms, Singapore, 5–8 December 2000; pp. 198–372.
57. Fischler, M.A.; Bolles, R.C. Random Sample Consensus: A Paradigm for Model Fitting with Applications to Image Analysis and Automated Cartography. *Commun. ACM* **1981**, *24*, 381–395. [CrossRef]
58. Rupnik, E.; Daakir, M.; Pierrot Deseilligny, M. MicMac—A Free, Open-Source Solution for Photogrammetry. *Open Geospat. Data Softw. Stand.* **2017**, *2*, 14. [CrossRef]
59. Deseilligny, M.; Paparodit, N. A Multiresolution and Optimization-Based Image Matching Approach: An Application to Surface Reconstruction from SPOT5-HRS Stereo Imagery. *Int. Arch. Photogramm. Remote Sens. Spat. Inf. Sci.* **2012**, *36*, 1–5.
60. Agisoft Metashape. Available online: https://www.agisoft.com/ (accessed on 28 August 2019).
61. Meshlab. Available online: http://www.meshlab.net/ (accessed on 27 May 2015).
62. Hong, S.; Jung, J.; Kim, S.; Cho, H.; Lee, J.; Heo, J. Semi-Automated Approach to Indoor Mapping for 3D as-Built Building Information Modeling. *Comput. Environ. Urban Syst.* **2015**, *51*, 34–46. [CrossRef]
63. Koutsoudis, A.; Vidmar, B.; Ioannakis, G.; Arnaoutoglou, F.; Pavlidis, G.; Chamzas, C. Multi-image 3D reconstruction data evaluation. *J. Cult. Herit.* **2014**, *15*, 73–79. [CrossRef]

MDPI

St. Alban-Anlage 66

4052 Basel

Switzerland

Tel. +41 61 683 77 34

Fax +41 61 302 89 18

www.mdpi.com

Sensors Editorial Office

E-mail: sensors@mdpi.com

www.mdpi.com/journal/sensors